SURFACE CHARACTERIZATION METHODS

SURFACTANT SCIENCE SERIES

FOUNDING EDITOR

MARTIN J. SCHICK
1918–1998

SERIES EDITOR

ARTHUR T. HUBBARD
Santa Barbara Science Project
Santa Barbara, California

ADVISORY BOARD

DANIEL BLANKSCHTEIN
Department of Chemical Engineering
Massachusetts Institute of Technology
Cambridge, Massachusetts

S. KARABORNI
Shell International Petroleum
  Company Limited
London, England

LISA B. QUENCER
The Dow Chemical Company
Midland, Michigan

JOHN F. SCAMEHORN
Institute for Applied Surfactant
  Research
University of Oklahoma
Norman, Oklahoma

P. SOMASUNDARAN
Henry Krumb School of Mines
Columbia University
New York, New York

ERIC W. KALER
Department of Chemical Engineering
University of Delaware
Newark, Delaware

CLARENCE MILLER
Department of Chemical Engineering
Rice University
Houston, Texas

DON RUBINGH
The Proctor & Gamble Company
Cincinnati, Ohio

BEREND SMIT
Shell International Oil Products B.V.
Amsterdam, The Netherlands

JOHN TEXTER
Strider Research, Incorporated
Rochester, New York

1. Nonionic Surfactants, *edited by Martin J. Schick* (see also Volumes 19, 23, and 60)
2. Solvent Properties of Surfactant Solutions, *edited by Kozo Shinoda* (see Volume 55)
3. Surfactant Biodegradation, *R. D. Swisher* (see Volume 18)
4. Cationic Surfactants, *edited by Eric Jungermann* (see also Volumes 34, 37, and 53)
5. Detergency: Theory and Test Methods (in three parts), *edited by W. G. Cutler and R. C. Davis* (see also Volume 20)
6. Emulsions and Emulsion Technology (in three parts), *edited by Kenneth J. Lissant*
7. Anionic Surfactants (in two parts), *edited by Warner M. Linfield* (see Volume 56)
8. Anionic Surfactants: Chemical Analysis, *edited by John Cross*
9. Stabilization of Colloidal Dispersions by Polymer Adsorption, *Tatsuo Sato and Richard Ruch*
10. Anionic Surfactants: Biochemistry, Toxicology, Dermatology, *edited by Christian Gloxhuber* (see Volume 43)
11. Anionic Surfactants: Physical Chemistry of Surfactant Action, *edited by E. H. Lucassen-Reynders*
12. Amphoteric Surfactants, *edited by B. R. Bluestein and Clifford L. Hilton* (see Volume 59)
13. Demulsification: Industrial Applications, *Kenneth J. Lissant*
14. Surfactants in Textile Processing, *Arved Datyner*
15. Electrical Phenomena at Interfaces: Fundamentals, Measurements, and Applications, *edited by Ayao Kitahara and Akira Watanabe*
16. Surfactants in Cosmetics, *edited by Martin M. Rieger* (see Volume 68)
17. Interfacial Phenomena: Equilibrium and Dynamic Effects, *Clarence A. Miller and P. Neogi*
18. Surfactant Biodegradation: Second Edition, Revised and Expanded, *R. D. Swisher*
19. Nonionic Surfactants: Chemical Analysis, *edited by John Cross*
20. Detergency: Theory and Technology, *edited by W. Gale Cutler and Erik Kissa*
21. Interfacial Phenomena in Apolar Media, *edited by Hans-Friedrich Eicke and Geoffrey D. Parfitt*
22. Surfactant Solutions: New Methods of Investigation, *edited by Raoul Zana*
23. Nonionic Surfactants: Physical Chemistry, *edited by Martin J. Schick*
24. Microemulsion Systems, *edited by Henri L. Rosano and Marc Clausse*
25. Biosurfactants and Biotechnology, *edited by Naim Kosaric, W. L. Cairns, and Neil C. C. Gray*
26. Surfactants in Emerging Technologies, *edited by Milton J. Rosen*
27. Reagents in Mineral Technology, *edited by P. Somasundaran and Brij M. Moudgil*
28. Surfactants in Chemical/Process Engineering, *edited by Darsh T. Wasan, Martin E. Ginn, and Dinesh O. Shah*
29. Thin Liquid Films, *edited by I. B. Ivanov*
30. Microemulsions and Related Systems: Formulation, Solvency, and Physical Properties, *edited by Maurice Bourrel and Robert S. Schechter*
31. Crystallization and Polymorphism of Fats and Fatty Acids, *edited by Nissim Garti and Kiyotaka Sato*

32. Interfacial Phenomena in Coal Technology, *edited by Gregory D. Botsaris and Yuli M. Glazman*
33. Surfactant-Based Separation Processes, *edited by John F. Scamehorn and Jeffrey H. Harwell*
34. Cationic Surfactants: Organic Chemistry, *edited by James M. Richmond*
35. Alkylene Oxides and Their Polymers, *F. E. Bailey, Jr., and Joseph V. Koleske*
36. Interfacial Phenomena in Petroleum Recovery, *edited by Norman R. Morrow*
37. Cationic Surfactants: Physical Chemistry, *edited by Donn N. Rubingh and Paul M. Holland*
38. Kinetics and Catalysis in Microheterogeneous Systems, *edited by M. Grätzel and K. Kalyanasundaram*
39. Interfacial Phenomena in Biological Systems, *edited by Max Bender*
40. Analysis of Surfactants, *Thomas M. Schmitt*
41. Light Scattering by Liquid Surfaces and Complementary Techniques, *edited by Dominique Langevin*
42. Polymeric Surfactants, *Irja Piirma*
43. Anionic Surfactants: Biochemistry, Toxicology, Dermatology. Second Edition, Revised and Expanded, *edited by Christian Gloxhuber and Klaus Künstler*
44. Organized Solutions: Surfactants in Science and Technology, *edited by Stig E. Friberg and Björn Lindman*
45. Defoaming: Theory and Industrial Applications, *edited by P. R. Garrett*
46. Mixed Surfactant Systems, *edited by Keizo Ogino and Masahiko Abe*
47. Coagulation and Flocculation: Theory and Applications, *edited by Bohuslav Dobiáš*
48. Biosurfactants: Production • Properties • Applications, *edited by Naim Kosaric*
49. Wettability, *edited by John C. Berg*
50. Fluorinated Surfactants: Synthesis • Properties • Applications, *Erik Kissa*
51. Surface and Colloid Chemistry in Advanced Ceramics Processing, *edited by Robert J. Pugh and Lennart Bergström*
52. Technological Applications of Dispersions, *edited by Robert B. McKay*
53. Cationic Surfactants: Analytical and Biological Evaluation, *edited by John Cross and Edward J. Singer*
54. Surfactants in Agrochemicals, *Tharwat F. Tadros*
55. Solubilization in Surfactant Aggregates, *edited by Sherril D. Christian and John F. Scamehorn*
56. Anionic Surfactants: Organic Chemistry, *edited by Helmut W. Stache*
57. Foams: Theory, Measurements, and Applications, *edited by Robert K. Prud'homme and Saad A. Khan*
58. The Preparation of Dispersions in Liquids, *H. N. Stein*
59. Amphoteric Surfactants: Second Edition, *edited by Eric G. Lomax*
60. Nonionic Surfactants: Polyoxyalkylene Block Copolymers, *edited by Vaughn M. Nace*
61. Emulsions and Emulsion Stability, *edited by Johan Sjöblom*
62. Vesicles, *edited by Morton Rosoff*
63. Applied Surface Thermodynamics, *edited by A. W. Neumann and Jan K. Spelt*
64. Surfactants in Solution, *edited by Arun K. Chattopadhyay and K. L. Mittal*
65. Detergents in the Environment, *edited by Milan Johann Schwuger*

66. Industrial Applications of Microemulsions, *edited by Conxita Solans and Hironobu Kunieda*
67. Liquid Detergents, *edited by Kuo-Yann Lai*
68. Surfactants in Cosmetics: Second Edition, Revised and Expanded, *edited by Martin M. Rieger and Linda D. Rhein*
69. Enzymes in Detergency, *edited by Jan H. van Ee, Onno Misset, and Erik J. Baas*
70. Structure–Performance Relationships in Surfactants, *edited by Kunio Esumi and Minoru Ueno*
71. Powdered Detergents, *edited by Michael S. Showell*
72. Nonionic Surfactants: Organic Chemistry, *edited by Nico M. van Os*
73. Anionic Surfactants: Analytical Chemistry, Second Edition, Revised and Expanded, *edited by John Cross*
74. Novel Surfactants: Preparation, Applications, and Biodegradability, *edited by Krister Holmberg*
75. Biopolymers at Interfaces, *edited by Martin Malmsten*
76. Electrical Phenomena at Interfaces: Fundamentals, Measurements, and Applications, Second Edition, Revised and Expanded, *edited by Hiroyuki Ohshima and Kunio Furusawa*
77. Polymer-Surfactant Systems, *edited by Jan C. T. Kwak*
78. Surfaces of Nanoparticles and Porous Materials, *edited by James A. Schwarz and Cristian I. Contescu*
79. Surface Chemistry and Electrochemistry of Membranes, *edited by Torben Smith Sørensen*
80. Interfacial Phenomena in Chromatography, *edited by Emile Pefferkorn*
81. Solid–Liquid Dispersions, *Bohuslav Dobiáš, Xueping Qiu, and Wolfgang von Rybinski*
82. Handbook of Detergents, *editor in chief: Uri Zoller*
     Part A: Properties, *edited by Guy Broze*
83. Modern Characterization Methods of Surfactant Systems, *edited by Bernard P. Binks*
84. Dispersions: Characterization, Testing, and Measurement, *Erik Kissa*
85. Interfacial Forces and Fields: Theory and Applications, *edited by Jyh-Ping Hsu*
86. Silicone Surfactants, *edited by Randal M. Hill*
87. Surface Characterization Methods: Principles, Techniques, and Applications, *edited by Andrew J. Milling*

ADDITIONAL VOLUMES IN PREPARATION

Interfacial Dynamics, *edited by Nikola Kallay*

Computational Methods in Surface and Colloid Science, *edited by Malgorzata Borowko*

SURFACE CHARACTERIZATION METHODS
Principles, Techniques, and Applications

edited by
Andrew J. Milling
University of Durham
Durham, England

 Marcel Dekker, Inc.   New York · Basel

ISBN: 0-8247-7336-5

This book is printed on acid-free paper.

Headquarters
Marcel Dekker, Inc.
270 Madison Avenue, New York, NY 10016
tel: 212-696-9000; fax: 212-685-4540

Eastern Hemisphere Distribution
Marcel Dekker AG
Hutgasse 4, Postfach 812, CH-4001 Basel, Switzerland
tel: 41-61-261-8482; fax: 41-61-261-8896

World Wide Web
http://www.dekker.com

The publisher offers discounts on this book when ordered in bulk quantities. For more information, write to Special Sales/Professional Marketing at the headquarters address above.

Copyright © 1999 by Marcel Dekker, Inc. All Rights Reserved.

Neither this book nor any part may be reproduced or transmitted in any form or by any means, electronic or mechanical, including photocopying, microfilming, and recording, or by any information storage and retrieval system, without permission in writing from the publisher.

Current printing (last digit):
10 9 8 7 6 5 4 3 2 1

PRINTED IN THE UNITED STATES OF AMERICA

Preface

During my career in chemistry, I have become increasingly aware that a multi-disciplinary approach is required to carry out research in the field of surface characterization, and that this field is certainly not the sole reserve of the physical chemist. A host of experimental methodologies are available to researchers interested in surface analysis per se, or as an adjunct to other applications. The current literature is on the whole highly specialized, dealing with specific topics in great detail. While such books are of use to the specialist, it is felt that there is certainly a need for a reference text that provides a more general appreciation of the surface characterization methods currently in use or being developed in modern laboratories. The ubiquity of surfaces entails that many fields, ranging from the processing of particulate materials (such as colloids) to molecular recognition processes, require a deeper understanding of surface and interfacial properties. Indeed, this is exemplified in the burgeoning interest in surface chemistry within the life sciences, and this is an important feature of the text.

The book, comprising a series of monographs by contemporary experts, outlines the underlying scientific principles and experimental techniques for a broad sample of discrete surface analysis techniques. The assembled material draws heavily from cornerstones of physical and analytical chemistry. Specific themes such as surface energies, electrokinetic characterization, van der Waals interactions, wetting behavior, self-assembly, adsorption behavior, mass spectroscopy, and scattering methodologies are described within the general context of surface analysis.

The book is intended to serve as a resource text and also to aid the active researcher in solving surface analysis problems that may arise in a variety of circumstances. The "techniques" aspect illustrates that there are often various approaches to characterizing a particular aspect of surface behavior, using in many cases equipment that either is commercially available or can be readily assembled. The scope of the material will appeal to researchers from final-year undergraduate students through senior researchers.

I would like to collectively thank several colleagues for their kind assistance in proofreading at various stages, for which I am extremely grateful. I would like to especially thank all the contributing authors, Professor A. Hubbard, for his great help in the genesis of this project, and Anita Lekhwani and Joseph Stubenrauch at Marcel Dekker, Inc., for their significant contribution and enthusiasm in the production of this text.

Andrew J. Milling

Contents

Preface iii
Contributors vii

1. Measurement of the Surface Tension and Surface Stress of Solids   1
   Hans-Jürgen Butt and Roberto Raiteri

2. Contact Angle Techniques and Measurements   37
   Daniel Y. Kwok and A. W. Neumann

3. Measurement of Ion-Mediated and van der Waals Forces Using Atomic Force Microscopy   87
   Ian Larson and Andrew J. Milling

4. Measurement of Electro-osmosis as a Method for Electrokinetic Surface Analysis   113
   Norman L. Burns

5. X-Ray Photoelectron Spectroscopy (XPS) and Static Secondary Ion Mass Spectrometry (SSIMS) of Biomedical Polymers and Surfactants   143
   Kevin M. Shakesheff, Martyn C. Davies, and Robert Langer

6. Evanescent Wave Scattering at Solid Surfaces   173
   Adolfas K. Gaigalas

7. Characterizing Colloidal Materials Using Dynamic Light Scattering   199
   Leo H. Hanus and Harry J. Ploehn

8. Light Scattering Studies of Microcapsules in Suspension 249
   Toshiaki Dobashi and Benjamin Chu

9. Three-Dimensional Particle Tracking of Micronic Colloidal Particles 269
   Y. Grasselli and Georges Bossis

10. Low-Mass Luminescent Organogels 285
    Pierre Terech and Richard G. Weiss

11. Chromatographic Methods for Measurement of Antibody–Antigen Association Rates 345
    Claire Vidal-Madjar and Alain Jaulmes

12. The Acid–Base Behavior of Proteins Determined by ISFETs 373
    Wouter Olthuis and Piet Bergveld

Index 405

Contributors

Piet Bergveld MESA Research Institute, University of Twente, Enschede, The Netherlands

Georges Bossis Department of Physics, CNRS–University of Nice, Nice, France

Norman L. Burns Institute for Surface Chemistry, Stockholm, Sweden

Hans-Jürgen Butt Institut für Physikalische Chemie, Universität Mainz, Mainz, Germany

Benjamin Chu Department of Chemistry, State University of New York at Stony Brook, Stony Brook, New York

Martyn C. Davies School of Pharmaceutical Sciences, University of Nottingham, Nottingham, England

Toshiaki Dobashi Department of Biological and Chemical Engineering, Gunma University, Kiryu, Gunma, Japan

Adolfas K. Gaigalas Biotechnology Division, National Institute of Standards and Technology, Gaithersburg, Maryland

Y. Grasselli Department of Physics, CNRS–University of Nice, Nice, France

Leo H. Hanus Department of Chemical Engineering, University of South Carolina, Columbia, South Carolina

Alain Jaulmes Laboratoire de Recherche sur les Polymères, CNRS, Thiais, France

Daniel Y. Kwok Department of Chemical Engineering, Massachusetts Institute of Technology, Cambridge, Massachusetts

Robert Langer Department of Chemical Engineering, Massachusetts Institute of Technology, Cambridge, Massachusetts

Ian Larson Ian Wark Research Institute, University of South Australia, Mawson Lakes, South Australia, Australia

Andrew J. Milling Department of Chemistry, University of Durham, Durham, England

A. W. Neumann Department of Mechanical and Industrial Engineering, University of Toronto, Toronto, Ontario, Canada

Wouter Olthuis MESA Research Institute, University of Twente, Enschede, The Netherlands

Harry J. Ploehn Department of Chemical Engineering, University of South Carolina, Columbia, South Carolina

Roberto Raiteri Institut für Physikalische Chemie, Universität Mainz, Mainz, Germany

Kevin M. Shakesheff School of Pharmaceutical Sciences, University of Nottingham, Nottingham, England

Pierre Terech Département de Recherche Fondamentale sur la Matière Condensée, UMR 5819, CEA–CNRS–Université J. Fourier, Grenoble, France

Claire Vidal-Madjar Laboratoire de Recherche sur les Polymères, CNRS, Thiais, France

Richard G. Weiss Department of Chemistry, Georgetown University, Washington, D.C.

SURFACE CHARACTERIZATION METHODS

1
Measurement of the Surface Tension and Surface Stress of Solids

HANS-JÜRGEN BUTT and ROBERTO RAITERI Institut für Physikalische Chemie, Universität Mainz, Mainz, Germany

I.	Introduction	2
II.	Basic Thermodynamics of Solid Surfaces	4
	A. The Gibbs–Duhem equation	4
	B. Elastic and plastic strain interrelationships	5
III.	Experimental Techniques to Measure the Surface Tension and the Surface Stress	7
	A. Zero creep method	7
	B. Change in lattice constant	8
	C. Cleavage experiments	9
	D. Adhesion measurements	11
	E. Adsorption measurements	14
	F. Calorimetric methods	16
	G. Solubility changes	19
	H. Contact angle measurements	21
	I. Bending plate methods	24
IV.	Electrocapillarity of Solid Electrodes	27
V.	Summary	28
	References	30

I. INTRODUCTION

The surface tension is a fundamental parameter of a solid since it depends directly on the binding forces of the material [1,2]. It is also of great practical interest. The adsorption of substances onto solids is determined by the surface tension, and it is of fundamental importance in biocompatibility [3]. The behavior of colloidal dispersions, adhesion, and friction are influenced by the surface tension. Since the surface tension enters the Young equation, it is important for contact angle phenomena such as detergency, wetting, water repellency, and flotation. When making micro- or nanoscopic structures, a knowledge of the surface tension is essential since, owing to the large surface-to-volume ratio, surface phenomena dominate the fabrication process [4]. The reconstruction of silicon and germanium surfaces [5,6] and the shape and structure of small particles [7,8,9] depends on the surface tension. In addition, it influences crystal growth [10,11].

The surface tension is equal to the reversible work per unit area needed to create a surface. For liquids this definition is sufficient. If a liquid (effects due to a possible curvature of the surface are ignored) is distorted, there is no barrier to prevent molecules from entering or leaving the surface. In the new equilibrium state each molecule covers the same area as in the original undistorted state. The number of molecules in the surface has changed, but the area per molecule remains the same. Such a deformation is called plastic.

The main difference between a solid and a liquid is that the molecules in a solid are not mobile. Therefore, as Gibbs already noted, the work required to create new surface area depends on the way the new solid surface is formed [12]. Plastic deformations are possible for solids too. An example is the cleavage of a crystal. Plastic deformations are described by the surface tension γ also called superficial work.* The surface tension may be defined as the reversible work at constant elastic strain, temperature, electric field, and chemical potential required to form a unit area of new surface. It is a scalar quantity. The surface tension is usually measured in adhesion and adsorption experiments.

New surface area of a solid can also be created elastically by stretching pre-existing surface. In this case molecules cannot migrate to the surface and therefore the number of molecules remains constant but the area occupied by

---

*Different authors use different symbols and different expressions for the surface tension. The term "superficial work" with the symbol σ was proposed by Linford [16]. The IUPAC recommends the symbol γ_π [15]. To avoid confusion with the surface charge density and for practical reasons we follow Lyklema (J. Lyklema, *Fundamentals of Interface and Colloid Science*, Vol. 1, Academic Press, London, 1991, p. 2.100) and Moy and Neumann (E. Moy and A. W. Neumann, in A. W. Neumann and J. K. Spelt (eds.), Surfactant Science Series 63, *Applied Surface Thermodynamics*, Marcel Dekker, New York, 1996, pp. 333–378), who used the symbol γ. Rusanov and Prokhorov [21] use "thermodynamic surface tension σ."

Surface Tension and Surface Stress of Solids

one molecule increases. Elastic deformations are described by the surface stress Υ_{ij}.* The surface stress can be defined as the reversible work needed to form a unit area of new surface by stretching a solid surface at constant temperature, electric field, and chemical potential with a linear stress. Since the response of a solid surface may depend on the direction the stress is applied, the surface stress is, in general, a tensorial quantity. For an isotropic material the directional dependence of the surface stress disappears, and it becomes a scalar quantity Υ. The surface stress is a surface excess property. However one should note that, in a real experiment, the measured mechanical work embodies terms that depend on the strain state of both bulk and surface [13]. The surface stress is mainly determined in mechanical experiments.

The change in the surface area of a solid is often described in terms of the surface strain. The total surface strain ε_{tot} is given by $d\Omega/\Omega = d\varepsilon_{\text{tot}}$, where Ω is the total area. The total strain may be divided into the plastic strain $d\varepsilon_p$ and the elastic strain $d\varepsilon_e$ so that $d\varepsilon_{\text{tot}} = d\varepsilon_p + d\varepsilon_e$.

In general the work required to form new surface area of a solid (plastic and elastic) is given by the expression [14]

$$\gamma^S = \frac{d\varepsilon_p}{d\varepsilon_{\text{tot}}} \cdot \gamma + \frac{d\varepsilon_e}{d\varepsilon_{\text{tot}}} \cdot \Upsilon \tag{1}$$

γ^S is called the "generalized surface intensive parameter" or "surface energy" [15,16]. The generalized surface intensive parameter depends on the path a certain state is reached while the surface tension and the surface stress are independent of the specific process. Therefore only γ and Υ are properties characterizing a solid surface, while γ^S is not a state function in a thermodynamic sense.

From a more fundamental point of view the difference between γ and Υ in an elastic solid arises from the inequality for immobile components of chemical potentials. In principle, for real solids such equalization is possible, but it proceeds very slowly so that always $\gamma \neq \Upsilon$ in practice.

Whether elastic or plastic behavior is observed depends not only on the process but also on the temperature. At very low temperature, real solids display mainly elastic properties. The higher the temperature, the larger the plasticity of the solid is. At temperatures close to the melting point, often methods developed for liquids can be employed to measure surface tension.

A determination of γ and Υ is often complicated in that solid surfaces are usually not in thermodynamic equilibrium. Even ideally pure solids have dislocations or vacancies that disturb the normal structure of the crystalline lattice.

---

*Most authors use the term "surface stress." Rusanov and Prokhorov [21] prefer the expression "mechanical surface tension γ." The symbol Υ was also used by Linford [16], and we do not see a reason to choose another symbol.

In addition, solid surfaces are usually rough. In practice it would take too much time to wait for spontaneous smoothing at temperature significantly lower than the melting point. On the other hand, annealing the surface does not help, for during the cooling process the surface structure might change. Also polishing and grinding change the properties of a surface significantly.

The surface tension of solids has been the subject of several reviews. Theoretical advances are reviewed in a paper of Linford [16] and more recently by Rusanov [17]. Also reviews about experimental techniques for determining the surface tension of solids in general [18] and of electrocapillary measurements [19], and a collection of experimental results [20], have appeared. Rusanov and Prokhorov provide a detailed review about the theoretical background of more classical experimental methods [21].

The aim of this paper is to review methods of measuring γ and Υ of solids from a practical viewpoint. It is mainly directed to readers who are not yet specialists concerning the surface tension of solids. It should provide a background of how γ and Υ are determined and how universal, technically demanding, and reliable the methods are. Some relatively specialized methods like the inert gas bubble method [22,23] are left out. Results are reported but not extensively discussed. In addition, a brief section about the Gibbs–Duhem equation of solid surfaces is given. The Gibbs–Duhem equation is the basis for interpreting some experimental methods. Also a section about elastic and plastic strain interrelationships is included. The intention is to give a more intuitive understanding of γ and Υ. Both sections can be skipped or read later.

II. BASIC THERMODYNAMICS OF SOLID SURFACES

A. The Gibbs–Duhem Equation

When dealing with surfaces, the Gibbs–Duhem relation is an important equation of chemical thermodynamics. It can be derived in the following way [24,25,16,26]. The differential of the internal surface energy, dU^σ, can be expressed as

$$dU^\sigma = TdS^\sigma + \sum \mu_i dn_i^\sigma + \varphi dQ^\sigma + \gamma^S d\Omega \qquad (2)$$

T is the temperature, S^σ is the entropy of the surface, φ is the surface electric potential, Q^σ is the total surface charge, μ_i is the chemical potential, n_i^σ denotes the excess amount of mobile substance i at the surface, and γ^S is the generalized surface intensive parameter as defined in Eq. (1). It is convenient to set the dividing plane so that the excess amount of the solid, immobile phase is zero. Hence the sum runs only over all mobile components. To derive the Gibbs–Duhem equation, the internal energy function is written in the integral form as a homogeneous first-order equation in the extensive variables. Such an equation involves only γ, since increasing the size of the system without changing its surface properties such the elastic strain ($d\varepsilon_e = 0$) excludes a contribution of the surface stress Υ (Ref. 24, p. 215; Ref. 27, p. 227; Ref. 21, p. 43). Hence, one can write

$$U^\sigma = TS^\sigma + \sum \mu_i n_i^\sigma + \varphi Q^\sigma + \gamma \Omega \tag{3}$$

The Gibbs–Duhem relation follows by differentiation and comparison with Eq. (2):

$$0 = S^\sigma \, dT + \sum n_i^\sigma \, d\mu_i + Q^\sigma \, d\varphi + \Omega \, d\gamma + \gamma \, d\Omega - \gamma^S \, d\Omega \tag{4}$$

Dividing by Ω leads to

$$0 = s^\sigma \, dT + \sum \Gamma_i \, d\mu_i + \sigma \, d\varphi + d\gamma + \gamma \, d\varepsilon_{\text{tot}} - \gamma^S \, d\varepsilon_{\text{tot}} \tag{5}$$

where s^σ is the surface entropy per unit area, Γ_i is the surface excess of the mobile component i, and σ is the surface charge density. The last two terms can be written as

$$\gamma \, d\varepsilon_{\text{tot}} - \gamma^S \, d\varepsilon_{\text{tot}} = \gamma \cdot (d\varepsilon_p + d\varepsilon_e) - \left(\gamma \cdot \frac{d\varepsilon_p}{d\varepsilon_{\text{tot}}} + \Upsilon \cdot \frac{d\varepsilon_e}{d\varepsilon_{\text{tot}}}\right) \cdot d\varepsilon_{\text{tot}}$$
$$= (\gamma - \Upsilon) \cdot d\varepsilon_e \tag{6}$$

Inserting this into Eq. (5) simplifies the Gibbs–Duhem equation to

$$0 = s^\sigma dT + \sum \Gamma_i \, d\mu_i + \sigma \, d\varphi + d\gamma + (\gamma - \Upsilon) \, d\varepsilon_e \tag{7}$$

One should note that Eq. (7) only contains the state functions γ and Υ instead of γ^S.

It is instructive to compare this relation with the Gibbs–Duhem equation for a liquid surface. For a liquid, no elastic strain can exist. Consequently $d\varepsilon_e$ must be zero and $\gamma^S = \gamma$. This leads to

$$0 = s^\sigma dT + \sum \Gamma_i \, d\mu_i + \sigma \, d\varphi + d\gamma \tag{8}$$

B. Elastic and Plastic Strain Interrelationships

A relation between the surface tension γ and the surface stress Υ can be directly derived from the Gibbs–Duhem equation, Eq. (7). At constant temperature, chemical potentials, and electric potential we have

$$\Upsilon = \gamma + \left.\frac{\partial \gamma}{\partial \varepsilon_e}\right|_{T, \mu_i, \varphi} \tag{9}$$

This equation was first presented in a classical paper by Shuttleworth [28]. If a surface area is increased by stretching, the increase in free energy (which in that case is ΥdA) is given by the surface tension plus the change of the surface tension with the elastic strain. The Shuttleworth equation states that in order

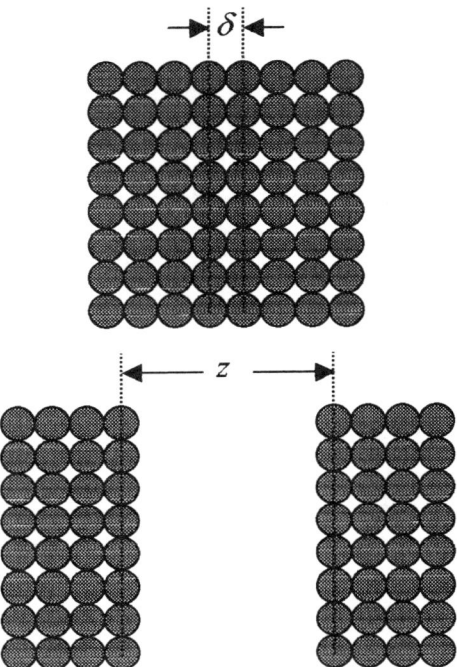

FIG. 1 Schematic drawing of splitting a body into two half-spaces.

to know the surface stress one needs to know the surface tension but also the dependence of γ on the elastic strain ε_e.

To get an idea of how γ depends on the elastic strain, we consider the formation of new surface by splitting a body into two halves (Fig. 1). The work (per unit area) to separate the two half-spaces from the initial molecular distance δ to an infinite distance is determined by the force acting between the two bodies (per unit area) Π. This force per unit area depends on the distance z. At the end of the process two new surfaces are formed. The total work is

$$2\gamma = \int_{\delta}^{\infty} \Pi(z)dz \tag{10}$$

Let us first consider molecular solids. There the force is the van der Waals force. The van der Waals force increases with increasing molecular density. When stretching a solid surface the number of molecules per unit surface area decreases. Hence, for molecular solids, it is reasonable to assume that γ decreases with increasing elastic strain and $\partial\gamma/\partial\varepsilon_e < 0$. It follows that $\Upsilon < \gamma$ for a completely elastic molecular solid. The same relation seems to hold for metals [29,30]. In contrast, calculations showed that for ionic crystals, $\partial \gamma / \partial \varepsilon_e > 0$ and consequently $\Upsilon > \gamma$ [31]. This at first sight surprising result is due to mutual neutralization of opposite charges at the surface of an ionic crystal. The neutralization is weakened upon stretching, which increases adhesion or cohesion. However, these relations give only tendencies. In general [32], the dependence of surface tension on surface strain is complicated and depends on a possible surface reconstruction.

III. EXPERIMENTAL TECHNIQUES TO MEASURE THE SURFACE TENSION AND THE SURFACE STRESS

A. Zero Creep Method

One of the oldest and theoretically most secure methods for obtaining surface energies of metal surfaces is the zero creep method. Thin metal foils and wires shrink when maintained at a constant high temperature because of the influence of surface tension. The method depends on finding the tensile load necessary to exactly counterbalance the effect of shrinkage, e.g., to produce a condition of zero creep. Therefore the length of a foil or wire is measured versus time for various loads. The first zero creep experiments were done in 1910 by Chapman and Porter, who measured the contraction of a gold foil at temperatures above 340°C [33].

A complication was observed by Udin et al.: when one heats copper wires to temperatures near the melting point, the wire developed a grainy structure [34,35]. After considering these grain boundary effects, the exact counterbalancing force, divided by the length along which it acts, can be equated to the reversible work required to form a unit area of surface [36]. In a related method, Maiya and Blakely observed the relaxation to flatness of a sinusoidal nickel surface [37]. The parameter measured is the surface stress. Because the shrinkage effect manifests itself only when the sample is heated to about 90% of its melting point, the strains are mainly plastic. Hence, close to the melting point the difference between γ and Υ becomes small.

Examples of applications are given in Table 1. The method is limited to materials that flow plastically at a reasonable time scale of minutes to hours and at temperatures significantly lower than their sharply defined melting points (so that they can still be treated as solids). This includes mainly metals.

A related technique is the multiple scratch method. A parallel set of scratches is made on a flat solid surface. When annealed at a temperature close to the melting point, the surface topography changes to a smooth sinusoidal profile. This is driven by the surface stress. The rate of the decay from the original sawtooth to the sinusoidal profile is measured, and the surface stress can be

TABLE 1 Surface Stress (Expressed in 10^{-3} N/m) Measured with the Zero Creep Method

Material	Conditions	Υ	Reference
Gold	1033°C	1450 ± 80	175
	1020°C	1400	176
	920°C	1680	176
Fe in δ-phase	1410–1450°C, UHV	2100–2400	177, 178
	1400–1535°C, in Ar	1950 ± 200	179
Fe in δ-phase with 0.1% phosphorus	1450°C, UHV	1600	177
Fe in δ-phase with adsorbed O_2	1410°C, 10^{-2} Pa O_2	1500	178
Cu	950°C, in He of H_2	1770	180
Cu with 0.26% Sn	950°C, in He or H_2	980	180
Ni(100)	1219°C, HV	1820 ± 180	37
Ni(110)	1219°C, HV	1900 ± 190	37
Sn	215°C, vacuum	685	181
Paraffin	29.5°C	65	181

calculated [38]. In this way Milles and Leak determined the surface stress of iron-3% silicon; in the temperature range of 1075°C to 1380°C its surface stress is 2000–2800 mN/m [39].

B. Change in Lattice Constant

In small crystals the surface tension significantly compresses the density of a solid due to the high surface-to-volume ratio. This surface tensional compression reduces the lattice constant. Such changes in lattice constant have mainly been observed by electron diffraction because, in contrast to x-rays, electrons only penetrate a few nanometers into solids. Therefore they are sensitive enough to give good diffraction patterns even if only one layer of scattering particles is used as a sample. Boswell [40] observed the reduction of the lattice constant of several alkali halide crystals with a size less than 10 nm and for gold crystals with a size less than 4 nm. He attributed this effect to the surface tension. This was confirmed for NaCl, KCl, CsCl, and aluminum by Rymer and Wright [41].

In these experiments small particles are normally made by evaporating the substance onto carbon or another support usually used in electron microscopy. Many substances form small, almost spherical particles rather than a thin homogeneous film. In the imaging mode of the electron microscope the mean radius of the particles is measured. Then the lattice constant is determined in the diffraction mode. Particles of different radii are made by changing the evap-

Surface Tension and Surface Stress of Solids

TABLE 2 Surface Stresses (in 10^{-3} N/m) as Determined by the Change in Lattice Constant of Small Particles

Material	Conditions	Υ	Reference
TaCl		5960	42
Gold	50°C	1150 ± 200	182, 183
	985°C	410	182
		3830	44
Silver	55°C	1415 ± 200	43
Pt	65°C	2570 ± 400	184
Pd	25°C	3150 ± 150	185

oration conditions. To keep contamination as low as possible, evaporation is done directly in the electron microscope.

De Planta et al. did the first quantitative analysis [42]. They derived the equation

$$\frac{\Delta d_o}{d_o} = \frac{2}{3} \cdot \Upsilon \cdot \frac{\chi}{R} \tag{11}$$

to obtain the surface stress Υ from the relative decrease of the lattice constant d_o, the coefficient of compressibility χ, and the radius of the particles R.

The surface energies of several materials have been determined by measuring the change of the lattice constant (Table 2). One problem of the technique lies in the preparation of the sample. Only a limited number of substances can be prepared as small spherical particles with a defined radius on a carbon support. Often the particles are not spherical, which limits the applicability of the above equation. The surface stress can only be determined for the solid/vacuum interface, not in gas or liquids. In addition, the interpretation of diffraction effects from small particles becomes increasingly difficult with diminishing particle size [43,44].

C. Cleavage Experiments

In cleavage experiments a solid block is split and the work required is measured (Fig. 2). Splitting of a solid creates two new surfaces. Hence, if the process is reversible, the work required to create the two new surfaces should be $W = 2\gamma\Omega$. Ω is the cleaved area. In addition, elastic work is needed to bend the cleaved parts of the solid away from each other. This elastic work depends on the specific geometry of the solid and on the specific method of cleavage applied. Usually this elastic work can be calculated. An equation is obtained, which relates the force F applied for splitting and the crack propagation length L with the Young

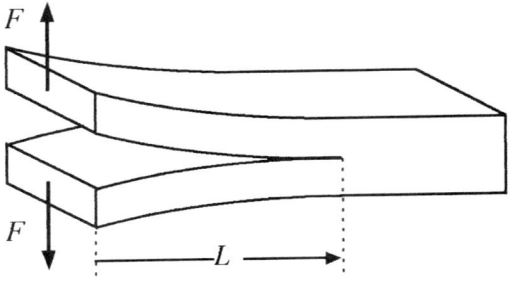

FIG. 2 Scheme of a typical cleavage experiment.

modulus and the surface tension [45,46]. Hence from measuring the force and the crack length the surface tension can be calculated. Practically cleavage is usually initiated by spark machining or by cutting with a blade. Once a crack is initiated it is propagated by pulling on either side with a tensile testing machine or by driving a wedge into the crack.

The first cleavage experiments to obtain surface energies were done by Obreimoff with mica [47]. Mica is a naturally occurring mineral consisting of aluminosilicate layers held together by ionic bonds to intercalated potassium. Due to its layered composition mica can be cleaved to atomically flat surfaces with a defined structure. Cleaving mica in vacuum yielded high surface energies of several J/m^2 [47]. In dry air the value is drastically reduced to roughly 1 J/m^2 [48]. The surface tension decreases with increasing relative humidity but does not appear to be affected by inert gases such as argon or nitrogen [49,50]. In liquid water the surface tension is further reduced to 0.107 J/m^2 [49].

With the cleavage technique, surface tensions of several materials have been measured, including salts [51], metals [52,53], and semiconductors (Table 3) [54].

Although the cleavage technique is conceptually simple, there are severe problems:

1. The chief problem is that plastic deformation effects at the crack tip usually obscure the process of surface formation. The work needed to propagate a crack is utilized not only to form two surfaces and to bend the cleaved material elastically but also for plastic deformation around the crack tip [16]. The energy needed for this plastic deformation is dissipated as heat. Hence the measured quantity, which is called the effective fracture surface tension γ_{EFES}, is numerically larger than γ. Reliable results can only be obtained at temperatures below ≈ 150 K when plasticity practically disappears [21,55].

2. Freshly cleaved surfaces often reconstruct. This surface reconstruction can drastically reduce the surface tension. Again, heat is dissipated.

TABLE 3 Effective Fracture Surface Energies (in 10^{-3} N/m) Determined by Cleavage Experiments. Below $\approx -15°C$ the Effective Fracture Surface Energy is Close to γ

Material	Conditions	γ_{EFES}	Reference
Ge(100)	$-196°C$, in N_2 gas	1835	54
Ge(110)	$-196°C$, in N_2 gas	1300	54
Si(100)	$-196°C$, in N_2 gas	2130	54
Si(110)	$-196°C$, in N_2 gas	1510	54
Si(111)	$-196°C$, in liq. N_2	1240	52
Tungsten(010)	$-196°C$	6300	53
Tungsten(100)	$-196°C$	1700	186
Zn(0001)	$-196°C$, in liq. N_2	105	52
KCl(100)	$25°C$	110	51
LiF(100)	$-196°C$, in liq. N_2	340	52
MgO(100)	$-196°C$, in liq. N_2	1200	52
	$20°C$	1150	187
CaF_2(111)	$-196°C$, in liq. N_2	450	52
BaF_2(111)	$-196°C$, in liq. N_2	280	52
$CaCO_3(10\bar{1}0)$	$-196°C$, in liq. N_2	230	52

3. Another problem was pointed out by Yudin and Hughes [56]. They explicitly considered that the interaction between the cleaved surfaces does not immediately fall to zero after cleavage but that, as long as the surfaces are only several 10 nm apart, surfaces forces are still active. This leads to a severe correction.

4. Finally, cleavage measurements cannot be done with all materials, since often brittle fracture occurs. In that case, once a flaw is initiated, a crack propagates quickly, often accelerating catastrophically under a small driving force. Examples are the fracture of glass or ceramic, displaying the characteristic sudden failure, low strength, unreliability, invisible defects, sensitivity to manufacturing conditions, and ease of damage by impact or by environmental agents [57].

It should also be pointed out that cleavage experiments were originally done to understand the fracture of solids under stress [58–60]. The fracture process is still far from being understood [61].

D. Adhesion Measurements

Closely related to cleavage experiments are adhesion measurements. Adhesion measurements have been done with mica in the so-called surface force apparatus (SFA) [62–64]. In the surface force apparatus two crossed mica cylinders of

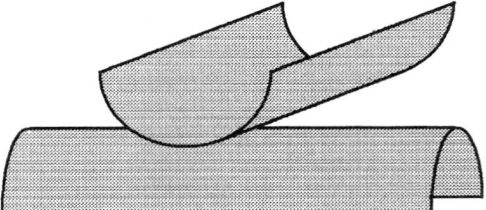

FIG. 3 Scheme of the interacting surfaces in the surface force apparatus.

typically 1 cm diameter are brought into contact, and the force necessary to separate them again is measured (Fig. 3).

When attempting to relate the adhesion force obtained with the SFA to surface energies measured by cleavage, several problems occur. First, in cleavage experiments the two split layers have a precisely defined orientation with respect to each other. In the SFA the orientation is arbitrary. Second, surface deformations become important. The reason is that the surfaces attract each other, deform, and adhere in order to reduce the total surface tension. This is opposed by the stiffness of the material. The net effect is always a finite contact area. Depending on the elasticity and geometry this effect can be described by the JKR [65] or the DMT [66] model. Theoretically, the pull-off force F between two ideally elastic cylinders is related to the surface tension of the solid and the radius of curvature by

$$F = n\pi R\gamma \tag{12}$$

with $n = 3-4$. The precise value of the prefactor n is still under debate [67,68]. With the SFA the surface tension of mismatched mica was measured to be 130–170 mN/m in dry air.

The main limit of the SFA is its restriction to mica. It is restricted to mica because atomically smooth surfaces are required over the contact area. Otherwise the interaction is determined by protrusions that would reduce the effective contact area. Typical contact areas in the SFA are several 100 μm^2. Only mica can be cleaved to obtain atomically flat homogeneous surfaces over such large areas.

The restriction to mica was overcome by a relatively recent technique: the atomic force microscope (AFM), sometimes also called the scanning force microscope [69]. AFMs are usually used to image solid surfaces. Therefore a sharp tip at the free end of a cantilever spring is scanned over a surface. Tip and cantilever are microfabricated. While scanning, surface features move the tip up and down and thus deflect the cantilever. By measuring the deflection of the cantilever, a topographic image of the surface can be obtained.

Surface Tension and Surface Stress of Solids 13

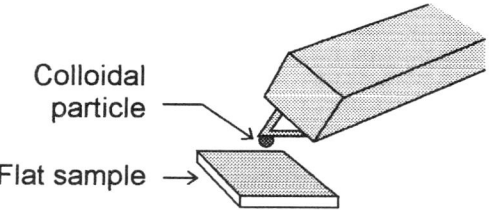

FIG. 4 Scheme of an atomic force microscope with a particle attached to the free end of the cantilever spring.

For an adhesion measurement a small spherical particle with a typical diameter of 5 μm is attached to the free end of the cantilever spring (Fig. 4). Then the particle is moved onto the flat sample and allowed to adhere. Finally the force to pull the particle off the sample is measured. From the pull-off force and the particle radius the surface tension can be calculated using Eq. (12) (Refs. 70 and 71 for introduction and review). The advantage of the AFM is that the contact area is reduced to typical values of a few 100 nm^2. As a consequence many more materials can be used. The main limits in obtaining quantitative results are the availability of spherical particles, the still present roughness of the interacting surfaces, and the difficulty of measuring the spring constant of the cantilever.

In some AFM adhesion measurements, integrated silicon or silicon nitride sharp tips are used instead of spherical particles. With integrated tips, the lateral distribution of the surface tension of a sample can be mapped or even imaged with a resolution of roughly 10 nm [72–75]. It is, however, difficult to obtain quantitative results because the size and shape of the tip in the 1–10 nm regime is unknown and practically impossible to determine [76–78].

A technique devised by Kendall et al. might be viewed as an extension of adhesion measurements [79]. They evaluated solid surface energies experimentally by measuring the elastic modulus of submicrometer powder assemblies and comparing it with the bulk elastic modulus. The idea is the following: neighboring particles not only have a point contact but also, as in the SFA, have a certain contact area. This contact area depends on the surface tension. A high surface tension leads to a large contact area. The larger the contact area, the closer the elastic modulus of the powder comes to that of the bulk. Knowing the size of the monodisperse particles and the volume fraction, a measurement of the difference in elastic modulus between bulk and powder gives the surface tension.

In adhesion measurements one is confronted with some of the problems of cleavage experiments. Deformation occurs close to the contact line, although in this case deformation is, to a large extent, elastic. In addition, the range of surface forces needs to be considered.

E. Adsorption Measurements

Adsorption measurements can be used to measure the change of the surface tension upon adsorption of a substance. The conceptually simplest method in this category is inverse chromatography. In standard chromatography a certain substance is characterized by its adsorption behavior to a solid matrix. The emphasis is usually on determining the properties of the solute or gas. Inverse chromatography, on the contrary, is a method of studying the properties of the solid, stationary phase. The interaction of the stationary phase and some well-known probe gases or solutes is determined by using the same procedure as in direct chromatography, but the results are used to derive properties of the stationary phase. An introduction to inverse gas chromatography is Ref. 80. Theory and the relation to immersion experiments is presented by Schröder [81] and Melrose [82].

Adsorption is described by the Gibbs–Duhem equation. Considering only a single vapor at constant temperature, electric potential, and elastic strain, the change in surface tension is given by

$$d\gamma = -\Gamma \cdot d\mu \tag{13}$$

Γ is the surface excess of the adsorbed vapor molecules, and μ is their chemical potential. For an ideal gas at pressure p the chemical potential is

$$\mu = \mu_0 + RT \cdot \ln p \tag{14}$$

In equilibrium the adsorbed vapor has the same chemical potential as in the gas phase. Inserting Eq. (14) into Eq. (13) yields the Gibbs adsorption isotherm

$$d\gamma = -\Gamma RT \cdot d\ln p \tag{15}$$

Integration leads to

$$\gamma_S - \gamma_{SV} = \pi = -\int_0^p d\gamma = RT \cdot \int_0^p \Gamma \cdot d\ln p' \tag{16}$$

π is the difference between the surface tension of the bare solid, γ_S, and the solid with adsorbed vapor, γ_{SV}. It can be interpreted as the spreading pressure. The surface excess is related to the number of moles adsorbed n and the specific surface area of the material Σ by $\Gamma = n/\Sigma$. It follows that

$$\pi = \gamma_S - \gamma_{SV} = \frac{RT}{\Sigma} \int_0^p n \cdot d\ln p' \tag{17}$$

From measuring adsorption isotherms one obtains $n(p)$, and the integral can be solved. The specific surface area is determined by measuring the adsorption isotherm $n(p)$ with a reference gas (e.g., nitrogen), where the area occupied by an adsorbed molecule is known. Therefore the adsorption isotherms are fitted

TABLE 4 Spreading Pressure (in 10^{-3} N/m) of Adsorbed Films on Solids Measured by Adsorption Experiments

	$\gamma_S - \gamma_{SV}$	Reference
Hydrocarbons adsorbing to cellulose and wood fibers	20–27	80
Water at saturating vapor pressure adsorbing to quartz at 25°C	103	188
Hydrocarbons adsorbing to silica	65–175[a]	189
	76	190
Water to α-Al$_2$O$_3$	2640	87
Water to γ-Al$_2$O$_3$	1670	87
Hydrocarbons adsorbing to water-coated silica	25–45	191

[a] Values range for various samples, which were heated to temperatures between 0 and 700°C.

with an appropriate model for the adsorption process, usually the BET model (Ref. 3, pp. 560–570 and 609–613; Ref. 83). With the knowledge of the specific surface area, the difference of the surface tension between a pure solid surface γ_S and the solid–gas interface γ_{SV} can be measured. Examples of applications are given in Table 4. Adsorption experiments can also be done in solution. In that case the change of γ upon adsorption of a solute is determined [84–86].

A limit of adsorption experiments is that only changes of the surface tension upon adsorption of a substance can be measured. In many practical applications this is not a severe limitation because $\gamma_S \gg \gamma_{SV}$. One example is alumina. For γ-Al$_2$O$_3$, McHale et al. measured a value of 1.67 J/m for $\gamma_S - \gamma_{SV}$ [87]. This is close to the surface tension of bare γ-Al$_2$O$_3$, so that 1.67 J/m $\approx \gamma_S$.

One way to obtain γ_S from the measured difference $\gamma_S - \gamma_{SV}$ is described by Weiler et al., who determined the surface tension of finely divided sodium chloride [88]. They measured $\gamma_S - \gamma_{SV}$ for ethanol and benzene up to saturating vapor pressure. Then they assumed that at saturating vapor pressure γ_{SV} is similar to the solid–liquid surface tension γ_{SL}. At saturating vapor pressure the surface of a solid is covered with a multilayer of the adsorbed vapor. Consequently the surface tension should not be significantly different from the solid–liquid interface. Taking $\gamma_{SL} = 171$ mN/m from solubility experiments [204] and measuring $\gamma_S - \gamma_{SV} = 56$ mN/m for ethanol, they obtained $\gamma_S = 227$ mN/m.

To deal with measurable amounts of adsorbed material, only solids with a high specific surface area can be used. In addition, all problems related to determining the specific surface area are present. A general problem is the inhomogeneity of the material. Different crystalline surfaces, edges, dislocations, or contaminated areas are exposed and contribute to the adsorption. Methods have been proposed to determine the energy distribution function, which results from surface heterogeneity, instead of a single surface tension [89,90].

F. Calorimetric Methods

Several calorimetric methods exist to obtain information about the energetics of solid surfaces. One is to measure the heat of immersion (Fig. 5, left) [91]. If a clean solid is immersed in a liquid, there is generally a liberation of heat. This heat is given by the change of enthalpy if the process is done at constant pressure. Since almost no volume changes occur, this equals the change in internal energy. At constant chemical potentials and negligible electric potential, the heat of immersion is derived from Eq. (2):

$$Q_{\text{imm}} = H_S^\sigma - H_{SL}^\sigma = \Delta H^\sigma = \Delta U^\sigma = \Omega \cdot (T \cdot \Delta s^\sigma + \Delta \gamma) \tag{18}$$

H_S^σ is the total enthalpy of the bare surface, and H_{SL}^σ is the total enthalpy of the solid–liquid interface. Since usually the total surface area does not change during immersion and consequently the surface stress remains constant, only the surface tension enters into the equation.

The surface entropy can be related to the surface tension. Therefore the Gibbs–Duhem equation is rearranged to

$$d\gamma = -s^\sigma dT - \sum \Gamma_i d\mu_i - \sigma d\varphi + (\Upsilon - \gamma) d\varepsilon_e \tag{19}$$

It follows that

$$s^\sigma = -\left. \frac{\partial \gamma}{\partial T} \right|_{\mu_i, E, \varepsilon_e} \tag{20}$$

At constant chemical potentials, electric potential, and elastic strain, the dependence of the surface tension on temperature is given by the surface entropy. Inserting into Eq. (19) yields [92,93]

$$\frac{Q_{\text{imm}}}{\Omega} = u_S^\sigma - u_{SL}^\sigma = \gamma_S - \gamma_{SL} - T \cdot \left(\frac{\partial \gamma_S}{\partial T} - \frac{\partial \gamma_{SL}}{\partial T} \right) \tag{21}$$

u_S and u_{SL} are the internal surface energies per unit area for the bare and the immersed surface, respectively. Hence from measuring the heat of immersion one can obtain information about the internal surface energy, but it is not possible to measure the surface tension directly.

The heat of immersion is determined calorimetrically by measuring the heat evolved on immersion of a clean solid in the liquid. In order to get detectable amounts of heat, only materials with high specific surface areas like fibers, powders, or porous materials can be used (Table 5).

The total surface area needs to be known to determine the change of internal surface energy from the heat of immersion. This is often done by adsorption measurements [94,83]. An alternative method was suggested by Harkins and Jura [92]. They proposed not to immerse a clean solid but to expose the solid first to the vapor of the liquid. If the liquid wets the solid at the saturating vapor

Surface Tension and Surface Stress of Solids

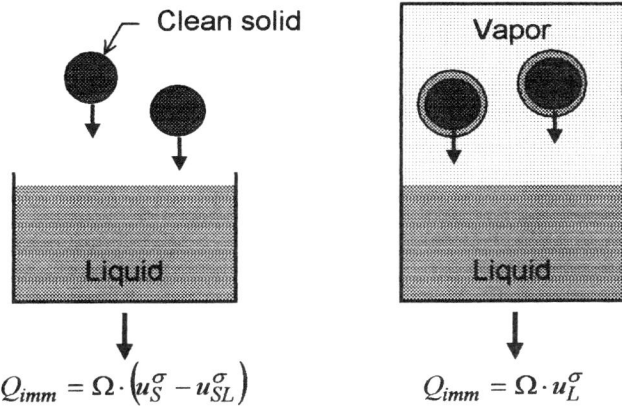

FIG. 5 Heat of immersion of clean solid particles and of a solid exposed to saturating vapor.

TABLE 5 Heats of Immersion (in 10^{-3} N/m) for Various Solids at 25°C

Material	$u_S^\sigma - u_{SL}^\sigma$	Reference
SiO$_2$ in water	600	192
	400–600[a]	193
SiO$_2$ in ethanol	520	192
TiO$_2$ in water	200–650[a]	194
	488	93
	510	92
Al$_2$O$_3$ in water	430	93
	630	92
NiO in water	780	195
NiO in heptane	110	195
ZnO in benzene	180–430[b]	196
ZnO in toluene	220–720	196
ZnO in chlorobenzene	190–550	196
Carbon in n-heptane	77	197
C (graphite) in water	80	198
C (graphite) in benzene	140	198

[a]There was a considerable variation of Q_{imm} depending on the origin of the sample, the particle size, and the outgassing conditions.
[b]Increasing with increasing degassing temperature (20–600°C) prior to immersion.

TABLE 6 Internal Surface Energies at 25°C (in 10^{-3} N/m) from Heats of Solution for Various Solids

Material	u_S^σ	Reference
NaCl in water	408	95
NaCl in water	276	199
CaO in water	1310	200
Ca(OH)$_2$ in water	1180	200
MgO in water	1040	201

pressure, the solid surface is covered with a thin layer of the liquid, and the surface tension becomes that of the liquid (Fig. 5, right). A layer two molecules thick seems to be enough to screen completely the solid surface [93]. Upon immersion, not the clean solid surface is replaced by a solid–liquid interface but a liquid surface, with roughly the same total area as that of the solid. Since the surface tension of liquids can easily be measured, the total surface area can be calculated from the heat of immersion: $\Omega = Q_{\text{imm}}/u_L$. u_L is the internal surface tension per unit area of the liquid.

Measuring heat of immersion is technically demanding. The amounts of heat are relatively small, and the surface area needs to be determined. In addition, results cannot directly be related to one of the surface parameters γ^S, γ, or Υ. Moreover, Eq. (20) is based on the assumption that the surface strain does not change upon immersion.

A variation of the technique is to measure the internal surface energies from the heat of solution. When the solid interface is destroyed, as by dissolving, the internal surface energy appears as an extra heat of solution. With accurate calorimetric experiments it is possible to measure the small difference between the heat of solution of coarse and of finely crystalline material (Table 6). The calorimetric measurements need to be done with high precision, since there is only a few joules' per mole difference between the heats of solution of coarse and those of finely divided material. A typical example is NaCl. For large crystals the heat of solution in water is 4046 J/mol. Lipsett et al. [95] measured with finely divided NaCl (specific surface area of 125 m^2/mol) a heat of solution that was 51 J/mol smaller.

Another method to determine the surface enthalpy and entropy and the internal surface energy was devised by Jura and Garland [96]. They measured the heat capacity for a powder versus temperature. The surface enthalpy per unit area is

$$h_S^\sigma = h_S^{\sigma\,298} + \frac{1}{\Omega} \cdot \int_{298K}^{T} \Delta C_p \cdot dT \tag{22}$$

ΔC_p is the difference of the heat capacity between bulk and powder. The surface entropy of the solid is

$$s_S^\sigma = \frac{1}{\Omega} \cdot \int_0^T \frac{\Delta C_p}{T} dT \tag{23}$$

assuming that the entropy of a small crystal approaches zero at 0 K. From this and with Eq. (4) the surface tension can be calculated as

$$\gamma = h_S^\sigma - T s_S^\sigma \tag{24}$$

To obtain ΔC_p they subtracted their measured values from heat capacities of the bulk. The surface enthalpy at 298 K $h_S^{\sigma\,298}$ was taken from solubility experiments. For MgO and using $h_S^{\sigma\,298} = 1040$ mN/m they obtained a surface tension of 957 mN/m and a surface entropy of 0.28 mNJm^{-1}K^{-1}.

G. Solubility Changes

The surface stress of some solids in a liquid might be determined by measuring solubility changes of small particles [97,98]. As small liquid drops have an increased vapor pressure in gas, small crystals show a higher solubility than larger ones. The reason is that, due to the curvature of the particles' surface, the Laplace pressure increases the chemical potential of the molecules inside the particle. This is described by the Kelvin equation, which can be written (Ref. 3, p. 380)

$$RT \cdot \ln \frac{a}{a_0} = \frac{2\Upsilon V_m}{r} \tag{25}$$

a is the activity in units a_0 and V_m is the molar volume of the dissolved material in the liquid. r denotes the inscribed radius of the solid particles. For details about the applicability of the Kelvin equation to solids see Ref. 99.

Most studies of solubility effects have been made with salts [100]. In the case of a sparingly soluble salt that dissociates into v^- anions A^- and v^+ cations B^+, the solubility S is given by $S = [B^+]/v^+ = [A^-]/v^-$. With the activity of the solute

$$a = [A^-]^{v^-} \cdot [B^+]^{v^+} = S^{v^-+v^+} \cdot (v^-)^{v^-} \cdot (v^+)^{v^+} \tag{26}$$

it follows that

$$RT \cdot (v^+ + v^-) \cdot \ln \frac{S}{S_0} = \frac{2\Upsilon V_m}{r} \tag{27}$$

where S_0 is the saturation solubility of large crystals. Hence by measuring the increase in solubility, the surface stress can be calculated. Some results are given in Table 7.

TABLE 7 Surface Stresses (in 10^{-3} N/m) from Solubility Changes

Substance	Conditions	Υ	Reference
$CaSO_4 \cdot 2H_2O$/water	25°C	370	202
CaF_2/water	30°C	2500	100
Silica/water	pH 2–8, 25°C	46	203
NaCl/ethanol	25°C	171	204
$SrSO_4$/water	25°C	84	205

Determining the surface stress of solids by solubility changes is certainly not a universal method. It is limited to materials that can be prepared as small particles with a relatively homogeneous size distribution and that only sparingly dissolve in the liquid. A major problem is to measure the correct size of the particles, and it should be kept in mind that solubility is largely determined by the smaller particles of the distribution [101]. In addition, instead of using the inscribed radius of the particles, their specific geometry has to be taken into account [102].

A somewhat related technique to measure the surface tension between a solid and its melt is described by Still and Skapski [103]. A thin wedge is filled with the liquid substance and then allowed to solidify. Slowly the temperature is raised. The solid substance starts to melt from the thin part of the wedge, even below its melting temperature (Fig. 6). As in solubility changes, this is caused by the Laplace pressure, which increases the chemical potential in the solid phase. The reduction in melting temperature is given by

$$\Delta T = \frac{T_m C \Upsilon_{SL}}{\rho_S Q_m} \tag{28}$$

T_m is the melting temperature, Q_m is the heat of fusion, and ρ_S denotes the density of the solid. C is the inverse effective radius of curvature of the solid–liquid interface. It is $2/r$ for a sphere and $1/r$ for the wedge.

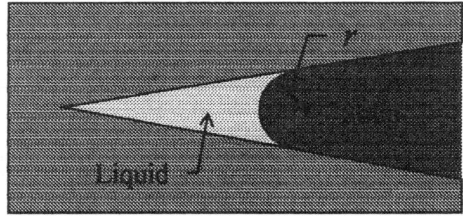

FIG. 6 Typical configuration of a solubility experiment.

With their method, Skapski et al. [104] measured the surface stress of the ice–water interface to be 120 mN/m. This agrees with a more recent result of Hansen et al., who report a value of 130 mN/m for the ice–water interface [105]. By using a combination of NMR and calorimetry to detect the melting, they extended the method to porous solids instead of a wedge. A surface stress around 120 mN/m is, however, surprisingly high. Earlier experimental and theoretical results were in the range of 10–35 mJ/m [106].

H. Contact Angle Measurements

Surface tensions of low-energy surfaces like many polymers are often determined from contact angle measurements. A review of the method and its application to polymer science was written by Koberstein [107]. In equilibrium, the contact angle of a liquid drop on a solid surface is given by the Young equation:

$$\gamma_L \cdot \cos\theta = \gamma_{SV} - \gamma_{SL} \tag{29}$$

The Young equation contains the surface tension of the liquid γ_L, which can easily be measured, and the difference of the surface tensions of the solid–vapor γ_{SV} and the solid–liquid interface γ_{SL}. That the surface tension enters the Young equation is not beyond doubt. Linford [16] inserted the generalized intensive surface parameter γ^S, arguing that at the three-phase contact line elastic deformations take place. In accordance with Rusanov [17] we use the surface tension, because the spreading of a liquid on a surface is a process similar to immersion or adsorption. Immersion is usually considered to effect the surface tension since no extension or contraction of the surface occurs.

In order to obtain the individual surface tensions and not only the difference $\gamma_{SV} - \gamma_{SL}$, an additional relation between γ_L, γ_{SV}, and γ_{SL} is required. To obtain such a relation it is appropriate to write the interfacial energy as

$$\gamma_{SL} = \gamma_L + \gamma_{SV} - W_a \tag{30}$$

W_a is the work of adhesion per unit area. γ_L is determined by the intermolecular force between liquid molecules. γ_{SV} is mainly caused by forces between solid molecules. The work of adhesion comes from the interaction between solid and liquid molecules. Hence the task of finding a relation $W_a = W_a(\gamma_L, \gamma_{SV})$ is somehow similar to that of determining the force between a liquid and a solid molecule from knowledge of the force between two solid and two liquid molecules. In a strict sense this is impossible, and one can only hope for empirically verified and theoretically sound approximations.

One such approximation was proposed by Good and Girifalco assuming that mainly van der Waals forces act between all molecules [108]. Based on the theory of van der Waals forces between macroscopic bodies they suggested

$$W_a = 2\phi\sqrt{\gamma_{SV}\gamma_L} \tag{31}$$

which leads to

$$\gamma_{SL} = \gamma_{SV} + \gamma_L - 2\phi\sqrt{\gamma_{SV}\gamma_L} \tag{32}$$

ϕ is an empirical interaction parameter that varies between 0.32 and 1.17 [109]. Inserting this into the Young equation one obtains

$$\gamma_L \cdot \cos\theta = -\gamma_L + 2\phi\sqrt{\gamma_{SV}\gamma_L} \tag{33}$$

Rearranging leads to

$$\cos\theta = 2\sqrt{\phi\frac{\gamma_{SV}}{\gamma_L}} - 1 \tag{34}$$

Hence a plot of $\cos\theta$ versus $1/\sqrt{\gamma_L}$ should yield a straight line with an inclination given by the surface tension of the solid. Good and Girifalco used several liquids to verify this behavior. As an example they determined the surface tension of polytetrafluoroethylene and octadecylamine monolayers to be roughly 28 and 30 mN/m, respectively.

Good and Girifalco assumed that the molecules in the liquid and in the solid interact only via dispersive van der Waals forces. One would expect that this is only a reasonable assumption for hydrocarbons or fluorocarbons. Fowkes, however, pointed out that even if one of the media has a polar nature and its molecules are held together by electrostatic forces and hydrogen bonding, the interaction between solid and liquid can still be of van der Waals type [110]. This is the case if the other medium is a hydrocarbon or fluorocarbon [111]. Assuming that the contributions of all intermolecular forces to the surface tensions are additive and

$$\gamma_{SV} = \gamma_{SV}^n + \gamma_{SV}^d \quad \text{and} \quad \gamma_L = \gamma_L^n + \gamma_L^d$$

he suggested that we use the equation

$$W_a = 2\sqrt{\gamma_{SV}^d \gamma_L^d} \tag{35}$$

γ_L^d and γ_{SV}^d are the contributions of the dispersive van der Waals forces. γ_L^n and γ_{SL}^n contain all nondispersive interactions, i.e., polar forces and hydrogen bonding. Since only dispersive van der Waals forces are allowed between the two media, only the dispersive contributions to the surface energies appear in Eq. (35). This leads to

$$\gamma_{SL} = \gamma_{SV} + \gamma_L - 2\sqrt{\gamma_{SV}^d \gamma_L^d} \tag{36}$$

TABLE 8 Surface Tension Components (in 10^{-3} N/m) of Various Test Liquids at 20°C

Liquid	γ_L	γ_L^d	γ_L^n	Reference
Water	72.8	21.8	51.0	110
Glycerol	63.3	20.2	43.1	122
Formamide	58.4	19.8	38.6	122
Diiodomethane	50.8	50.4	0.4	117
Ethylene glycol	48.2	18.9	29.3	122
Dimethylsulfoxide	44	35	9	114

Owens and Wendt proposed an even more general equation for the work of adhesion between two arbitrary media [112]. In analogy to dispersion forces they used the geometric mean also for the other force components:

$$\gamma_{SL} = \gamma_{SV} + \gamma_L - 2\sqrt{\gamma_{SV}^d \gamma_L^d} - 2\sqrt{\gamma_{SV}^n \gamma_L^n} \tag{37}$$

For two media with relatively low energy surfaces often the alternative expression

$$\gamma_{SL} = \gamma_{SV} + \gamma_L - \frac{4\gamma_{SV}^d \gamma_L^d}{\gamma_{SV}^d + \gamma_L^d} - \frac{4\gamma_{SV}^n \gamma_L^n}{\gamma_{SV}^n + \gamma_L^n} \tag{38}$$

is applied. A discussion of the different components contributing to the surface tension is given in Refs. 113–118.

In order to obtain γ_{SL} from γ_{SV} and γ_L a knowledge of the dispersive and nondispersive components is necessary. For various test liquids these components have been determined in the following way:

1. First, it is assumed that for liquid hydrocarbons only dispersive forces contribute to the surface tension. Hence from a surface tension measurement of the liquid hydrocarbon, γ_h, its dispersive component is obtained, since $\gamma_h = \gamma_h^d$.

2. Second, the interfacial tensions of various polar test liquids with liquid hydrocarbons are determined. This is only possible with liquids that are immiscible with liquid hydrocarbons. It is assumed that the interaction between the polar liquid and the liquid hydrocarbon is only due to dispersive forces, and hence the equation $\gamma_{L,h} = \gamma_h + \gamma_L - 2\sqrt{\gamma_h^d \cdot \gamma_L^d}$ is valid. $\gamma_{L,h}$ is the interfacial tension between the fluid hydrocarbon and the polar test liquid. Like the surface tension of the test liquid, $\gamma_{L,h}$ can be precisely measured. Once $\gamma_{L,h}$ and $\gamma_h = \gamma_h^d$ are known, γ_L^d can be calculated. The nondispersive contribution to the surface tension is simply $\gamma_L^n = \gamma_L - \gamma_L^d$. Dispersive and nondispersive surface tensions for various liquids are tabulated. Examples are given in Table 8. The

errors of the dispersive and polar components are in the order of 10%. The error arises because results of measurements done with different hydrocarbons, such as n-hexane, n-heptane, n-decane, and cyclohexane, were slightly different.

3. Third, contact angles of the solid with at least two different liquids of known γ_L^n and γ_L^d are measured. Then, combining the Young equation with Eq. (37) or (38), γ_{SV}^n, γ_{SV}^d, and consequently γ_{SV} can be calculated. If the geometric mean equation is used for the work of adhesion, the contact angle is given by

$$\frac{\gamma_L}{2} \cdot (1 + \cos\theta) = \sqrt{\gamma_{SV}^d \gamma_L^d} + \sqrt{\gamma_{SV}^n \gamma_L^n} \qquad (39)$$

Several problems of contact angle experiments limit the applicability and reliability of the method to measure surface tensions:

1. Usually a hysteresis between advancing and receding contact angles is observed.

2. Beside the Young equation the additional relation between γ_L, γ_{SV}, and γ_{SL} is necessary as discussed. The choice of this semiempirical relation is somehow arbitrary and influences the result.

3. It is sometimes difficult to find liquids that do not spread completely over a solid surface. In that case one might use two immiscible liquids and instead of measuring the contact angle at the solid–liquid–gas contact line, determine the contact angle at the solid–liquid A–liquid B contact line [119–121].

Owing to the last problem the contact angle method has mainly been applied to low-energy surfaces like polymers [112,122–126] or silicas grafted with alkyl chains [127,128]. Some results are summarized in Table 9. In addition, it is often used to determine the surface tension of materials with a high specific surface area like powders or porous substances [129–133]. In that case the rise of the test liquids in a capillary filled with the substance is measured.

I. Bending Plate Methods

Cahn and Hanneman [134] and Finn and Gatos [135] observed that thin annealed wafers of several III-V compounds, like GaAs, bend spontaneously. This effect can be attributed to a difference in surface stress of the two faces. The III-V compounds have, due to their crystal structure, different opposite faces. If, for instance, the surface stress of the top side is smaller than that of the bottom side, the wafer bends downwards. The degree of bending is related to the difference in surface stress $\Delta\Upsilon$ by Stoney's formula [136]:

$$\Delta\Upsilon = \frac{Ed^2}{6r(1-v)} \qquad (40)$$

TABLE 9 Surface Tension Components (in 10^{-3} N/m) of Various Solid Surfaces at 20°C Determined by Contact Angle Measurements

Solid	γ_{SV}	γ_{SV}^d	γ_{SV}^n	Reference
Paraffin	25.4	25.4	0.0	112
	25.1	25.1	0.0	122
Polyethylene	33.1	32.0	1.1	112
	32.8	32.1	0.7	122
	23–33	22–33	0–1	206
Polymethyl methacrylate	40.2	35.9	4.3	112
	44.9	39.0	5.9	122
	23–48	14–34	9–14	206
Polystyrene	42.0	41.4	0.6	112
	23–41	17–34	6–7	206
Polytetrafluoroethylene	14.0	12.5	1.5	112
	21.8	21.7	0.1	122
	16–29	15–28	1	206
	17			207
Polyvinyl chloride	41.5	40.0	1.5	112
Carbon fibers	35–38[a]	28–38	0–9	208
Mica	120	30	90	121

[a]Different carbon fibers were analyzed, which explains the spread of results.

E is the modulus of elasticity, d is the thickness of the plate, r is the radius of curvature, and v is the Poisson ratio (typically 0.3). For a discussion of Eq. (40) see Refs. 137 and 138. Spontaneous bending was used to calculate the difference in surface stress between opposite faces of several crystals, including InSb, GaAs, InAs, GaSb, and AlSb [134,139] and aluminum nitride [140].

Martinez et al. started to use the bending of a plate deliberately to determine changes in surface stress differences (Fig. 7) [5]. Therefore they deposited a monolayer of gallium atoms onto a silicon plate, and they measured the related surface stress change. In a similar way, Sander and Ibach determined the change in surface stress on silicon upon adsorption of oxygen [141] and of Ni(100) upon adsorption of sulphur, oxygen, and carbon [142]. Rao et al. determined changes in surface stress of silicon(111) after adsorption of lead [143]. Some results are listed in Table 10. These experiments gave insights into the molecular origins of adsorbate-induced surface stress [144].

Recently the bending plate technique was significantly improved by using microfabricated cantilevers made for atomic force microscopy [145]. Such cantilevers are typically 100 μm long, 20 μm wide, and 0.5 μm thick. Due to their

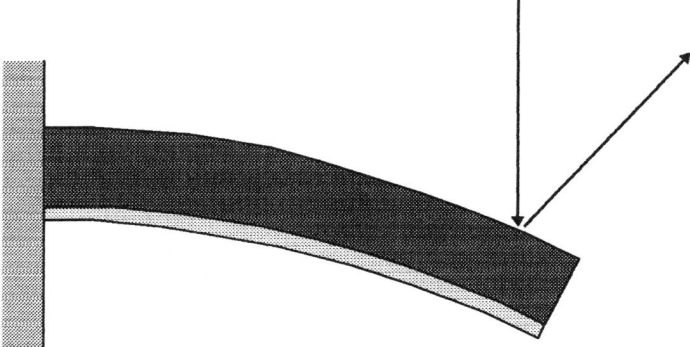

FIG. 7 Schematic of the bending plate technique. In this case a cantilever with different faces is used. The bending is often detected by measuring the angle of a light beam that is reflected off the cantilever.

small size and their high resonance frequency they are highly sensitive and at the same time relatively immune to external vibrations and turbulence. In this way changes in surface stress at the solid–liquid interface could be measured [146]. The unspecific binding of proteins to a hydrophobic surface in aqueous electrolyte was for instance detected.

With the bending plate technique no absolute values of the surface stress can be measured. Only changes in the difference between the surface stresses of the two faces are detectable. In the adsorption experiments done in UHV this was achieved by directing the beam of adsorbing molecules to one side only. In liquid the two sides of the cantilever need to be of different composition.

TABLE 10 Surface Stress Changes (in 10^{-3} N/m) Determined with the Bending Plate Method. ML Stands for Monolayer

Material	Conditions	$\Delta \Upsilon$	Reference
1 ML CO on Ni(100)	UHV	−1000	144
1 ML C on Ni(100)	UHV	−8500	142
1 ML O_2 on Si(111)	UHV	−7200	141
1 ML O_2 on Si(100)	UHV	+260	141
1 ML Ge on Si(001)	UHV	+800	209
Unspecific binding of protein to hydrophobic surface	Water	10	146

IV. ELECTROCAPILLARITY OF SOLID ELECTRODES

In electrocapillarity experiments the change of surface tension with changing surface potential is measured. With mercury such measurements were already done in the nineteenth century [147]. Mercury has the advantage that the surface tension can easily be measured, since it is a liquid. Such measurements yielded important information about the behavior of electrified surfaces. In particular, the relation between surface charge and surface potential or the adsorption of ions to metal surfaces was investigated by electrocapillary experiments.

The important quantity that can be derived is the surface charge density. At constant temperature and chemical potentials the surface charge density σ is simply given by the Lippmann equation:

$$\left.\frac{\partial \gamma}{\partial \varphi}\right|_{T,\mu_i} = -\sigma \tag{41}$$

At the surface of a solid electrode the molecules are not free to move and the interface is in general not in thermodynamic equilibrium with the bulk. In addition, any real solid electrode is at least partially polarizable. These two effects lead to an additional superficial charge σ_e, and the Lippmann equation becomes [26]

$$\left.\frac{\partial \gamma}{\partial \varphi}\right|_{T,\mu_i} = -(\sigma + \sigma_e) \tag{42}$$

The additional charge and the corresponding additional surface tension are time-dependent quantities in which the equilibrium between the bulk and the interface is not established. The irreversible contribution can be separated from the reversible by considering the time dependence, if the experimental time scale allows for such a test. Time-dependent effects can be observed by impedance measurements at different frequencies. For gold, as an example, impedance measurements showed spectra characteristic for equilibration processes at least over a time scale of 0.1 ms to 100 s. Gold also shows a surface reconstruction depending on the potential [148]. Fortunately, the variation of the interfacial strain with potential is usually so small that the original Lippmann equation (41) for a solid is practically the same as for a liquid electrode [149].

Electrocapillary experiments with solid electrodes have been done with a wide variety of techniques. Some of these techniques are unique for electrocapillary; others are already described. All methods applied measure the surface stress. A review of early experiments is Ref. 150.

A technique unique for electrocapillarity measurements is the extensometer [151,152]. In the extensometer the change in length of a ribbon or wire due to changes in surface stress is measured. The ribbon is mounted axially into a glass tube and extended by a weak spring. The glass tube is filled with the electrolyte of interest. Via a counterelectrode a potential is applied and the change in length is measured versus the potential. Lin and Beck used this technique to study the adsorption of various anions to gold [153,154]. A disadvantage is that the method is seriously affected by temperature fluctuations, since the glass tube and the metal electrode have different expansion coefficients.

The contact angle of an electrolyte on a solid electrode changes when changing the potential. This effect was used by Morcos and Fischer, who measured the potential-dependent capillary rise of an electrolyte meniscus at the surface of a partially immersed metal plate [155]. The interfacial energy at a solid electrode/solution interface can also be measured by the Wilhelmy plate method [156,157].

Electrocapillary curves of various metals under different conditions have been determined with the bending plate technique [158,159]. The deflection of the plate is usually measured by an optical lever. Pangarov and Kolarov used an interferometric method to detect the bending of the plate [160]. Recently the method was improved by using microfabricated cantilevers [161–163]. Bard et al. glued flat metal [164,165] or semiconductor [166] electrodes onto piezoelectric disks. When applying a potential to the electrode the resultant bending caused a potential on the piezo, which was detected.

Heusler et al. observed that the resonance frequency of a quartz oscillator changes when one changes the surface stress [167,168]. They used this effect for electrocapillary measurements by coating the quartz plate with the metal and detecting changes in surface stress by measuring the accompanying change in resonance frequency.

In a somewhat related experiment, surface stress changes due to an applied electric potential were determined by measuring the bending of a circular plate. This bending was measured interferometrically [169–172] or with a STM [173,174].

V. SUMMARY

There is no simple, reliable, and universal method of measuring the surface tension or the surface stress of a solid. Depending on the material, different techniques are applied. The cleavage technique is important for measuring the surface tension of metals and salts. Close to the melting point, the surface stress of metals can be measured with the zero-creep method. With contact angle measurements the surface tension of polymers and other low-energy surfaces is determined. For porous materials or powders, adsorption experiments are suitable for measuring the surface tension. Adhesion measurements with the atomic force microscope reveal the lateral distribution of the surface tension at a high resolution. Changes in surface stress can be detected with high sensitivity

TABLE 11 Summary of Methods for Measuring the Surface Stress Υ, the Surface Tension γ, the Internal Surface Energies of the Solid Surface u_S^σ or of the Solid–Liquid Interface u_{SL}^σ, and Changes in Surface Stress $\Delta\Upsilon$

Method	Quantity measured	Application	Main problem
Zero creep	$\Upsilon (\approx \gamma)$	Metals	Limited to temperatures close to the melting point
Reduction of lattice constant for small particles	Υ	Salts, metals	Production of small enough particles with a defined size is necessary
Cleavage	γ	Salts, semiconductors, mica	Plastic deformation at cleavage line above a temperature of 150 K
Adhesion	γ	Mica	
Inverse chromatography	$\gamma_S - \gamma_{SL}$	Porous material or powders	The specific surface area needs to be known
			Only concerns the difference between the bare and gas exposed surface
Heat of immersion	$u_S^\sigma - u_{SL}^\sigma$	Porous material or powders	The specific surface area needs to be known
			Not directly related to γ or Υ
Heat of solution	u_S^σ	Powders	The specific surface area needs to be known
			Not directly related to γ or Υ
Solubility changes	Υ	Sparingly soluble salts in the form of powder	The particle size needs to be known and the size distribution must be narrow
Contact angle	γ	Polymers, fluoro or carbonhydrates	Restricted to low energy surfaces
			Conceptual problems since an additional relation between γ_L, γ_{SV}, and γ_{SL} needs to be assumed
Bending plate method	$\Delta\Upsilon$		Only changes in surface stress differences can be measured

with the bending plate method. All methods depend to a certain degree on special assumptions or on theoretical models. A summary of the techniques is given in Table 11.

REFERENCES

1. J. N. Israelachvili, *Intermolecular and Surface Forces*, Academic Press, London, 1992, pp. 201–205.
2. A. W. Adamson, *Physical Chemistry of Surfaces*, John Wiley, New York, 1990, pp. 298–305.
3. E. Ruckenstein and S. V. Gourisankar. J. Colloid Interface Sci. *101*:436 (1984).
4. R. C. Cammarata. Progress in Surface Sci. *46*:1 (1994), and references therein.
5. R. E. Martinez, W. M. Augustyniak, and J. A. Golovchenko. Phys. Rev. Lett. *64*:1035 (1990).
6. F. K. Men, W. E. Packard, and M. B. Webb. Phys. Rev. Lett. *61*:2469 (1988).
7. T. Wang, C. Lee, and L. D. Schmidt. Surf. Sci. *163*:181 (1985).
8. A. Yanase and H. Komiyama. Surf. Sci. *264*:147 (1992).
9. J. M. McHale, A. Auroux, and A. Navrotsky. Science *277*:788 (1997).
10. R. Kern and P. Müller. Bulgarian Research Communications *27*:24 (1994).
11. R. Kern and P. Müller. J. Crystal Growth *146*:193 (1995).
12. J. W. Gibbs, in *The Collected Works of J. Willard Gibbs*, Vol. I, Yale University Press, New Haven, reprinted 1957, pp. 219 ff.
13. C. Herring, in *Structure and Properties of Solid Surfaces* (R. Gomer and C. W. Smith, eds.), University of Chicago Press, Chicago, 1952, p. 5.
14. P. R. Couchman, W. A. Jesser, D. Kuhlmann-Wilsdorf, and J. P. Hirth. Surf. Sci. *33*:429 (1972).
15. S. Trasatti and R. Parsons. Pure Appl. Chem. *58*:437–454 (1986).
16. R. G. Linford. Chem. Rev. *78*:81 (1978).
17. A. I. Rusanov. Surface Science Reports *23*:173 (1996).
18. V. Levy and M. P. Regnier. J. de Physique Colloque C1 *suppl. 4*:159 (1970).
19. I. Morcos. J. Electroanal. Chem. *62*:313 (1975).
20. V. K. Kumikov and K. B. Khokonov. J. Appl. Phys. *54*:1346 (1983).
21. A. I. Rusanov and V. A. Prokhorov, in *Interfacial Tensiometry, Studies in Interface Science 3*, Elsevier, Amsterdam, 1996, pp. 351–385.
22. R. S. Nelson, D. J. Mazey, and R. S. Barnes. Phil. Mag. *11*:91 (1965).
23. W. A. Miller, G. J. C. Carpenter, and G. A. Chadwick. Phil. Mag. *19*:305 (1969).
24. P. R. Couchman and W. A. Jesser. Surf. Sci. *34*:212 (1973).
25. P. R. Couchman, D. H. Everett, and W. A. Jesser. J. Colloid Interface Sci. *52*:410 (1975).
26. G. Lang and K. E. Heusler. J. Electroanal. Chem. *377*:1 (1994).
27. J. C. Eriksson. Surf. Sci. *14*:221 (1969).
28. R. Shuttleworth. Proc. Roy. Soc. London A *63*:444 (1950).
29. J. S. Vermaak and D. Kuhlmann-Wilsdorf, in *Structure and Chemistry of Solid Surfaces* (G. A. Samorjai, ed.), John Wiley, New York, 1969, p. 75.

30. G. J. Ackland and M. W. Finnis. Phil. Mag. A *54*:301 (1986).
31. G. C. Benson and K. S. Yun, in *The Liquid-Gas Interface, Vol. 1* (E. A. Flood, ed.), Marcel Dekker, New York, 1967, p. 172.
32. D. Wolf. Surf. Sci. *226*:389 (1990).
33. J. C. Chapman and H. L. Porter. Proc. Roy. Soc. London A *83*:65 (1910).
34. H. Udin, A. J. Shaller, and J. Wulff. Trans. Amer. Inst. Metall. Engrs. *185*:186 (1949).
35. H. Udin. Trans. Amer. Inst. Metall. Engrs. *191*:63 (1951).
36. C. Herring. J. Appl. Phys. *21*:437 (1950).
37. P. S. Maiya and J. M. Blakely. J. Appl. Phys. *38*:698 (1967).
38. W. W. Mullins. J. Appl. Phys. *30*:77 (1959).
39. B. Mills and G. M. Leak. Acta Metallurgica *16*:303 (1968).
40. F. W. C. Boswell. Proc. Phys. Soc. London A *64*:465 (1951).
41. T. B. Rymer and K. H. R. Wright. Proc. Royal Soc. London A *215*:550 (1952).
42. T. de Planta, R. Ghez, and F. Piuz. Helvetica Physica Acta *37*:74 (1964).
43. H. J. Wasserman and J. S. Vermaak. Surf. Sci. *22*:164 (1970).
44. C. Solliard and P. Buffat. J. de Physique C2 *7*:167 (1977).
45. P. P. Gillis and J. J. Gilman. J. Appl. Phys. *35*:647 (1964).
46. S. M. Wiederhorn, A. M. Shorb, and R. L. Moses. J. Appl. Phys. *39*:1569 (1968).
47. J. W Obreimoff. Proc. Roy. Soc. London A *127*:290 (1930).
48. A. I. Bailey. J. Appl. Phys. *32*:1407 (1961).
49. A. I. Bailey and S. M. Kay. Proc. Roy. Soc. A *301*:47 (1967).
50. K. T. Wan and B. R. Lawn. Acta Metall. Mater. *38*:2073 (1990).
51. A. R. C. Westwood and T. T. Hitch. J. Appl. Phys. *34*:3085 (1963).
52. J. J. Gilman. J. Appl. Phys. *31*:2208 (1960).
53. D. Hull, P. Beardmore, and A. P. Valintine. Philos. Mag. *12*:1021 (1965).
54. R. J. Jaccodine. J. Electrochem. Soc. *110*:524 (1963).
55. J. A. Williams and I. G. Palmer. Phil. Mag. *23*:1155 (1971).
56. M. Yudin and B. D. Hughes. Phys. Rev. B. *49*:5638 (1994).
57. D. Y. C. Chan, B. D. Hughes, and L. R. White. J. Colloid Interface Sci. *115*:240 (1987).
58. A. A. Griffith. Philos. Trans. Roy. Soc. London A *221*:163 (1920).
59. K. Kendall. J. Adhesion Sci. Technol. *8*:1271 (1994).
60. D. Maugis. J. Materials Sci. *20*:3041 (1985).
61. A. Hellemans. Science *281*:943 (1998).
62. D. Tabor, F. R. S. Winterton, and R. H. S. Winterton. Proc. Royal Soc. London A *312*:435 (1969).
63. J. N. Israelachvili and D. Tabor. Proc. Royal Soc. London A *331*:19 (1972).
64. P. M. Claesson, T. Ederth, V. Bergeron, and M. W. Rutland. Adv. Colloid Interf. Sci. *67*:119 (1996).
65. K. L. Johnson, K. Kendall, and A. D. Roberts. Proc. Roy. Soc. London A *324*:301 (1971).
66. B. V. Derjaguin, V. M. Muller, and Y. P. Toporov. J. Colloid Interface Sci. *53*:314 (1975).

67. J. N. Israelachvili, E. Perez, and R. K. Tandon. J. Colloid Interface Sci. 78:260 (1980).
68. H. K. Christenson, J. Phys. Chem. 97:12034 (1993).
69. G. Binnig, C. F. Quate, and C. Gerber. Phys. Rev. Lett. 56:930 (1986).
70. F. Creuzet, G. Ryschenkow, and H. Arribart. J. Adhesion 40:15 (1992).
71. H.-J. Butt, M. Jaschke, and W. Ducker. Bioelectrochem. Bioenergetics 38:191 (1995).
72. H. A. Mizes, K. G. Loh, R. J. D. Miller, S. K. Ahuja, and E. F. Grabowski. Appl. Phys. Lett. 59:2901 (1991).
73. K. O. van der Werf, C. A. J. Putman, B. G. de Grooth, and J. Greve. Appl. Phys. Lett. 65:1195 (1994).
74. M. Radmacher, J. P. Cleveland, M. Fritz, H. G. Hansma, and P. K. Hansma. Biophys. J. 66:2159 (1994).
75. M. P. L. Werts, E. W. van der Vegte, and G. Hadziioannou. Langmuir 13:4939 (1997).
76. A. Kawai, H. Nagata, and M. Takata. Jpn. J. Appl. Phys. 31:L977 (1992).
77. G. Toikka, R. A. Hayes, and J. Ralston. J. Colloid Interface Science 180:329 (1996).
78. P. Siedle, H.-J. Butt, E. Bamberg, D. N. Wang, W. Kühlbrandt, J. Zach, and M. Haider. Inst. Phys. Conf. Ser. 130:361 (1992).
79. K. Kendall, N. McN. Alford, and J. D. Birchall. Nature 325:794 (1987).
80. G. M. Dorris and D. G. Gray. J. Colloid Interface Sci. 71:93 (1979).
81. J. Schröder. Progr. Org. Coatings 12:159 (1984).
82. J. C. Melrose. J. Colloid Interface Sci. 24:416 (1967).
83. S. J. Gregg and K. S. W. Sing, in *Adsorption, Surface Area and Porosity*, Academic Press, London, 1982.
84. I. Dekany, T. Marosi, Z. Kiraly, and L. G. Nagy. Colloids Surfaces 49:81 (1990).
85. Z. Kiraly and I. Dekany. Langmuir 12:423 (1996).
86. T. Rheinländer, H. Dropsch, and G. H. Findenegg. Progr. Colloid Polym. Sci. 83:59 (1990).
87. J. M. McHale, A. Auroux, A. J. Perrotta, and A. Navrotsky. Science 277:788 (1997).
88. R. R. Weiler, J. Beekmans, and R. McIntosh. Can. J. Chem. 39:1360 (1961).
89. J. Roles and G. Guichon. J. Phys. Chem. 95:4098 (1991).
90. H. Balard. Langmuir 13:1260 (1997).
91. A. C. Zettlemoyer. Ind. Eng. Chem. 57:27 (1965).
92. W. D. Harkins and G. Jura. J. Am.. Chem. Soc. 66:1362 (1944).
93. S. Partyka, F. Rouquerol, and J. Rouquerol. J. Colloid Interface Sci. 68:21 (1979).
94. See Ref. 2, pp. 561–570.
95. S. G. Lipsett, F. M. G. Johnson, and O. Maas. J. Am. Chem. Soc. 49:925 (1927).
96. G. Jura and C. W. Garland. J. Phys. Chem. 74:6033 (1952).
97. A. Hulett. Z. physik. Chemie 37:385 (1901); 47:357 (1904).
98. M. L. Dundon. J. Am. Chem. Soc. 45:2658 (1923).
99. D. E. Wolf and P. Nozières. Zeitschr. f. Physik B 70:507 (1988).

100. M. L. Dundon. J. Am. Chem. Soc. *45*:2658 (1923).
101. B. V. Enüstün and J. Turkevich. J. Am. Chem. Soc. *82*:4501 (1960).
102. J. W. Cahn. Acta Metallurgica *28*:1333 (1980).
103. R. C. Sill and A. S. Skapski. J. Chem. Phys. *24*:644 (1956).
104. A. Skapski, R. Billups, and A. Rooney. J. Phys. Chem. *26*:1350 (1957).
105. E. W. Hansen, H. C. Gran, and E. J. Sellevold. J. Phys. Chem. B *101*:7027 (1997).
106. W. M. Ketcham and P. V. Hobbs. Philosophical Mag. *19*:1161 (1969).
107. J. T. Koberstein, in *Encyclopedia of Polymer Science and Engineering*, Vol. 8 (J. I. Kroschwitz, ed.), John Wiley, New York, 1987, pp. 237–279.
108. R. J. Good and L. A. Girifalco. J. Phys. Chem. *64*:561 (1960).
109. L. A. Girifalco and R. J. Good. J. Phys. Chem. *61*:904 (1957).
110. F. M. Fowkes. J. Phys. Chem. *67*:2538 (1963).
111. A. C. Zettlemoyer. J. Colloid Interf. Sci. *28*:343 (1968).
112. D. K. Owens and R. C. Wendt. J. Appl. Polym. Sci. *13*:1741 (1969).
113. C. J. van Oss, M. K. Chaudhury, and R. J. Good. Chem. Rev. *88*:927 (1988).
114. C. J. van Oss, L. Ju, M. K. Chaudhury, and R. J. Good. J. Colloid Interface Sci. *128*:313 (1989).
115. H. J. Jacobasch, K. Grundke, S. Schneider, and F. Simon. J. Adhesion *48*:57 (1995).
116. W. Wu, R. F. Giese, and C. J. van Oss. Langmuir *11*:379 (1995).
117. N. T. Correia, J. J. M. Ramos, B. J. V. Saramago, and J. C. G. Calado. J. Colloid Interface Sci. *189*:361 (1997).
118. C. D. Volpe, and S. Siboni. J. Colloid Interface Sci. *195*:121 (1997).
119. Y. Tamai, K. Makuuchi, and M. Suzuki. J. Phys. Chem. *71*:4176 (1967).
120. Y. Tamai and H. Kobayashi. J. Colloid Interface Sci. *32*:369 (1970).
121. J. Schultz, K. Tsutsumi, and J. B. Donnet. J. Colloid Interface Sci. *59*:272 (1977).
122. B. Janczuk. T. Bialopiotrowicz, and W. Wojcik. J. Colloid Interface Sci. *127*:59 (1989).
123. J. R. Dann. J. Colloid Interface Sci. *32*:302 (1970).
124. J. R. Dann. J. Colloid Interface Sci. *32*:321 (1970).
125. H. Kamusewitz, W. Possart, and D. Paul, in *Polymer Surfaces and Interfaces* (K. L. Mittal and K. W. Lee, eds.), VSP, 1997, pp. 125–143.
126. J. Höpken and M. Möller. Macromolecules *25*:1461 (1992).
127. Z. Kessaissia, E. Papirer, and J. B. Donnet. J. Colloid Interface Sci. *82*:526 (1981).
128. J. M. Park and J. H. Kim. J. Colloid Interface Sci. *168*:103 (1994).
129. D. T. Hansford, D. J. W. Grabt, and J. M. Newton. J. Chem. Soc. Faraday Trans. I *76*:2417 (1980).
130. E. Papirer, J. Schultz, and C. Turchi. Eur. Polym. J. *20*:1155 (1984).
131. L. Holysz and E. Chibowski. J. Colloid Interface Sci. *164*:245 (1994).
132. A. Siebold, A. Walliser, M. Nardin, M. Oppliger, and J. Schultz. J. Colloid Interface Sci. *186*:60 (1997).
133. A. Siebold, A. Walliser, M. Nardin, M. Oppliger, and J. Schultz. J. Colloid Interface Sci. *186*:60 (1997).

134. J. W. Cahn and R. E. Hanneman. Surf. Sci. *1*:387 (1964).
135. M. C. Finn and H. C. Gatos. Surf. Sci. *1*:361 (1964).
136. G. G. Stoney. Proc. Roy. Soc. London A *82*:172 (1909).
137. P. Müller and R. Kern. Surf. Sci. *301*:386 (1994).
138. R. J. von Preissig. J. Appl. Phys. *66*:4262 (1989).
139. A. Taloni and D. Haneman. Surf. Sci. *8*:323 (1967).
140. C. M. Drum. Phil. Mag. *13*:1239 (1966).
141. D. Sander and H. Ibach. Phys. Rev. B. *43*:4263 (1991).
142. D. Sander, U. Linke, and H. Ibach. Surf. Sci. *272*:318 (1992).
143. K. Rao, R. E. Martinez, and J. A. Golovchenko. Surf. Sci. *277*:323 (1992).
144. H. Ibach. J. Vac. Sci. Technol. A *12*:2240 (1994).
145. G. Y. Chen, T. Thundat, E. A. Wachter, and R. J. Warmack. J. Appl. Phys. *77*:3618 (1995).
146. H.-J. Butt. J. Colloid Interface Sci. *180*:251 (1996).
147. See Ref. 2, pp. 226–237.
148. D. M. Kolb, A. S. Dakkouri, and N. Batina, in *Nanoscale Probes of the Solid/Liquid Interface* (A. A. Gewirth and H. Siegenthaler, eds.), Kluwer Academic Publishers, Dordrecht, 1995, pp. 263–284.
149. J. Lipkowski, W. Schmickler, D. M. Kolb, and R. Parsons. J. Electroanal. Chem. *452*:193 (1998).
150. I. Morcos. J. Electroanal. Chem. *62*:313 (1975).
151. T. R. Beck. J. Phys. Chem. *73*:466 (1968).
152. D. M. Mohilner and T. R. Beck. J. Phys. Chem. *83*:1160 (1979).
153. K. F. Lin and T. R. Beck. J. Electrochem. Soc. *123*:1145 (1976).
154. K. F. Lin. J. Electrochem. Soc. *125*:1077 (1978).
155. I. Morcos and H. Fischer. J. Electroanal. Chem. *17*:7 (1968).
156. O. J. Murphy and J. S. Wainright. Langmuir *5*:519 (1989).
157. J. A. M. Sondag-Huethorst and L. G. J. Fokkink. Langmuir *8*:2560 (1992).
158. R. A. Fredlein, A. Damjanovic, and J. O'M. Bockris. Surface Science *25*:261 (1971).
159. R. A. Fredlein and J. O'M. Bockris. Surface Science *46*:641 (1974).
160. N. Pangarov and G. Kolarov. J. Electroanal. Chem. *91*:281 (1978).
161. R. Raiteri and H.-J. Butt. J. Phys. Chem. *99*:15728 (1995).
162. T. A. Brunt, E. D. Chabala, T. Rayment, S. J. O'Shea, and M. E. Welland. J. Chem. Soc. Faraday Trans. *92*:3807 (1996).
163. T. A. Brunt, T. Rayment, S. J. O'Shea, and M. E. Welland. Langmuir *12*:5942 (1996).
164. R. E. Malpas, R. A. Fredlein, and A. J. Bard. J. Electroanal. Chem. *98*:171 (1979).
165. R. E. Malpas, R. A. Fredlein, and A. J. Bard. J. Electroanal. Chem. *98*:339 (1979).
166. L. J. Handley and A. J. Bard. J. Electrochem. Soc. *127*:338 (1980).
167. K. E. Heusler, A. Grzegorzewski, L. Jäckel, and J. Pietrucha. Ber. Bunsenges. phys. Chem. *92*:1218 (1988).
168. K. E. Heusler and J. Pietrucha. J. Electroanal. Chem. *329*:339 (1992), and references therein.

169. T. J. Lewis, J. P. Llewellyn, and M. J. van der Sluijs. Polymer *33*:2636 (1992).
170. T. J. Lewis, J. P. Llewellyn, and M. J. van der Sluijs. IEE Proc. A *140*:385 (1993).
171. L. Jaeckel, G. Láng, and K. E. Heusler. Electrochimica Acta *39*:1031 (1994).
172. G. Láng and K. E. Heusler. J. Electroanal. Chem. *391*:169 (1995).
173. W. Haiss and J. K. Sass. J. Electroanal. Chem. *386*:267 (1995).
174. W. Haiss and J. K. Sass. Langmuir *12*:4311 (1996).
175. F. H. Buttner, E. R. Funk, and H. Udin. J. Phys. Chem. *56*:657 (1952).
176. B. H. Alexander, M. H. Dawson, and H. P. Kling. J. Appl. Phys. *22*:439 (1951).
177. E. D. Hondros. Proc. Roy. Soc. London *286*:479 (1965).
178. E. D. Hondros. Acta Metallurgica *16*:1377 (1968).
179. A. T. Price, H. A. Holl, and A. P. Greenough. Acta Metallurgica *12*:49 (1964).
180. M. C. Inman, D. McLean, and H. R. Tipler. Proc. Roy. Soc. London A *273*:538 (1963).
181. E. B. Greenhill and S. R. McDonald. Nature *171*:37 (1953).
182. J. S. Vermaak and Kuhlmann-Wilsdorf. J. Phys. Chem. *72*:4150 (1968).
183. C. W. Mays, J. S. Vermaak, and D. Kuhlmann-Wilsdorf. Surf. Sci. *12*:134 (1968).
184. H. J. Wasserman and J. S. Vermaak. Surf. Sci. *32*:168 (1972).
185. E. Salomons, R. Griessen, D. G. de Groot, and A. Magerl. Europhys. Lett. *5*:449 (1988).
186. J. E. Cordwell and D. Hull. Phil. Mag. *19*:951 (1969).
187. A. R. C. Westwood and D. L. Goldheim. J. Appl. Phys. *34*:3335 (1963).
188. J. W. Whalen. J. Phys. Chem. *65*:1676 (1961).
189. G. Ligner, A. Vidal, H. Balard, and E. Papirer. J. Colloid Interface Sci. *133*:200 (1989).
190. B. Bilinski and E. Chibowski. Powder Technol. *35*:39 (1983).
191. G. M. Dorris and D. G. Gray. J. Phys. Chem. *85*:3628 (1981).
192. W. D. Harkins and G. E. Boyd. J. Am. Chem. Soc. *64*:1195 (1942).
193. A. C. Makrides and N. Hackerman. J. Phys. Chem. *63*:594 (1959).
194. W. H. Wade and N. Hackerman. J. Phys. Chem. *65*:1681 (1961).
195. M. Topic, F. J. Micale, H. Leidheiser, and A. C. Zettlemoyer. Rev. Sci. Instrum. *45*:487 (1974).
196. T. Morimoto, Y. Suda, and M. Nagao. J. Phys. Chem. *89*:4881 (1985).
197. A. J. Groszek. Carbon *27*:33 (1989).
198. F. Rodriguez-Reinoso, M. Molina-Sabio, and M. T. González. Langmuir *13*:2354 (1997).
199. G. C. Benson, H. P. Schreiber, and F. van Zeggeren. Can. J. Chem. *34*:1553 (1956).
200. S. Brunauer, D. L. Kantro, and C. H. Weise. Can. J. Chem. *37*:714 (1959).
201. G. Jura and C. W. Garland. J. Am. Chem. Soc. *75*:1006 (1953).
202. M. L. Dundon and E. Mack. J. Am. Chem. Soc. *45*:2479 (1923).
203. G. B. Alexander. J. Phys. Chem. *61*:1563 (1957).
204. F. van Zeggeren and G. C. Benson. Can. J. Chem. *35*:1150 (1957).
205. B. V. Enüstün, M. Enuysal, and M. Dösemeci, J. Colloid Interface Sci. *57*:143 (1976).

206. H. J. Busscher, A. W. J. van Pelt, H. P. de Jong, and J. Arends. J. Colloid Interface Sci. *95*:23 (1983).
207. K. Grundke, T. Bogumil, T. Giezelt, H.-J. Jacobasch, D. Y. Kwok, and A. W. Neumann. Progr. Colloid Polym. Sci. *101*:58 (1996).
208. H.-J. Jacobasch, K. Grundke, P. Uhlmann, F. Simon, and E. Mäder. Composite Interfaces *3*:293 (1996).
209. A. J. Schell-Sorokin and R. M. Tromp. Phys. Rev. Lett. *64*:1039 (1990).

2
Contact Angle Techniques and Measurements

DANIEL Y. KWOK   Department of Chemical Engineering, Massachusetts Institute of Technology, Cambridge, Massachusetts

A. W. NEUMANN   Department of Mechanical and Industrial Engineering, University of Toronto, Toronto, Ontario, Canada

I.	Introduction	37
II.	Contact Angle Measurements	39
	A. Flat surface	39
	B. Nonflat surface and particles	58
III.	Preparation of Solid Surfaces	73
	A. Heat pressing	74
	B. Solvent casting	74
	C. Dip-coating	75
	D. Langmuir–Blodgett film deposition	75
	E. Self-assembled monolayers (SAMs)	76
	F. Vapor and molecular deposition techniques	76
	G. Siliconization	77
	H. Surface polishing	78
	I. Preparation of powders for contact angle measurements	78
	References	79

I. INTRODUCTION

In many areas of applied surface thermodynamics, measurement of contact angles plays an important role. The range of applications of contact angle measurement is remarkable. It can be used as a simple tool to assess, for example, the cleanliness of the surfaces, or it can be a highly sensitive scientific measurement aimed at obtaining information on the solid surface tension and the physical state of the surface. When first encountered, the measurement of contact angles appears to be quite straightforward. This apparent simplicity is, however, misleading. Experience has shown that the acquisition of thermodynamically significant contact angles requires substantial effort.

Contact angles provide a unique means of determining solid–vapor and solid–liquid interfacial tensions because of the Young equation

$$\gamma_{lv} \cos\theta_Y = \gamma_{sv} - \gamma_{sl} \tag{1}$$

where γ_{lv}, γ_{sv}, and γ_{sl} are, respectively, the interfacial tensions of the liquid–vapor, solid–vapor, and solid–liquid interfaces; θ_Y is the Young contact angle, i.e., a contact angle that can be used in conjunction with the Young equation. If we measure contact angles with the intention of drawing conclusions with respect to surface energetics of the solid phase involved, the all-important question arises whether these contact angles can be used in conjunction with the Young equation. The importance of this question lies in that all contact angle approaches of current interest [1–3] to estimate surface energetics make use of this equation.

A detailed discussion of the Young contact angle θ_Y is available elsewhere [4,5]. While the equilibrium contact angle θ_e on a smooth and homogeneous solid surface (no contact angle hysteresis) is indeed a Young contact angle, there are certain other contact angles that are also Young contact angles. An important example is the advancing contact angle on a heterogeneous solid surface [4,5]. Unfortunately, roughness also gives rise to contact angle hysteresis, but the observed advancing and receding contact angles cannot, at this time, be linked unambiguously to surface energetics. Worse still, there is no independent means of distinguishing between contact angle hysteresis due to heterogeneity and that due to roughness. It is therefore of particular interest to produce solid surfaces that are so smooth and homogeneous that there will be small or negligible contact angle hysteresis. In addition to these complexities, static and dynamic conditions, penetration of the liquid into the solid, swelling of the solid by the liquid [6], and physio-chemical reactions can all play a role.

This chapter discusses the commonly used experimental techniques and practical issues pertaining to the measurement of contact angles, including the preparation of suitable solid surfaces for energetics calculations.

Many different techniques have been developed for the measurement of contact angles [7,8]. Of these, the three most useful methods are the Wilhelmy technique, the technique of capillary rise at a vertical plate, and the drop shape methods. These techniques require the solid surface to be flat and smooth. Direct measurement of contact angles on fibers (of uniform thickness) can also be performed using the Wilhelmy technique. For nonflat surfaces or particles, indirect methods such as capillary penetration into columns of powders, sedimentation volume of particles, and solidification fronts of particles can be used. These methods are indirect because they provide *a priori* quantities different from the contact angle, in some cases the solid surface tension. The choice of technique depends largely on the specific geometry of the surfaces. Therefore the following sections are organized according to the geometric form of the solid of interest: (1) flat surface; and (2) nonflat surface and particles.

II. CONTACT ANGLE MEASUREMENTS

A. Flat Surface

As mentioned above, many contact angle measurement techniques have been described in the literature [7,8]. Although only a few have found wide application, we shall nevertheless briefly describe some of the old, largely forgotten, techniques for the sake of completeness.

Historically, Adam and Jessop [9] originated a tilting plate method that is attractive in its simplicity. A flat plate of the material of interest (about 2 cm wide) is gripped at one end and immersed in the measuring liquid. It is then tilted until the meniscus becomes horizontal on one side of the plate. The angle the plate makes with the liquid surface is then the contact angle. Improvements have been made by Fowkes and Harkins [10], by providing means for keeping the surface of the liquid clean, as in film pressure investigations. The method has been used for the measurement of low contact angles (less than 10°) by suspending the plate at both ends and immersing it in the liquid so that the plane of the plate is only slightly inclined with respect to the liquid surface. A major disadvantage of the method is the difficulty in establishing a proper advancing contact angle θ_a; the contact angle may lie somewhere between the advancing and receding angle, θ_a and θ_r. The method has lost most of its former popularity.

Langmuir and Schaeffer [11] used the specular reflection from a drop surface to measure the contact angle in a method that was later refined by Fort and Patterson [12]. A light source is pivoted about the three-phase line to observe the angle at which reflection from the drop surface just disappears. This is the contact angle, and it can usually be measured in this way to within 1°. The method has been used with sessile drops and with menisci on flat plates and inside tubes [7].

Two rather specialized methods are available to measure contact angles less than approximately 60°. The method of interference microscopy makes use of fringe patterns reflected from the drop surface to calculate the contact angle [13]. Fisher [14] obtained contact angles less than 30° by simultaneously measuring the mass of the drop and the radius of the three-phase line. The contact angle was then derived from a semiempirical relationship involving these two quantities.

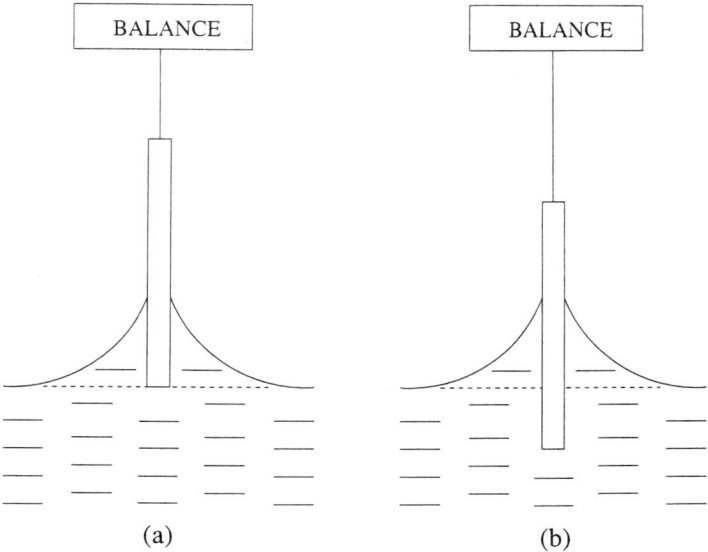

FIG. 1 Vertical plate methods; plate extends perpendicular to plane of paper; (a) zero net depth of immersion, (b) finite depth of immersion.

In the following subsections, the three most commonly used contact angle techniques, the Wilhelmy technique, capillary rise at a vertical plate, and drop shape methods, are described in detail.

1. Wilhelmy Method

If a smooth vertical plate is brought into contact with a liquid, as indicated in Fig. 1a, the liquid will exert a downward force on the plate given by

$$f = p\gamma_{lv} \cos\theta \tag{2}$$

where p is the perimeter of the plate. If the depth of immersion is not zero, so that a volume V of liquid is displaced (see Fig. 1b), the force exerted on the plate becomes

$$f = p\gamma_{lv} \cos\theta - V \Delta\rho\, g \tag{3}$$

where $\Delta\rho$ is the difference in density between the two fluids and V is the displaced liquid volume. These two equations are general and rigorous [15–19]. In order to evaluate the contact angle, the surface tension γ_{lv} must be known.

If a plate is not effectively smooth, it will not be suitable for determination of the contact angle of the liquid on that solid. However, a well-known useful technique for obtaining a zero contact angle, in order to use the Wilhelmy method for measuring surface tension, is to sandblast a glass plate or roughen it with an abrasive [20].

In this method, several experimental setups for contact angle measurements have been described in the literature [15,21,22]. The basic apparatus is shown schematically in Fig. 2. The plate (1), the perimeter of which should be constant, is suspended by a thin rod (2) from an electrobalance (3). The downward force is recorded on a recording device (4). The liquid (6) is contained in a beaker (5), which is partially covered by a lid (11) to minimize evaporation. This lid may conveniently consist of two halves so that it can be put on after the plate has been suspended from the electrobalance. The table (7) carrying the beaker can be moved up and down by means of a screw (8), driven by an electric motor (9) or by hand, in order to establish advancing and receding contact angles. Motion of the table (7) at speeds of the order of 1 mm/min appears to be convenient. To facilitate a rough positioning of the table (7) and the motor (9) it is convenient to mount them on a heavy vertical rod and to bring them into a desired position using a clamp (10). Vibration insulation may also be desirable, since the motor and screw could be sources of appreciable vibration.

Advancing and receding conditions are established as follows: After suspending the plate from the electrobalance, the table is raised from an initial low position to such a height that the surface of the liquid is approximately 1 mm below the lower edge of the plate. The table, together with the motor, is fixed in this position by means of the clamp. Then the table is raised slowly by means of the screw. After contact is established between the plate and the liquid, the motor is stopped immediately. The resulting configuration is one of advancing contact angle. The depth of immersion is zero, so that Eq. (3) may be used.

Receding conditions can be established by immersing the plate to some depth and subsequently withdrawing the plate through exactly the same distance. This can conveniently be done by timing the immersion and the withdrawal or by measuring the depth of immersion and withdrawal by means of a cathetometer. If the latter technique is used, a simpler setup without an electric motor can be employed. To obtain the true receding contact angle, it is imperative to immerse the plate to such a depth that motion of the three-phase line with respect to the plate occurs. The upper limit for that height h can be calculated from Eq. (4), which assumes zero contact angle.

$$h = \left(\frac{2\gamma_{lv}}{\Delta \rho \, g} \right)^{1/2} \tag{4}$$

While the Wilhelmy method at first sight seems to be an ideal method, it has several drawbacks that are not immediately obvious and that can restrict the usefulness of the method. The high sensitivity of the electrobalance employed in the Wilhelmy method can be exploited only if the perimeter is constant.

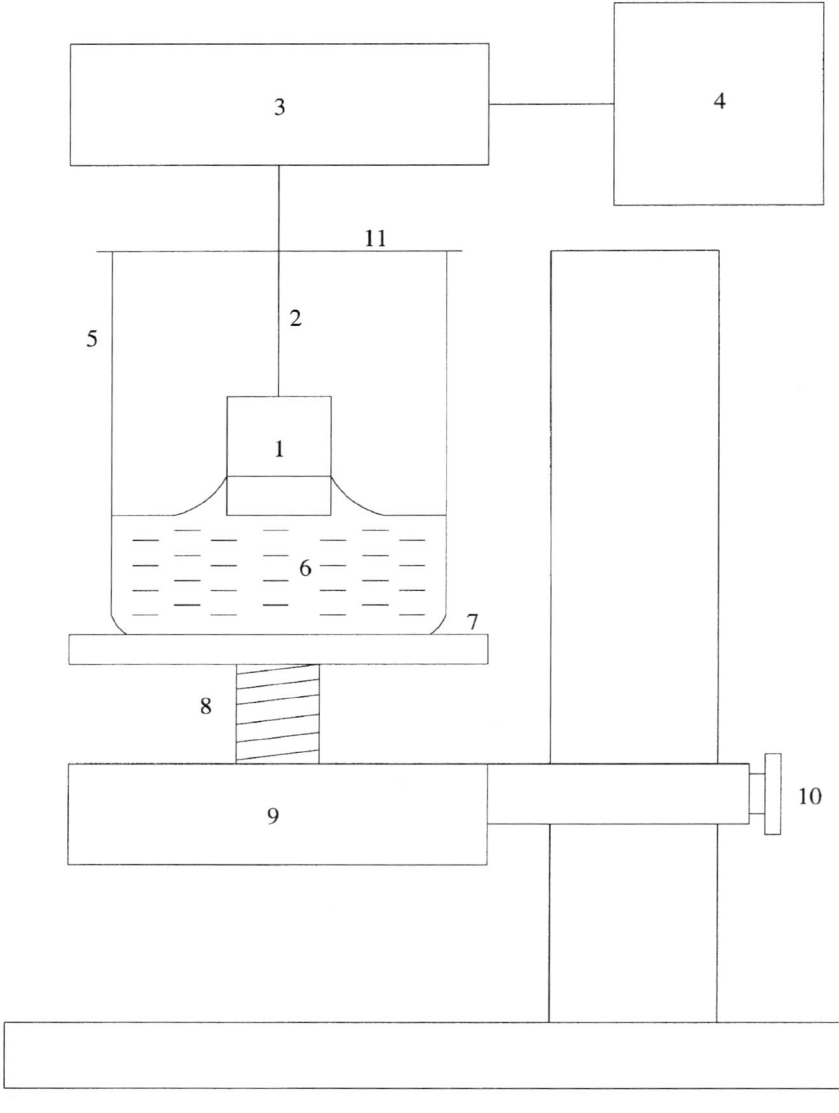

FIG. 2 Apparatus for Wilhelmy technique: (1) measuring plate, (2) glass fiber or rod, (3) electrobalance, (4) recording device, (5) measuring cell, (6) liquid, (7) movable platform, (8) screw or gear mechanism to raise or lower the platform, (9) motor, (10) clamp and support, (11) lid.

Furthermore, the plate must have the same composition and morphology at all surfaces: front, back, and both edges. This condition may be difficult to meet, particularly if one wants to investigate films deposited in vacuum, polished surfaces, or anistropic systems. In measurements that extend over appreciable time intervals, swelling or solution of the solid may become a problem. Swelling of solution may change the volume V and the mass of the displaced liquid, see Eq. (3), in an uncontrollable manner. Also, adsorption of the vapor of the liquid at various parts of the gravimetric system other than the plate may change the balance output. This is particularly serious for measurements of the temperature dependence of contact angles. Lastly, on surfaces that are not completely uniform, the interface does not intersect the solid in a perfectly smooth curve of constant shape, and it can only be hoped that an average or effective contact angle can be obtained.

All these difficulties are virtually absent in the capillary rise method, which works on a very similar geometric constellation.

2. Capillary Rise at a Vertical Plate

In this method, the solid surface is also aligned vertically and brought into contact with the liquid. Instead of the capillary pull as in the Wilhelmy method, the capillary rise h at the vertical surface is measured (see Fig. 3). This method has been found particularly effective for measuring contact angles as a function of the rate of advance and retreat and for determining the temperature coefficient of θ.

For an infinitely wide plate, a straightforward integration of the Laplace equation [16,23,24] yields

$$\sin \theta = 1 - \frac{\Delta \rho \, g h^2}{2 \gamma_{lv}} \tag{5}$$

It is by rearranging this equation and assuming θ to be zero that we obtain Eq. (4). For practical purposes, plates that are about 2 cm wide satisfy the theoretical requirement of "infinite" width. For such a plate (assumed to have a uniform surface), the line of contact is straight in the central part of the plate for all liquids of moderate surface tension, including water. If we assume g, $\Delta \rho$, and γ_{lv} to be known, the task of determining a contact angle is reduced to the measurement of a length (the capillary rise h), which can be determined optically, e.g., with a cathetometer.

Since this method is broadly useful, and since it has proved to be particularly suitable for measuring the temperature dependence of contact angles [4,25–29], we describe the apparatus in detail, including the temperature control system. The essential parts of the equipment (Fig. 4) are the following:

1. The plate (A) (about 2 × 3 cm and as thin as is practical), is aligned vertically; in certain cases, it may be tilted (see below). It is attached to a

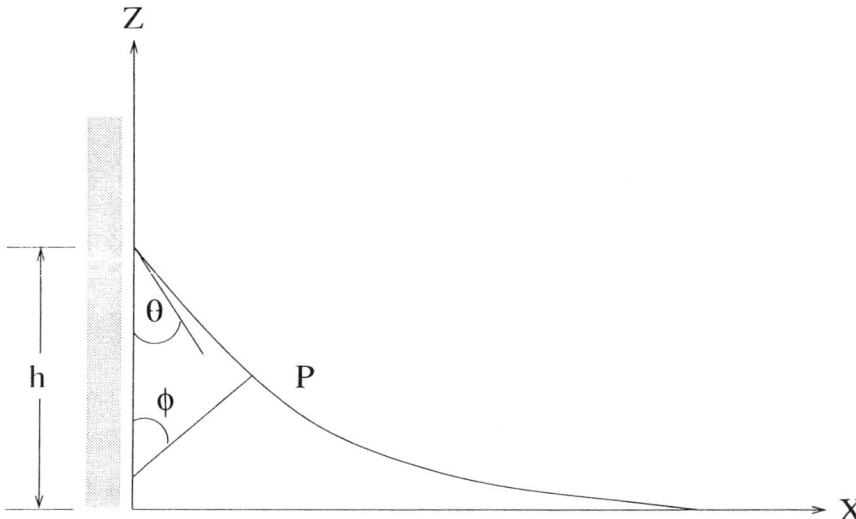

FIG. 3 Schematic of capillary rise at a vertical plate, where ϕ is the angle between the vertical axis and the normal at a point on the liquid–vapor surface, P; θ is the contact angle; and h is the capillary rise at a vertical plate.

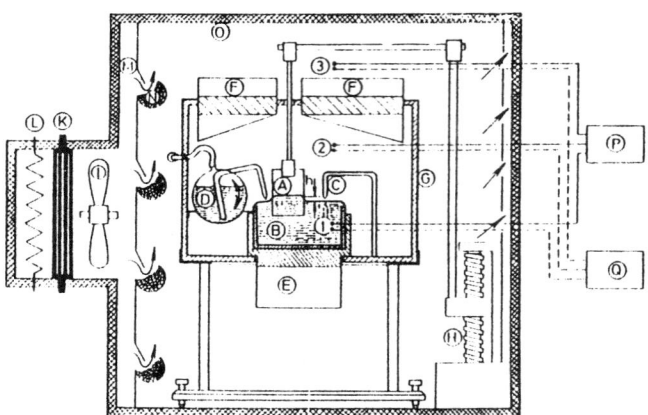

FIG. 4 Apparatus for measuring the temperature dependence of contact angles, based on the optical measurement of the capillary rise of the liquid at a vertical plate.

support that can be raised or lowered by a screw (H). The screw may be driven by a variable-speed reversible motor.

2. The liquid (B) is contained in an optical-quality glass cell whose top is horizontal and that is about 10×10 cm or larger. This relatively large area is needed to ensure that the liquid surface is horizontal in the region where the reference height is measured. It also helps in the attainment of equilibrium between liquid and vapor. In principle, a smaller cell can be used if it can be filled so that the liquid level is exactly at the top of the cell, so that the only source of curvature is the capillary rise at the plate. However, if the liquid surface area is small, the lowering of the plate into the liquid will raise the liquid level appreciable, particularly if the plate is not extremely thin. This will interfere with the determination of the reference level from which the capillary rise h is measured; thus it is desirable to use a cell with the largest area that is practical. The plate dips into the liquid, which rises as indicated in Fig. 3. If the contact angle is greater than $90°$, there is, of course, a capillary depression instead of a rise. The optical quality of the front of the cell is necessary in order for accurate observations to be made below the liquid level when there is a capillary depression.

3. A cathetometer, not shown in Fig. 4.

4. A point (C) located in the same plane as the plate, for use in determining the level of the horizontal liquid surface. It is difficult to see the surface directly to determine its elevation, so a vertical fiber (which may be of glass) is mounted above and close to the surface. The location of the liquid surface can be determined by observing the tip of the pointer and its mirror image in the surface.

5. A pneumatic device (D) for transferring liquid into the cell to maintain the level of the liquid is a considerable convenience; for high-precision work, it is a necessity. The bottle shown (D) is mounted on a horizontal axis so that it can be rotated and the tip brought in contact with the liquid; a vertical bottle supported on a platform that can be raised or lowered can serve the same purpose.

The line of contact of the liquid on the solid appears as a sharp light–dark border when the specimen is illuminated with light from the direction of the cathetometer telescope (illumination from within the telescope is best). The contrast may not be great, but it is usually possible to observe the boundary quite distinctly.

The optical measurements are made with the cathetometer. Readings can conveniently be made with commercial equipment to a precision of about 2×10^{-4} cm. Since three position readings are required, the overall precision in the measurement of h is no better than about 3×10^{-4} cm. The corresponding uncertainty in the value of θ for a typical liquid (assuming $\gamma_{lv} = 30$ mJ/m^2, density $\Delta\rho = 1.00$ g/cm^3, $\theta = 30°$) is about $0.1°$.

Dynamic advancing and receding contact angles can be measured by moving the plate up or down. It may be possible to choose a rate in a range such that the position of the line of contact is essentially independent of rate, about 1 mm/min or less. The motion can, of course, be stopped and the angle measured after a specified time.

If high-precision measurements are to be made, or measurements other than at room temperature, then for the purpose of temperature control the apparatus is doubly enclosed, with separate controls on the temperatures of three regions. First, the liquid can be heated and cooled from a battery of Peltier elements (E, Fig. 4). To establish vapor and thermal equilibrium, the interior of the box (G) can be thermostatted with the Peltier elements (F). Condensation at the walls of (G) would be a serious problem at temperatures different from ambient. Therefore, the box (G) is installed inside a climate chamber (O), which can be thermostatted by means of a resistance heater (L) and an external refrigerating unit with a cooling element (K). Heat exchange is enhanced by means of a ventilator (I). The hot or cold air enters through the slits (M) and is directed over troughs containing a desiccant such as calcium chloride. The temperature control is effected by resistance thermometers or thermocouples in conjunction with external electronic control units (P). The temperature is controlled in the three positions 1, 2, and 3. To avoid condensation, the temperature in G is always kept slightly (about $0.2°C$) higher than the temperature in B, and the temperature outside B in the climate chamber somewhat higher still. Peltier elements and the heating and cooling device of the climate chamber (O) are connected in such a way that each heating pulse is followed by a cooling pulse. By this means, the temperature can be kept constant in B to $\pm 0.1°C$, in G to $\pm 0.2°C$, and in O to $\pm 1.5°C$. This apparatus may be used in the temperature interval from $-80°C$ to $+80°C$.

Actual measurements can be performed in the following way: The specimen (A) is installed vertically in its holder inside the box (G), and the liquid is transferred into the cell (B) and the bottle (D). The glass tip is adjusted in such a way that it is at the same distance from the axis of the cathetometer as the specimen. Then the whole apparatus is closed, and when temperature equilibrium is reached at the first temperature to be investigated, the cell (B) is filled slightly higher than level from the bottle (D). The specimen is lowered until it dips about 0.2 cm into the liquid. Since each reading for an advancing angle measurement should be taken at a region of the surface that has not been previously in contact with the liquid in that run, the plate (A) is lowered very slowly, typically in the range of 0.01 cm/min, while a number of readings are taken at that particular temperature. The three-phase line usually remains stationary with respect to the telescope of the cathetometer, while the depth of immersion of the plate increases steadily. This procedure has an additional advantage of minimizing effects of small random vibrations. By establishing that the angle is independent of the rate of immersion of the specimen, it may be ensured that dynamic contact angle

Contact Angle Techniques and Measurements

effects are absent. The procedure just described yields advancing contact angles. Receding contact angles can be obtained by withdrawing the plate (A) from the liquid in the same manner.

As already mentioned, a slightly different arrangement is necessary when the contact angles to be measured are larger than 90°, i.e., when there is a capillary depression rather than a capillary rise. The cell (B) in Fig. 4 is then filled slightly less than level; the glass tip (C) should be in the liquid, supported from below and pointed upward, but again, close to the liquid surface. In other words, the tip and its mirror image are interchanged. Observations are then made through the wall of the cell (B) and the liquid.

The accuracy and reproducibility of this method is such that, for example, the solid–solid phase transition of cholesterol acetate at 40°C has been detected by means of the temperature dependence of the contact angle of water [26]. The discontinuity in the contact angle curve was about 0.3° of arc. For water on siliconized glass plates, the deviations of individual points in the plot of contact angle versus temperature were found to be about 0.1° [28], in good agreement with the error limits estimated above.

The technique of capillary rise at a vertical plate has been used in various contact angle [24–35] and electrocapillary [36–39] studies. The method has been automated [40,41] and is used to perform various dynamic advancing and receding contact angle measurements at immersion speeds ranging from 0.08 to 0.9 mm/min [6,41,42]. Sedev et al. [6] have employed the technique to study both advancing and receding dynamic contact angles on dry, prewetted, and soaked fluorocarbon FC-722 (3M) surfaces. Such measurements reveal that the behavior of a polymer surface, although chemically stable, can be affected by prolonged contact with liquids as inert as alkanes. The method has also been employed to study surface quality of various solids from different preparations [42].

Despite the many advantages that methods such as the capillary rise or the Wilhelmy method have over other methods, they have the disadvantage that the surface tension γ_{lv} of the liquid must be known. While this restriction does not usually present severe problems with most pure liquids, there may be a serious uncertainty for solutions of surface active substances. In such systems, adsorption of the surface active material at various interfaces may change the surface tension and the contact angle simultaneously. The joint use of the capillary rise method and the Wilhelmy method makes it possible to determine contact angle and liquid surface tension at the same time [19,31,43]. Eliminating γ_{lv} from Eqs. (2) and (5) and with the aid of the identity $\sin^2 \theta + \cos^2 \theta = 1$, we obtain

$$\cos \theta = \frac{4 \, \Delta M \, \Delta \rho \, h^2 p}{4(\Delta M)^2 + p^2 (\Delta \rho)^2 h^4} \tag{6}$$

Measuring the change in mass ΔM and the capillary rise h as described above and knowing the density ρ of the liquid and the perimeter p of the plate, we

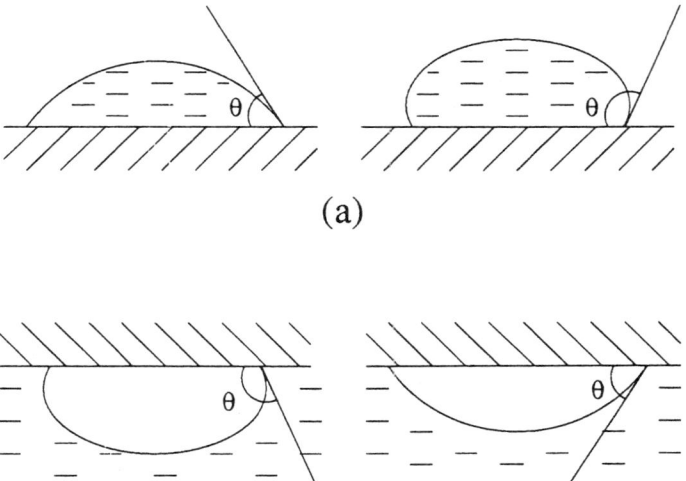

FIG. 5 Examples of systems with nonzero contact angles; (a) sessile drops, (b) adhering bubbles.

can calculate the contact angle θ from Eq. (6) without explicit knowledge of the surface tension of the liquid; γ_{lv} can then, of course, be calculated from Eq. (2) or Eq. (5). Alternatively, we may eliminate θ from Eqs. (2) and (5) to obtain the liquid surface tension

$$\gamma_{lv} = \left(\frac{\Delta M g}{p}\right)^2 \frac{1}{\Delta \rho g h^2} + \frac{\Delta \rho g h^2}{4} \qquad (7)$$

The contact angle can then be calculated from either Eq. (2) or Eq. (5). Details, particularly referring to work with solutions of surface active substances, have been given elsewhere [31,43].

3. Sessile Drop or Adhering Bubble

Another widely used method is the measurement of contact angles from sessile drops or adhering bubbles. Because of its simplicity, various procedures have been developed, as we shall describe below.

(*a*) *Direct Measurement of Angle from Drop Profile.* The method of the sessile drop or, alternatively, of the adhering gas bubble (Fig. 5) is at present the most widely used technique. The most obvious and widely used procedure is simply to align a tangent with the sessile drop profile at the point of contact with the solid surface. This is most frequently done directly using a telescope

Contact Angle Techniques and Measurements 49

equipped with a goniometer eyepiece, but it is perhaps more accurate to work with a photograph of the drop profile. In either case, the results are somewhat subjective and dependent on the experience of the operator, although certain training procedures can be used to improve both the accuracy and the precision [7]. The accuracy is enhanced by the use of relatively high magnifications (up to ×50) that permit the detailed examination of the intersection of the drop profile and the solid surface. The telescope should be tilted downward slightly (1 or 2°) out of the horizontal so that the near edge of the solid surface is out of focus while a portion of the profile reflected by the surface is in view. The effect of this for smooth specularly reflecting solid surfaces is to create a drop "tip" or cusp at the point of contact with the solid, out of the profile and its mirror image. The tangent is aligned to the profile at this point. Observations are facilitated by the use of a brightly lit diffuse background against which the drop appears as a black silhouette. A 150 W fiber-optic light source illuminating a heavily frosted glass diffuser plate works well and reduces any unwanted heating of the liquid.

In order to establish an advancing contact angle, it is best to grow the sessile drop slowly to a radius of approximately 5 mm using a micrometer syringe and a narrow-gage stainless steel or Teflon needle. The contact angles of smaller drops may be unduly influenced by the tension of the three-phase line at the drop perimeter. The needle must not be removed from the drop, as this may cause vibrations that can decrease the advancing contact angle to some lower metastable state. Provided that the needle diameter is not more than approximately one-fifth of the drop diameter, the distortion of the profile due to the needle will not interfere with the contact angle measurement. Contact angles should be measured on both sides of the drop profile unless one side of the drop is observed to "hinge" on a surface scratch and not advance. By repeatedly adding small amounts of liquid to the drop and advancing it over fresh areas of the solid, many contact angles can be measured to give an average that is representative of a relatively large area of the surface.

Another procedure that has been used to form sessile drops is to dip a fine platinum wire in the liquid and then gently flick it to create a pendant drop hanging from the tip of the wire. When the drop is slowly brought into contact with a solid surface, it flows from the wire and forms a sessile drop. This technique was used in the laboratory of Zisman, whose pioneering work did much to advance the use and interpretation of contact angle measurements [44]. Although it is claimed that the method gives reproducible contact angles, and although it is attractive because of its simplicity, there is concern that the kinetic energy associated with the flowing liquid and the detachment of the wire can vibrate the drop and lead to a lower metastable contact angle than the true advancing contact angle.

In general, placing a tangent to the sessile drops manually, e.g., by using a goniometer, is the most convenient method if high accuracy is not required. There are two advantages of using this method: only very small quantities of liquid are required, and solid surfaces as small as a few square millimeters can be used. It is commonly claimed that the accuracy of this method is ±2°. It would be better to call this a matter of precision; agreement between independent laboratories using this method has in the past been as poor as 10°. A recent study [45] has shown that the procedure can produce misleading contact angle results, in terms of surface energetics.

(b) *Measurements of Contact Angles from Drop Dimensions.* For a spherical drop [46,47], we have

$$\frac{2h}{\Delta} = \tan\frac{\theta}{2} \tag{8}$$

connecting the contact angle with the base diameter Δ and height of the drop h. There is also another equation [46,47],

$$\frac{\Delta^3}{V} = \frac{24\sin^3\theta}{\pi(2 - 3\cos\theta + \cos^3\theta)} \tag{9}$$

which connects the angle θ with the base diameter and V, the volume of the drop. However, care must be used with these methods. The size of the drop must be kept small so that the drop is indeed a spherical cap. It should be noted that, for drops that are small (less than 2–3 mm radius), line tension could play a significant role on the dependence of contact angles on drop size [48,49].

Otherwise, the limitations of this method are similar to those of the direct measurement of the contact angle at the sessile drop. The drop may not have perfect axial symmetry, and the contact angle may vary somewhat from point to point; the measurement of two drop dimensions takes more time, of course, than one measurement; it can best be done on a photograph of the drop. This further diminishes the suitability of the method for measuring advancing and receding angles. Perhaps even worse, it fosters the illusion that an equilibrium angle has been measured.

Measurements using drops that depart appreciably from spherical shape have the advantage that the surface tension γ_{lv} of the wetting liquid can be determined simultaneously with the contact angle. These methods consist, essentially, of measuring parameters that determine the entire profile of an axially symmetrical drop. Most of these methods start with the Laplace equation describing the shape of fluid interfaces [50]:

$$\gamma\left(\frac{1}{R_1} + \frac{1}{R_2}\right) = \Delta P \tag{10}$$

where R_1 and R_2 are the two principal radii of curvature and ΔP is the pressure difference across the interface. The earliest efforts in the analysis of axisymmetric drops were those of Bashforth and Adams [51]. They generated sessile-drop profiles for different values of surface tension and radius of curvature at the apex of the drop. The determination of the interfacial tension and contact angle of an actual drop was accomplished by interpolation of tabulated profiles. Hartland and Hartley [52] also collected numerous solutions for determining the interfacial tensions of axisymmetric liquid–fluid interfaces of different shapes. A FORTRAN computer program was used to integrate the appropriate form of the Laplace equation, and the results were presented in tables. The major shortcoming of these methods is in data acquisition. The description of the surface of the drop is accomplished by the measurement of a few preselected points. These points are critical, since they correspond to special features, such as inflection points on the interface, and must be measured with a high degree of accuracy. Also, for the determination of the contact angle, the point of contact with the solid surface, where the three phases meet, must be established. However, these measurements are not easily obtained. In addition, the use of these tables is limited to drops of a certain size and shape range.

Several graphical curve-fitting techniques have been developed (see Padday [53] for details) that can be used in conjunction with the numerical integration of the Laplace equation by Bashforth and Adams (and by subsequent workers) to determine θ and to obtain γ_{lv}. Smolders [54,55] used a number of coordinate points of the profile of the drop for curve fitting. If the surface tension of the liquid is known and if $\theta > 90°$, a perturbation solution of the Laplace equation derived by Ehrlich [56] can be used to determine the contact angle, provided the drop is not far from spherical. Input data are the maximum radius of the drop and the radius at the plane of contact of the drop with the solid surface. The accuracy of this calculation does not depend critically on the accuracy of the interfacial tension.

Another method of determining contact angles from the dimensions of sessile drops is by Staicopolus [57–59], who obtained a digital computer solution of the Laplace equation, from which he developed approximate equations and nomographs permitting the computation of surface tension and contact angle, provided the latter is larger than 90°. Parvatikar [60] has introduced a relation connecting z/r and V/r to the contact angle, which can be used in conjunction with the tables of Bashforth and Adams [51]; here, z is the vertical coordinate of the drop, r is the maximum radius of the drop, and V is its volume. The main disadvantage of these and similar curve-fitting techniques is that they can be used to determine the contact angle only if $\theta > 90°$.

Maze and Burnet [61,62] developed a more satisfactory scheme for the determination of contact angle and interfacial tension from the shape of sessile drops both above and below 90°. They utilized a numerical nonlinear regression procedure in which a calculated drop shape is made to fit a number of arbitrary selected and measured points on the drop profile. In other words, the measured drop shape (one-half or the meridian section) is described by a set of coordinate points, and no particular significance is assigned to any one of the points. In order to start the calculation, reasonable estimates of the drop shape and size are required, otherwise the calculated curve will not converge to the measured one. The initial estimates are obtained, indirectly, using values from the tables of Bashforth and Adams [51]. Despite the progress in strategy, there are several deficiencies in this algorithm. The error function is computed by summing the squares of the horizontal distances between the measured points and the calculated curve. This measure may not be adequate, particularly for sessile drops whose shapes are strongly influenced by gravity. For example, large drops of low surface tension tend to flatten near the apex. Therefore, any data point that is near the apex may cause a large error, even if it lies very close to the best-fitting curve, and lead to considerable bias of the solution. In addition, the identification of the apex of the drop is of paramount importance, since it acts as the origin of the calculated curves.

(c) Axisymmetric Drop Shape Analysis (ADSA). Rotenberg et al. [63] developed a technique called axisymmetric drop shape analysis-profile (ADSA-P), which is superior to the above-mentioned drop shape methods and does not suffer from their deficiencies. The ADSA-P technique fits the measured profile of a drop to a Laplacian curve, i.e., a theoretical curve that is calculated from the Laplace equation of capillarity, Eq. (10), by assuming that the drop is axisymmetric and that the gravitational force is the only external force. An objective function is formed that describes the deviation of the experimental profile from the theoretical one as the sum of the squares of the normal distances between the experimental points and the calculated curve. This function is minimized by a nonlinear regression procedure yielding the interfacial tension and the contact angle in the case of a sessile drop. The location of the apex of the drop is assumed to be unknown, and the coordinates of the origin are regarded as independent variables of the objective function. Thus the drop shape can be measured from any convenient reference frame, and any measured point on the surface is equally important. A specific value is not required for the surface tension, the radius of curvature at the apex, or the coordinates of the origin. The program requires as input several coordinate points along the drop profile, the value of the density difference across the interface, the magnitude of the local gravitational constant, and the distance between the base of the drop and the horizontal coordinate axis. An initial guess of the location of the apex and the radius of curvature at the apex are not required. The solution of the ADSA-P program yields not only the interfacial tension and contact angle but also the volume, surface area, radius of curvature, and contact radius of the drop. Essentially, ADSA-P employs a numerical procedure that unifies the method for both the sessile drop and the pendant drop. There is no need for any table, nor is there any drop-size restriction on the applicability of the method. The technique has also been automated [64–66].

Recently, a new version of ADSA-P has been developed [67,68] that is superior to the original program in terms of computation time and range of applicability. The new version is written in the C language (rather than FORTRAN) and utilizes more efficient and accurate numerical methods. For example, the new algorithm uses the curvature at the apex (rather than the radius of curvature), it permits an additional optimization parameter (the vertical misalignment of the camera), and it gives improved initial estimates of the apex location and shape.

ADSA-P has been employed in various surface tension and contact angle studies, including static (advancing) contact angles [69,70], dynamic (advancing) contact angles at slow motion of the three-phase contact line [45,71–74], and contact angle kinetics of surfactant solutions [75]. A schematic of the experimental setup for ADSA-P sessile drops is shown in Fig. 6. More details are available elsewhere [66].

It has been found that a contact angle accuracy of better than $\pm 0.3°$ can be obtained on well-prepared solids. In the static contact angle experiments of Li et al. [69,70], the procedures for measuring advancing contact angles are as follows: Static (advancing) contact angles are measured by supplying test liquids from below a flat surface into the sessile drop, using a motor-driven syringe device; see Fig. 7. A hole of about 1 mm in the center of each solid surface is required to facilitate such procedures. Liquid was pumped slowly into the drop from below until the three-phase contact radius was about 0.4 cm. After stopping the motor, the sessile drop was allowed to relax for approximately 30 s to reach equilibrium. Then three pictures of this sessile drop are taken successively at intervals of 30 s. More liquid is then pumped into the drop until it reaches another desired size, and the above procedure is repeated. These procedures ensure that the measured static contact angles are indeed the advancing contact angles. It should be noted that the procedure to measure advancing contact angles for a goniometer-sessile drop technique cannot be used for ADSA-P, since ADSA determines the contact angles and surface tensions based on the complete and undisturbed drop profile, so that having a syringe or a needle immersed into the drop would render ADSA-P inapplicable.

Alternatively, Kwok et al. [45,71,73,74] have suggested to measure dynamic (advancing) contact angles at very slow motion of the three-phase contact line. In these low-rate dynamic contact angle experiments, images of an advancing drop (and hence information such as surface tension and contact angle) are recorded continuously as drop volume is steadily increased from below the surface. The procedures described here are different from those described above, in that the contact angles measured are not static angles. Since ADSA-P determines the contact angle and the three-phase contact radius simultaneously for each picture, the advancing dynamic contact angles as a function of the three-phase contact radius (i.e., location on the surface) can be obtained. In addition, the change in the contact angle, surface tension, drop volume, drop surface area, and three-

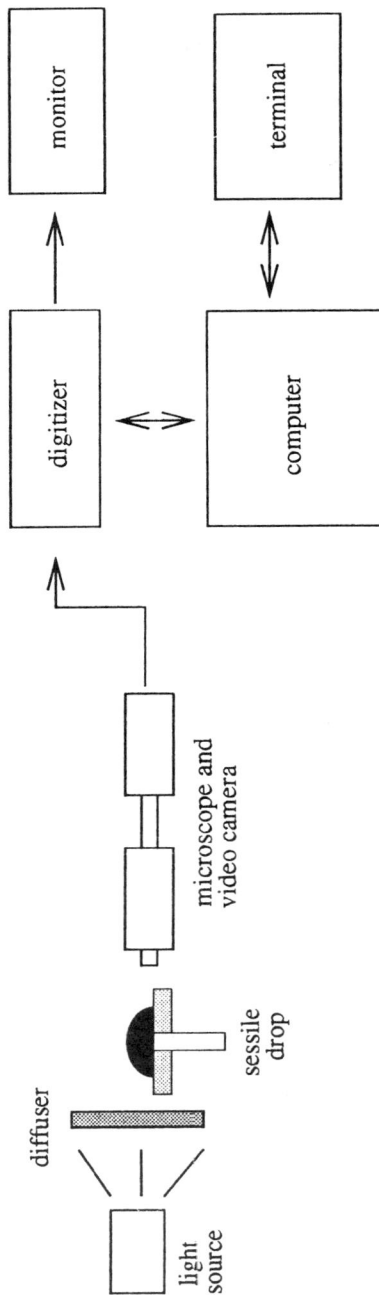

FIG. 6 Schematic of the ADSA-P sessile drop setup.

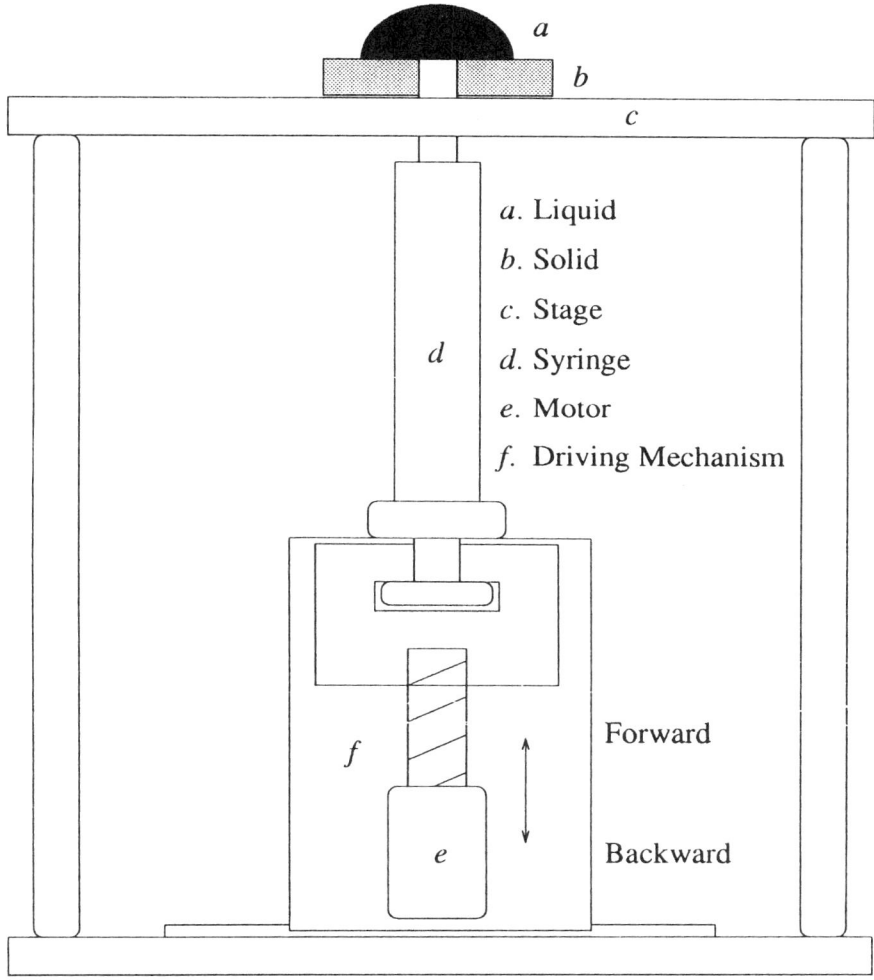

FIG. 7 Schematic of a motorized syringe mechanism for static and dynamic contact angle experiments.

phase contact radius can also be studied as a function of time. By stopping the motor, static contact angles can of course also be obtained. Typically, they are equal to low-rate dynamic angles.

In actual experiments, an initial liquid drop of about 0.3 cm radius is carefully deposited, covering the hole on the surface. This is to ensure that the drop will increase axisymmetrically in the center of the image field when liquid is supplied from the bottom of the surface and will not hinge on the lip of the hole. The motor in the motorized syringe mechanism is then set to a specific speed, by adjusting the voltage from a voltage controller. Such a syringe mechanism pushes the syringe plunger, leading to an increase in drop volume and hence the three-phase contact radius. A sequence of pictures of the growing drop is then recorded by the computer typically at a rate of 1 picture every 2–5 seconds, until the three-phase contact radius is about 0.5 cm or larger. For each low-rate dynamic contact angle experiment, at least 50 and up to 200 images are normally taken. The actual rate of advancing can be determined by linear regression, by plotting the three-phase contact radius over time. Different rates of advancing of the three-phase line can be achieved by adjusting the speed of the pumping mechanism [45,71].

The advantage of this procedure is that the quality of the surface is observed indirectly in the measured contact angles. If a solid surface is not very smooth, irregular and inconsistent contact angle values will be seen as a function of the three-phase contact radius. Contact angle complexities, such as slip/stick contact angle behavior, which affects the contact angle interpretation in terms of surface energetics, can also be identified [45,71].

We show in Fig. 8 the contact angle results [74] for water and diiodomethane on a poly(n-butyl methacrylate) PnBMA-coated silicon wafer surface. In Fig. 8a, the water contact angles are essentially constant, as the drop volume V increases and hence the three-phase contact radius R. Increasing the drop volume in this manner ensures the measured θ to be an advancing contact angle. These contact angles can be averaged to give a mean contact angle, i.e., $90.31 \pm 0.06°$.

Another contact angle result [74] is shown in Fig. 8b for diiodomethane on the same polymer. It can be seen that initially the apparent drop volume, as perceived by ADSA-P, increases linearly, and θ increases from $55°$ to $95°$ at essentially constant R. Suddenly, the drop front jumps to a new location as more liquid is supplied into the sessile drop. The resulting θ decreases sharply from $95°$ to $45°$. As more liquid is supplied into the sessile drop, the contact angle increases again. Such slip/stick behavior could be due to noninertness of the surface. Phenomenologically, an energy barrier for the drop front exists, resulting in sticking, which causes θ to increase at constant R. However, as more liquid is supplied into the sessile drop, the drop front possesses enough energy to overcome the energy barrier, resulting in slipping, which causes θ to decrease suddenly. It should be noted that as the drop front jumps from one location to the next, it is unlikely that the drop is or will remain axisymmetric. Such a nonaxisymmetric drop will obviously not meet the basic assumptions underlying ADSA-P, causing possible errors, e.g., in the apparent surface tension and drop volume. This can be seen from the discontinuity of the apparent surface tension and drop volume with time as the drop front sticks and slips. Obviously, the observed angles in Fig. 8b cannot all be the Young contact angles; since γ_{lv}

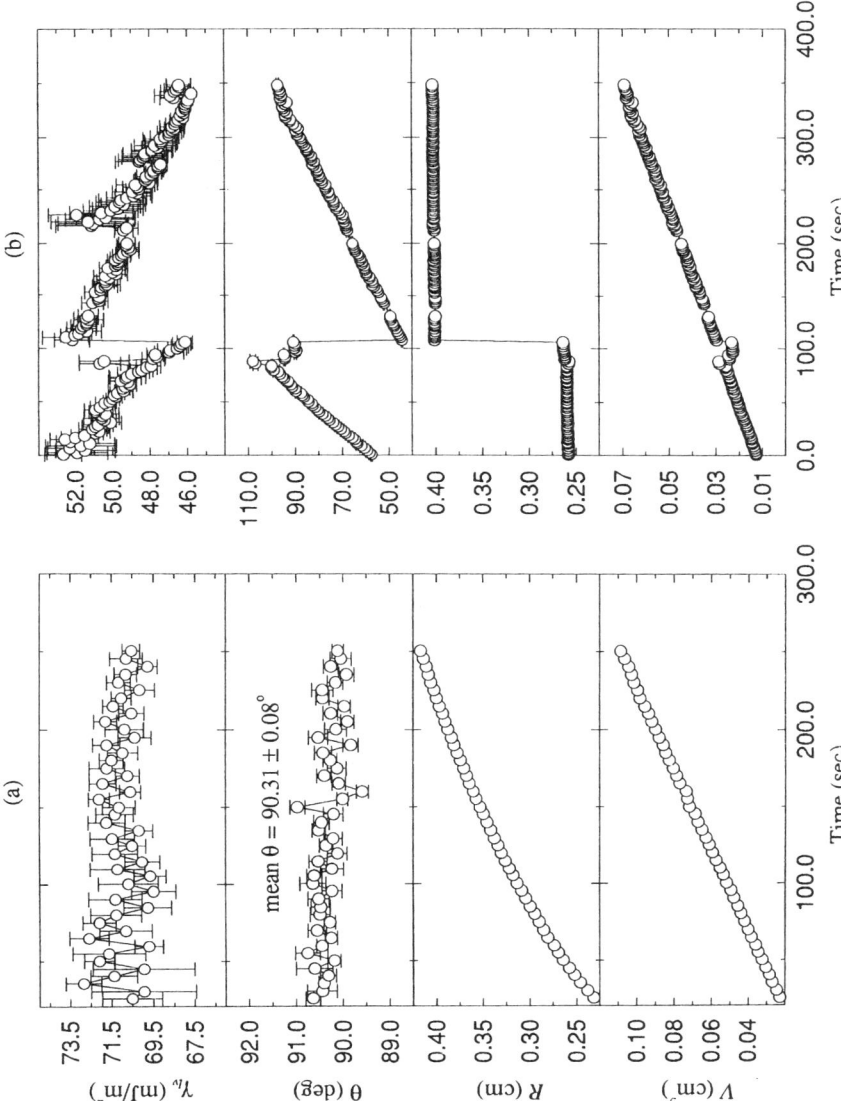

FIG. 8 Low-rate dynamic contact angles using ADSA-P on a poly(n-butyl methacrylate) PnBMA-coated silicon wafer surface [74] for (a) water and (b) diiodomethane. The slip/stick contact angles for diiodomethane cannot be used for the interpretation in terms of surface energetics.

TABLE 1 Comparison of Measured Contact Angles θ (deg) Using the Capillary Rise at a Vertical Plate Technique and ADSA-P for the Two Solid Surfaces, FC-721 and Teflon (FEP)

	FC-721		FEP	
Liquids	ADSA-P[a]	Capillary rise[c]	ADSA-P[a]	Capillary rise[c]
Tetradecane	73.31 ± 0.14	73.5 ± 0.1	52.51 ± 0.23	52.5 ± 0.1
	72.96 ± 0.21[b]			
Hexadecane	75.32 ± 0.27	75.6 ± 0.1	53.75 ± 0.22	53.9 ± 0.1
Dodecane	69.82 ± 0.25	70.4 ± 0.1	47.96 ± 0.21	47.8 ± 0.1
Dimethyl formamide			68.52 ± 0.21	68.6 ± 0.2

The error limits are 95% confidence limits.
[a] *Source:* Ref. 69.
[b] *Source:* Ref. 70.
[c] *Source:* Ref. 42.

and γ_{sv} (and γ_{sl}) are constants, θ ought to be a constant because of the Young equation. In addition, it is difficult to decide unambiguously at this moment whether the Young equation is applicable at all because of lack of understanding of the slip/stick mechanism. Therefore these contact angles should not be used for the interpretation in terms of surface energetics.

In several instances, ADSA-P contact angle measurements [69] are available for the same systems for which capillary rise measurements [42] were performed. Results for both techniques for FC-721 and FEP are summarized in Table 1. The capillary rise results presented are the average of 10 contact angles measured at 10 different velocities ranging from 0.08 to 0.49 mm/min. The contact angle results from ADSA-P were measured at zero velocity of the three-phase contact line. It is apparent that there is excellent agreement among the contact angles from the two techniques. The choice of methods depends on the specific application and is largely a matter of convenience and equipment availability.

B. Nonflat Surface and Particles

In many instances, the solids of interest may not be in the form of a flat surface. This section describes briefly the techniques to measure contact angle of solids of various geometries; a detailed discussion can be found elsewhere [7,8].

1. Nonflat Surface

(a) Individual Fiber. The Wilhelmy method, as described above, is a good technique for measuring the contact angle on individual fibers of known diame-

ter. The fiber diameter can be determined by first using the method with a liquid of known surface tension for which the contact angle is zero. Equation (3) can be used to calculate the perimeter.

Several direct procedures have also been developed. Schwartz and coworkers [76] suspended a fiber horizontally and used a goniometer to measure the contact angle of drops deposited on the fiber. The drops had a diameter slightly larger than that of the fiber, which could be rotated about its longitudinal axis to generate advancing and receding contact angles. Of related interest is the work of Roe [77], who computed equilibrium shapes for drops resting on fibers of different diameters. It was shown that the apparent contact angle can become misleading if the drop diameter is much larger than that of the fiber.

Bascom and Romans [78] passed a fiber vertically through the center of a liquid drop held within a small-diameter ring. The advancing and receding contact angles were measured with a microscope as the fiber was pulled through the stationary drop.

In principle, the capillary depression caused by a fiber floating on a liquid surface can be used to calculate the contact angle [7]. In practice, however, the relatively small depth of immersion can make it difficult to obtain sufficient accuracy. Recently, a technique utilizing image analysis has been developed to analyze the capillary rise profile around fibers or cylinders [79].

(b) *Capillary Tube.* If the inside and outside surfaces of the tubes are of the same material, the Wilhelmy balance can be used to measure the contact angle [7].

For tubes of small enough diameter, the meniscus may be considered to be spherical, and the capillary rise h is given by

$$h = \frac{2\gamma_{lv} \cos\theta}{rg\,\Delta\rho} \tag{11}$$

where r is the capillary radius and h is the capillary rise. In many cases, both h and r can be determined optically. If r is too small, it can be calculated from the length of capillary occupied by a known mass of mercury. The Langmuir–Schaeffer reflection method can also be used to measure directly the contact angle in this case [80].

(c) *Mat and Woven Fabric.* The apparent contact angle of a sessile drop on a mat of fibers or on a woven fabric should not be interpreted as the contact angle that the liquid would make on an individual fiber of the same material. The local geometry of the three-phase line will be affected by the positions of the fibers and will have a strong influence on the apparent contact angle. For this reason, only qualitative tests of fabric wettability are available. One such test is to put drops of liquids of different surface tension on the material and to note the surface tension required to cause repellency [81]. A second approach, known as the Draves test, involves recording the time required for a disk of fabric placed on a liquid surface to sink [7]. The disadvantage of both of these tests is that comparisons can only be made among fabrics with identical pore structure and size. In addition, it should be noted that if a series of liquid mixtures is used to generate a range of liquid surface tensions, care must be taken to ensure that preferential adsorption of one liquid does not occur. In general, the thermodynamic interpretation of contact angles of liquid mixtures is difficult.

(*d*) *Rough Surface.* As described above, the thermodynamic interpretation of contact angles in terms of the Young equation for energetic calculations is not possible for rough surfaces. However, if the goal is to study wettability as in the case of mat and woven fabric, rather than energetics, such measurements are still worth pursuing.

Prime examples of such surfaces are biological materials. It should be noted that measuring contact angles on rough surfaces such as biological materials is generally difficult. This is because such surfaces usually present not only small contact angles but also morphological and energetic imperfection, leading to irregularities of the three-phase contact line, as seen in Fig. 9. It will be difficult and dubious to measure contact angles on such drops, e.g., by finding a tangent of the drop profile at a three-phase contact point. Axisymmetric drop shape analysis–diameter (ADSA-D), another branch of ADSA, avoids such problems and provides accurate and consistent contact angles by viewing the drop from above [82,83]. A schematic of this setup is shown in Fig. 10.

In the ADSA-D program, the contact angle is determined from a numerical integration of the Laplace equation of capillarity, Eq. (10), when the contact diameter, liquid surface tension, and volume of the drop are known. It was first developed especially for situations where the contact angle is small, e.g., less than $20°$, and later found to be useful for measuring contact angles on imperfect solid surfaces [84–90]. The input information is drop volume, liquid surface tension, contact diameter of the drop, density difference between the liquid–fluid interface, and the local acceleration of gravity. A computer-assisted digitization procedure of the drop perimeter was used to measure the effective contact diameter of the drop on an image taken from above.

A video image of the drop contact area can be digitized semiautomatically, using a "mouse" on the video screen. The pattern of selection of points on the three-phase contact line is characterized schematically in Fig. 11. An average drop diameter can then be calculated and used to determine the contact angle. As an illustration, contact angle measurements on layers of several bacterial strains have been measured: Fig. 9 illustrates the typical shape of drops of double-distilled water formed on a layer of *Thiobacillus ferrooxidans* cells. For these drops, points at the drop perimeter were selected to estimate the average drop diameter, as schematically shown in Fig. 11. The results of such

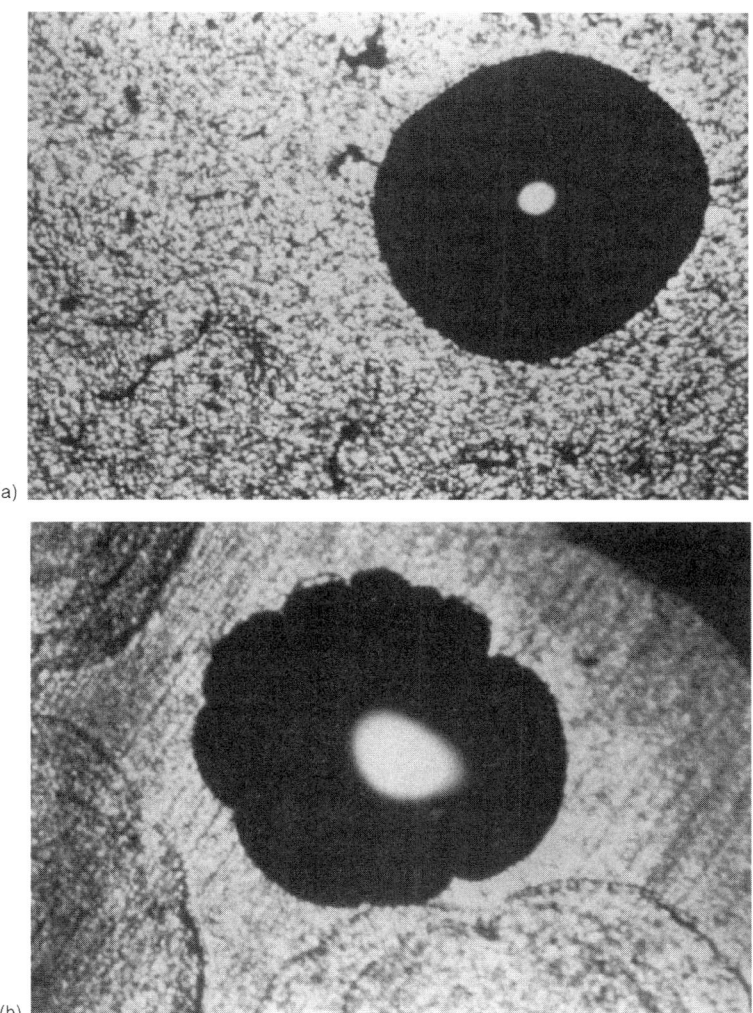

FIG. 9 Images of sessile drops (top view) of water on a layer of *Thiobacillus ferrooxidans* cells. The contact angles calculated using ADSA-D are (a) 12.7° and (b) 11.3°.

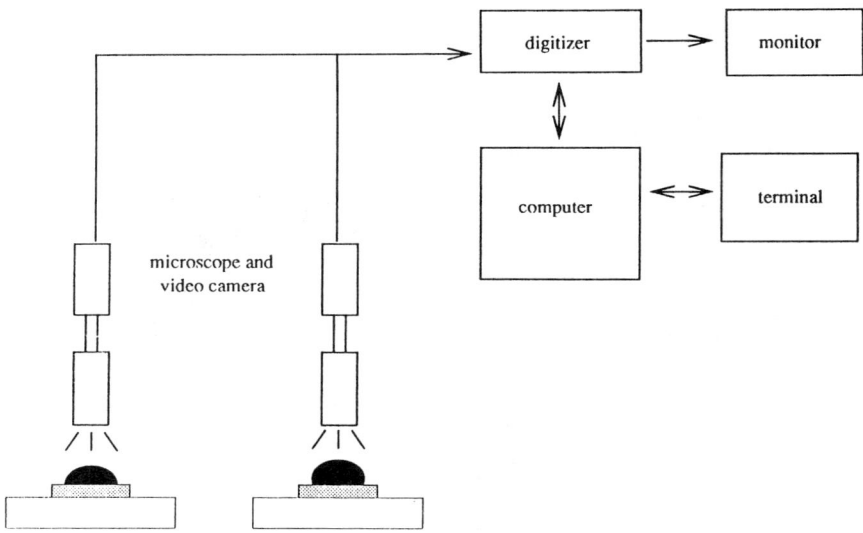

FIG. 10 Schematic of the ADSA-D setup.

measurements on three different species of bacteria (*T. thiooxidians*, *Staphylococcus epidermidis*, and two strains of *T. ferrooxidans*) are illustrated in Fig. 12 [84]. It is apparent that the measured contact angles remain in a fairly narrow range as the cell layers dry slowly over approximately 2 h. The average contact angle of 16 drops for *Staph. epidermidis* and *T. thiooxidans* are found to be

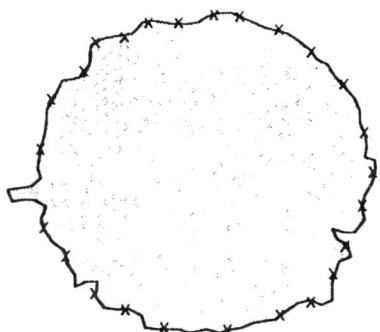

FIG. 11 Schematic of the determination of the perimeter of a fictitious sessile drop on a video screen using a cursor controlled by a mouse to select perimeter points.

Contact Angle Techniques and Measurements

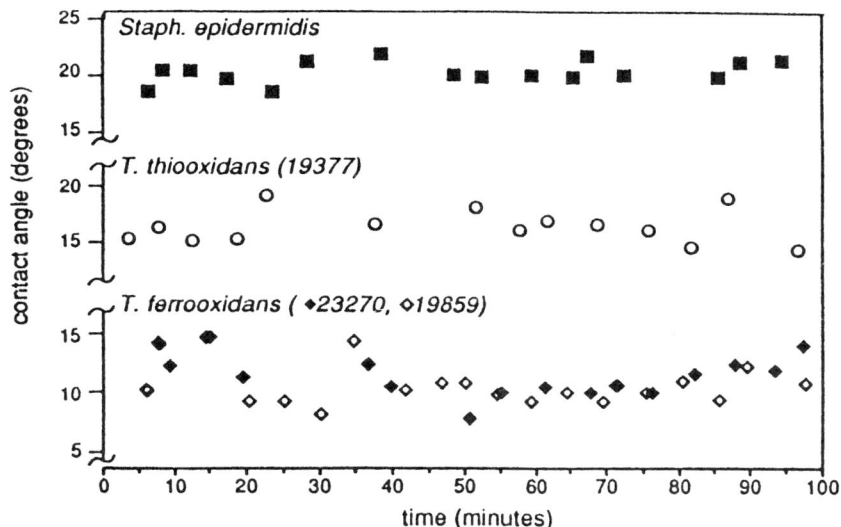

FIG. 12 The contact angles of sessile drops of water on layers of different bacteria measured using ADSA-D. The horizontal axis represents that time after formation of the layer at which drops were deposited. The mean contact angles (±95% confidence limits) are (■) 20.6 ± 0.9° (*Staphylococcus epidermidis*); (○) 16.9 ± 0.9° (*Thiobacillus thiooxidans*); (◆) 11.7 ± 1.0° (*T. ferrooxidans* strain 23270); and (◇) 10.5 ± 0.9° (*T. ferrooxidans* strain 19859).

20.6 ± 0.9° and 16.9 ± 0.9°, respectively, where the errors represent the 95% confidence limits. It should be noted that a relatively high degree of accuracy is attained from a relatively small number of measurements. The two strains of *T. ferrooxidans* do not show a statistically significant difference in their contact angles. While these angles cannot be employed for energetic calculations with any degree of certainty, they provide qualitative means to characterize the wettability of rough surfaces. ADSA-D has also been applied for various imperfect surfaces, including metals [91] and wood [92].

2. Particles

The measurement of contact angles on solids in powder or particle form is a challenging task, and numerous methods have been proposed [7]. In this section, three successful indirect approaches are discussed: capillary penetration into columns of powders, sedimentation volume of particles, and solidification fronts of particles. These methods are indirect because they provide the solid surface tension rather than the contact angle. For the sake of completeness, various direct approaches are also summarized briefly.

A direct and simple method is to measure the contact angle of a sessile liquid droplet on a compressed powder cake (tablet) of rigid particles. This method is often used to characterize the wettability of pharmaceutical powders [92–97] and organic pigments [98], since these substances usually do not exist as smooth, flat surfaces. Such data, however, must be viewed with caution, since they are affected by surface roughness [99,100], possible plastic deformation of the powder particles while forming the tablet [100], particle swelling [94], and tablet porosity [99,101]. If the powder is slightly soluble in the measuring liquids, contact angles have been measured using drops of solution saturated with the powder instead of using pure liquids [102]. Whenever possible, contact angles measured on compressed cakes should be confirmed by other means, such as measurements on films of the substance prepared by solvent casting or dip coating; see later.

Contact angles on compressed cakes have also been determined by the $h - \epsilon$ method introduced by Kossen and Heertjes [103,104]. This method is based on the assumption that the powder consists of identical spheres. After the powder is compressed into a cake, it is saturated with the measuring liquid and a drop is placed onto the surface. The contact angle is calculated from the height of the drop (e.g., as measured by a cathetometer), a knowledge of the cake porosity, the liquid density, and the surface tension.

The relative wettabilities of hydrophobic pharmaceutical powders have been determined using a vacuum balance technique to measure the rates of water vapor uptake and the equilibrium moisture levels of the powder samples [105].

Indirect methods such as film flotation [106–115] can be used to determine the solid surface tension, based on floatability of particles. In a film flotation experiment, closely sized particles are sprinkled onto the surface of the wetting liquid, and the fraction of particles that sink into the liquid is determined. Depending on wetting characteristics of the material and the surface tension of the test liquid, the particles either remain at the liquid–vapor interface or are immediately engulfed into the liquid. At a particular surface tension, those particles that do not sink into the wetting liquid are considered to be hydrophobic, while those that are imbibed into the liquid are considered to be hydrophilic. After performing a film flotation test, the hydrophobic and hydrophilic fractions are recovered, dried, and weighed. The percentage by weight of the hydrophobic fraction of the particles for each solution is plotted as a function of the surface tension of the solution. From this curve, four parameters for defining the wetting characteristics of the particulate samples can be determined: the mean critical wetting surface tension (the critical wetting surface tension is the liquid surface tension at which the solid surface is completely wetted), the minimum and maximum wetting surface tensions, and the standard deviation of the wetting surface tension [110,116].

This method has been employed to measure the critical wetting surface tensions of particles of sulfur, silver iodide, methylated glass beads, quartz, paraffin-wax–coated coal, and surfactant-coated pyrite. Generally, Fuerstenau and coworkers [106–115] found that the film flotation technique is sensitive to the surface hydrophobicity and the heterogeneity of the particles. It was found that particle size, particle shape, particle density, film flotation time, and the nature of the wetting liquids have negligible effects on the results of film flotation. But the liquid and the solid particles used in the experiments must not have any chemical interactions.

(*a*) *Capillary Penetration.* Early approaches to characterize the wettability of powders were introduced by Washburn [117] and by Bartell and Osterhof [118], who measured such quantities as the rate at which a liquid penetrates into the cake of compressed powder, and the gas pressure required to prevent further penetration of the liquid into the powder cake. The contact angle can then be estimated from these data. There is one important point to keep in mind. While the contact angles obtained on such surfaces are not normally the Young contact angles (i.e., it will not be possible to use it for the determination of the solid surface tension γ_{sv}), it is indeed this phenomenological contact angle that will, together with the liquid surface tension, determine the Laplace pressure ΔP and hence capillary penetration [50].

One method, therefore, is to measure the pressure necessary just to balance the Laplace pressure, which drives liquid into a capillary bed, i.e., the limiting pressure necessary to prevent capillary penetration:

$$\Delta P = \frac{2\gamma_{lv} \cos\theta}{r} \tag{12}$$

where r is the radius of the capillary. It has been shown [119] that if Eq. (12) is used in conjunction with the Hagen–Poiseuille equation for steady flow, the following equation can be derived:

$$K\gamma_{lv} \cos\theta = \left(\frac{\eta}{\rho^2}\right)\left(\frac{M^2}{t}\right) \tag{13}$$

where K is an unknown parameter that depends on the geometry of the capillary; η and ρ are respectively the viscosity and density of the liquid; and M is the weight of the liquid penetrated into the capillary at time t. Thus the values of $K\gamma_{lv} \cos\theta$ can be obtained indirectly when the liquid properties and the amount of capillary penetration are known. Equation (13) is referred to as the modified Washburn equation with the following assumptions: (1) laminar flow predominates in the pore spaces; (2) gravity can be neglected; and (3) the geometry of the porous solid is constant.

The basic precautions/procedures for capillary penetration experiments can be summarized as follows: in actual experiments, powder is packed in a glass tube of which the lower end is closed with a glass filter. Considerable care is necessary to obtain a constant and homogeneous powder packing; a precisely weighted quantity of powder is also required to fill up to the same height as the glass tube, by manually tapping the powder. The filled columns are attached to an electrobalance and brought into contact with several test liquids. Their penetration velocities are determined by measuring the weight gain with the electrobalance as a function of time. A tensiometer can be used for these gravimetric experiments.

There are two basic requirements that the test liquids must satisfy in order to be used for capillary penetration experiments: (1) they should be chemically inert with respect to the powder of interest; and (2) the surface tension γ_{lv} range of the test liquids should cover the anticipated surface tension of the powder. For a series of pure liquids of different surface tension, Grundke et al. [119] found that there exists an extremum (maximum) when the values $K\gamma_{lv}\cos\theta$ are plotted against the surface tension γ_{lv} of the liquids, at a value of $\gamma_{lv} = \gamma_{lv}^*$. It has been argued [119] that the surface tension of the powder can then be determined as $\gamma_{sv} = \gamma_{lv}^*$. For example, in the case of the PTFE particles, it was found [119] that a maximum occurs at $\gamma_{lv}^* = 20.4$ mJ/m^2. Thus the γ_{sv} value of the PTFE particles would be 20.4 mJ/m^2. When this γ_{sv} value and $\gamma_{lv} = 72.5$ mJ/m^2 for water are used in the equation of state approach for solid–liquid interfacial tensions [1], it predicts a contact angle of 104°. Remarkable, this is exactly what one would observe on a smooth Teflon surface. Although the packed powder bed certainly does not present a flat and smooth surface, the derived values for γ_{sv} are those obtained by contact angle measurements of a flat and smooth surface. It is well known that even highly compacted hydrophobic powder, presenting a seemingly flat and smooth solid surface, does not yield the same contact angle as truly smooth and coherent solid surfaces. It appears that this indirect method may provide more relevant information than direct contact angle measurements on imperfect solid surfaces.

At present, many authors [120–126] follow another concept: From the plot of $[\eta/\rho^2][M^2/t]$ versus the surface tension γ_{lv} of the liquids, the geometric factor K is calculated for those liquids that should wet the solid completely. By inserting this K value and $[\eta/\rho^2][M^2/t]$ for these liquids into Eq. (13), their contact angles θ are calculated and used for the interpretation of the solid–vapor surface tension of the porous material. This procedure is dubious, because it can be expected that the contact angles, calculated from the Washburn equation, are affected by roughness and porosity. If we apply this procedure to the PTFE powder for hexadecane, a contact angle $\theta = 88°$ would be obtained. However, it is well known that the contact angle of hexadecane on a flat and smooth surface should be about 46°. Obviously, the former value only reflects the contact angle/wettability of hexadecane on rough PTFE powder, which is meaningless for energetics calculations in conjunction with the Young equation.

(b) *Sedimentation Volume.* Sedimentation experiments are a well-established technique to study the stability of dispersions of powders in liquids [127–129]. While in many cases the behavior of such systems is governed by van der Waals and electrostatic interactions, it is to be expected that for polymer particles, particularly in nonaqueous media, the effect of the electrostatic interactions can be considered negligible. It is also of importance to note that van der Waals interactions can be related to surface tensions [130–132]. The van der Waals interaction between two parallel infinitely extended flat surfaces in a liquid medium was first calculated by Hamaker [131]. For the work done in bringing these surfaces from infinity to a distance d_0, Hamaker obtained

$$W = -\frac{A_{123}}{12\pi d_0^2} \qquad (14)$$

where the coefficient A_{123} has subsequently been called the "Hamaker coefficient." If we assume that d_0 is so small that we in fact have contact between the two solid phases, this work is the thermodynamic free energy of adhesion

$$\Delta F^{adh} = \gamma_{12} - \gamma_{13} - \gamma_{23} \qquad (15)$$

where the indices 1, 2, and 3 refer respectively to the solid (1), solid (2), and liquid (3).

It has been shown that the free energy of adhesion can be positive, negative, or zero, implying that van der Waals interactions can be attractive as well as repulsive [130,133,134]. While Eq. (14) can, strictly speaking, be expected to hold only for systems that interact by means of dispersion forces only, there are no restrictions on Eq. (15). Since this equation describes very well the fundamental patterns of the behavior of particles, including macromolecules, independent of the type of molecular interactions present, it was found to be convenient to define an "effective Hamaker coefficient" that reflects the free energy of adhesion [130].

While van der Waals interactions between unlike solids in a liquid can be attractive as well as repulsive, it is clear from the underlying thermodynamics [133,135] that like particles can only attract each other, with zero interaction in the limiting case. For interaction between particles of the same kind embedded in a liquid, this interaction is governed by the free energy of cohesion

$$\Delta F^{coh} = -2\gamma_{sl} \qquad (16)$$

Because solid–liquid interfacial tensions are always positive or zero, as a limiting case, it follows that $\Delta F^{coh} \leq 0$, implying that, in the absence of electrostatic forces, there will always be an attraction between like particles suspended in a liquid (with no interaction, as a limiting case, at $\gamma_{sl} = 0$).

It can be expected that the sedimentation volume of particles will show extrema in the special case of $\Delta F^{coh} = 0$. There are at least two possible patterns of behavior, depending on whether agglomeration of the particles at the early stages of sedimentation is possible.

1. If there is no agglomeration at finite values of the free energy of cohesion, then, for zero free energy of cohesion, least close packing of the sediment and hence a maximum in the sedimentation volume V_{sed} is expected.

2. If there is agglomeration at finite values of the van der Waals attraction in the early stages of sedimentation, then this agglomeration will cease when the van der Waals attraction approaches zero. Since the irregularly shaped aggregates resulting from agglomeration do not pack well, one would expect minimum sedimentation volume at zero van der Waals attraction.

In view of the possibility that the sedimentation volume of particles may show extrema when $\Delta F^{coh} = 0$, such an extremum in the sedimentation volume may provide a means to determine the solid–vapor surface tension of the particles. The solid–vapor surface tension γ_{sv} of particles would be equal to γ_{lv}, the surface tension of the suspending liquid at which the sedimentation volume extremum occurs.

Therefore to determine the particle surface tension by using the sedimentation volume technique, the required basic procedures in the experiments can be summarized as follows:

1. Prepare a series of liquids with a surface tension range covering the surface tension of the particles of interest in suitable graduated cylinders.
2. Put an equal amount of the particles into each liquid.
3. Determine the liquid surface tension γ_{lv}^* at which an extremum in the sedimentation volume occurs.

The surface tension of the particles can then be determined as $\gamma_{sv} = \gamma_{lv}^*$.

In practice, sedimentation volume experiments are performed with binary liquid mixtures as the suspending liquids, in order to have a sufficiently large range of surface tension and to be able to adjust the liquid surface tension to any specific value.

There are several requirements that the suspending liquids must satisfy in order to be used for the sedimentation experiments: (1) they should be chemically inert with respect to the solid particles; (2) the boiling temperature should be reasonably high to minimize evaporation; (3) the density of the liquid should be less than that of the particles; (4) liquids with zero or nearly zero dipole moment as well as those with higher dipole moment should be used in order to cover a wide polarity range; (5) the liquid components should be miscible in all ratios

Contact Angle Techniques and Measurements 69

with each other; and (6) the surface tension γ_{lv} of the mixtures should cover the surface tension range required.

In actual experiments, graduated micro(test)tubes (100 mm high with an inner diameter of 3 mm) were used. The total volume, 1.00 ml, of these tubes is divided into 100 graduations, and the sedimentation volume was read (or estimated) to 0.1 of a graduation, i.e., to 0.001 ml. Each of the dispersions prepared in the centrifuge tubes was then transferred into a microtube using Pasteur micropipets. The selected liquid mixture was used as rinsing liquid in each case, so that the polymer powder was totally and homogeneously suspended in the liquid medium. The reading of the sedimentation volume V_{sed} of the polymer powder was taken every day for 3 days to 1 week, depending on the sample, until no further change in V_{sed} occurred [127–129,136].

The method has been employed for various polymer powders [127–129,136]. The results obtained are in good agreement with those determined from contact angles on smooth polymers [8]. The method has also been used for various pharmaceutical powders [137–139].

(c) *Solidification Fronts.* The behavior of inert particles at the solid–liquid interface of an advancing solidification front is a multifaceted phenomenon. Consider a particle in a channel that contains a liquid. One end of the channel is cooled, causing a solidification front to move through the liquid as it solidifies. When the solidification front meets the particle, one of the following occurs [140,141]:

1. The particle is pushed along by the solidification front. This occurs when the free energy of engulfment is positive. The particle is not, however, pushed for all velocities of the solidification front. For each particle that is pushed, there is a maximum, or critical, velocity above which the particle will be engulfed by the advancing solidification front.

2. The particle is engulfed. This occurs for all velocities when the free energy of engulfment is negative, and for velocities greater than the critical velocity when the free energy of engulfment is positive.

This phenomenon is of practical importance in metallurgy [142], in materials science and crystal growth [143], in phagocytosis [144,145], in soil mechanics, soil genesis, and the behavior of vegetation in permafrost and cold regions [146], and in the separation of various types of solid particles from one another by particle chromatography [147,148].

The behavior of small particles at an advancing solidification front has been investigated experimentally for both vertical [149–154] and horizontal [140,141, 144,155–158] motion of the solidification front.

Thermodynamically, the process of particle engulfment by the advancing solidification front can be modeled by the net free energy change of the system

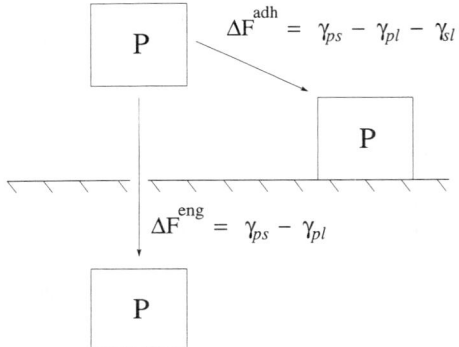

FIG. 13 Free energy changes during particle–surface adhesion and particle engulfment processes.

during the engulfing process. As illustrated in Fig. 13, the net free-energy change per unit surface area for the engulfment process of a square particle is given by

$$\Delta F^{eng} = \gamma_{ps} - \gamma_{pl} \tag{17}$$

where γ_{ps} and γ_{pl} are the particle–solid and particle–liquid interfacial tensions, respectively. Equation (17) describes how, for a particle of unit surface area, 1 cm^2 of particle–solid interface is generated and 1 cm^2 of particle–liquid interface is annihilated.

The condition for particle engulfment is that the net change in the free energy of the system, ΔF^{eng}, be negative, i.e.,

$$\Delta F^{eng} < 0 \tag{18}$$

If ΔF^{eng} is positive, i.e.,

$$\Delta F^{eng} > 0 \tag{19}$$

there will be particle rejection. An intermediate step in the process of particle engulfment is particle adhesion. The associated free energy change, the free energy of adhesion, is given by

$$\Delta F^{adh} = \gamma_{ps} - \gamma_{pl} - \gamma_{sl} \tag{20}$$

The process of particle adhesion involves generation of particle–solid interface and annihilation of particle–liquid interface as well as solid–liquid interface. It is not clear *a priori* which of the two free energies, ΔF^{eng} and ΔF^{adh}, should be used to study the interaction between a particle and a solidification front. A discussion of this point has been given elsewhere [159]. Fortunately, γ_{sl} is

Contact Angle Techniques and Measurements

usually quite small, so that there will not normally be a large difference between ΔF^{eng} and ΔF^{adh}.

As an experimental fact, particle engulfment depends on the rate of the advancing solidification front. Therefore it is necessary to study rate-dependent phenomena. It has been shown [141] that, at even the lowest rates, Teflon particles and siliconized glass spheres were engulfed by the advancing solid–liquid interfaces of biphenyl and naphthalene melts, while nylon and acetal particles were pushed. Polystyrene latex spheres were just reoriented without being pushed. As a result of these observations, nylon and acetal particles were chosen for rate studies. Rates of solidification were measured by timing the progress of the solidification front over distances measured with a micrometer eyepiece graticule. With the aid of a temperature programmer, it was possible to adjust the velocity of the interface at will.

Conventionally, such experiments were performed with the aim of determining what is called the critical velocity V_c, i.e., the velocity that separates pushing and engulfing of particles. However, from experimental observations, the transition from pushing to engulfing is not sharp, and three modes of particle behavior can be defined. At relatively high rates of solidification, the particles are engulfed instantly on contact with the solidification front (mode 1). At intermediate rates, the particles are pushed through various distances before being engulfed (mode 2). At relatively low rates of solidification, steady-state pushing of any individual particle can be observed without engulfment (mode 3). As typical examples, these three modes of engulfing and pushing for various particle diameters are represented in Fig. 14 for biphenyl (matrix)/nylon (particle) systems.

The three modes of particle pushing are clearly shown in the figure. The area between the broken lines represents mode 2. These transition velocities are clearly particle-size dependent. This was to be expected, since dynamic effects (e.g., viscous drag, wall friction) retard particle motion more strongly the larger the particles. The critical velocity V_c is defined as the central line through the area representing mode 2.

The critical velocity V_c can be understood as follows. If the free energy of adhesion between the solidification front and the solid particle is positive, then repulsion occurs, and the particle will be pushed along at low rates of solidification. This motion, however, sets up a viscous drag force acting on the particle and opposing the thermodynamic or van der Waals type of repulsion. As the rate of solidification increases, the viscous drag will increase and finally overpower the van der Waals type of repulsion; engulfing will take place, and the corresponding rate of solidification is the critical velocity V_c. The latter is thus a measure of the balance between van der Waals repulsion and viscous drag. Thus, in essence, measuring the critical velocity of engulfing V_c and knowing the viscous drag on the particle would allow one to determine the van der Waals interaction of the particle with the solidification front, and hence the free

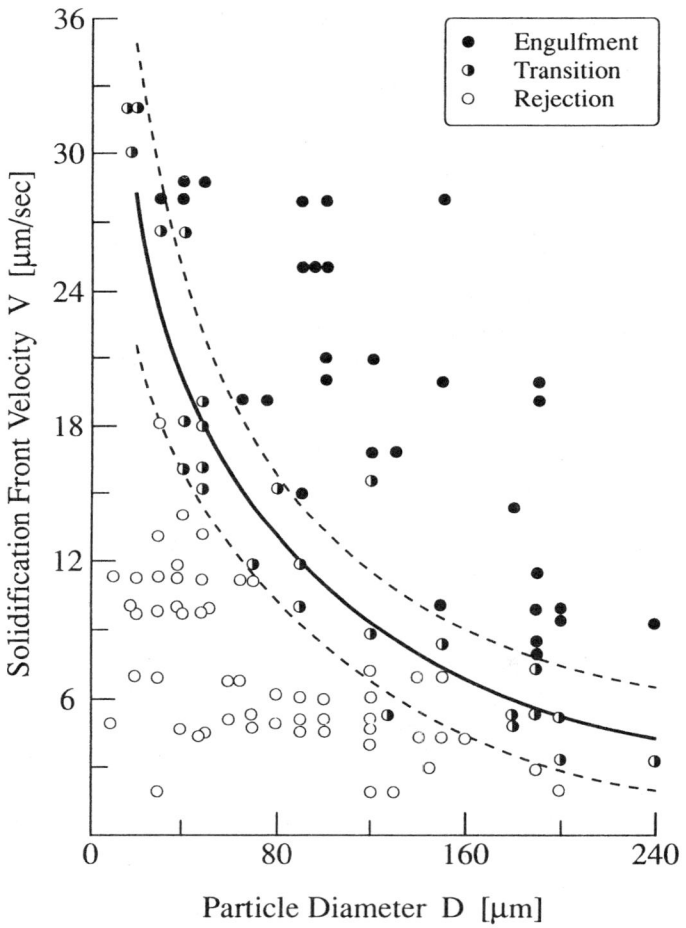

FIG. 14 Pushing and trapping velocities as a function of the particle diameters for the biphenyl (matrix)–nylon (particle) systems.

energy of adhesion ΔF^{adh} and other interfacial free-energy quantities, such as the solid–melt interfacial tension. Thus, through the appropriate dimensional analysis [157,159], the critical velocity V_c, the free energy of adhesion ΔF^{adh}, and material properties can be related. Having thus obtained a value of ΔF^{adh}, one can, through the equation-of-state approach [1], obtain the various relevant interfacial tensions and in this manner determine the surface tension of small particles, γ_{pv}. Thus the sequence of steps required to obtain, for example, γ_{pv} may be summarized as follows:

$V_c \to$ dimensional analysis $\to$ equation of state for interfacial tensions

$\to \gamma_{pv}$

This method has been applied successfully in the determination of the surface tensions of polymer particles [140,141,159], coal particles [160,161], and biological cells [162–168]. Contact angles predicted from these surface tensions involving solids agree well with experimental contact angles [159].

III. PREPARATION OF SOLID SURFACES

Contact angles can be used to characterize the fundamental surface properties of a solid material or to study the effective properties of the material in its natural or as-manufactured state. The acquisition of thermodynamically significant contact angle data for fundamental studies is largely dependent on the quality of the substrate surface. The effects of roughness and heterogeneity can easily overshadow the influences of interfacial energetics. It is therefore important, in fundamental scientific studies, to produce solid surfaces of sufficient quality to ensure that the contact angles accurately reflect the interaction between the solid and the liquid as given by the Young equation.

Contact angle hysteresis, the difference between advancing and receding contact angles, is a measure of surface nonideality, although it is somewhat dependent on the rate at which the receding angle is measured as well as the time of contact between the solid and the liquid. Hysteresis may be due to roughness, heterogeneity in the chemical composition of the surface, liquid dissolution of the surface, and adsorption on the solid. It has been shown that if hysteresis is due to heterogeneity consisting of high- and low-energy surface patches, then both the advancing and the receding contact angles should be "Young contact angles" in the sense that they may be used in conjunction with the Young equation [4,5]. In this case, the advancing contact angle represents the equilibrium angle on an ideal surface composed entirely of the low-energy component, and the receding angle similarly corresponds to the higher energy patches. Between these two angles are a range of metastable contact angles that are meaningless in terms of the Young equation. As a practical consequence of this, since most real surfaces are heterogeneous on a microscopic scale, contact angles must always be measured in the advancing mode if the goal is to infer the solid surface tension. Furthermore, it can be demonstrated that the receding contact angle is only a manifestation of the impurities of a surface and is difficult to measure reproducibly [169]. More recently, contact angle hysteresis has been linked to line tension [5,48].

If surface roughness is the primary cause of hysteresis, it may be impossible to obtain Young contact angles. In such cases, the advancing contact angle is influenced more by the microscopic geometry than it is by interfacial energetics.

Very smooth surfaces are required in order to eliminate all roughness effects. For example, on layers of organic pigment, contact angle hysteresis due to roughness was absent if roughness was less than 0.1 μm [98]. The experiments of Oliver et al. [170] indicated that scratches significantly less than 0.5 μm can result in a "slip/stick" behavior as a liquid spreads over a solid surface. There are also indications that the sensitivity to roughness effects increases with the surface tension of the solid [28].

In a practical sense, it is not necessary to eliminate completely all evidence of roughness and heterogeneity. If the average contact angle and liquid surface tension measurements for a variety of liquids all give the same solid surface tension via the equation of state [1], then effectively the surface is sufficiently smooth and homogeneous. The required level of accuracy will determine the acceptable solid surface quality.

The other major application of contact angles is the characterization of real surfaces as they are received or manufactured. These may not be Young contact angles and may therefore not be thermodynamically significant. Nevertheless, they can serve to assess solid surfaces qualitatively and detect relatively gross changes in surface energy.

The rest of this section deals with the techniques that can be used to improve the quality of solid surfaces in order to obtain more meaningful and reproducible contact angles.

A. Heat Pressing

This method involves pressing thermoplastic polymer (particles, film, or block) such as fluorinated ethylene propylene (FEP), polyethylene, polypropylene, or polysulfone between two very clean smooth surfaces at an elevated temperature [7,171]. Typically, glass plates are used as the platens, which are squeezed together with clamps in an oven or between the heated steel surfaces of a hydraulic heat press. The pressure and temperature employed during the heat-pressing process must be sufficient to cause the polymer to conform to the glass, but not so high as to cause any chemical changes in the polymer. In practice, this means that a thermoplastic must be heated to (or near) the glass transition temperature and then pressed to produce smooth surfaces. The required temperature and pressure must be established empirically for each system. Care should be taken to avoid the fracture of polymer surfaces, which may adhere to the platens. It has been found that immersing the polymer and platens together in warm water for 24 h will facilitate the separation without damaging the surface.

B. Solvent Casting

When the solid material can be dissolved in a volatile solvent, then solvent casting becomes an option for the preparation of smooth substrates. The method has been applied to a wide variety of materials including, for example, polymers [45,72–74,172], pharmaceuticals [173], and bitumen [174].

Although many variations of the basic procedure have been used, the following is a typical method. A solution of the substrate material with a concentration of the order of 1% is prepared with a volatile solvent. Glass microscope slides, cleaned with chromic acid, are mounted on a specially designed flat-bed centrifuge rotor head, which holds the slides in a horizontal position during centrifugation. A 500 μL drop of solution is deposited on each 75×25 mm^2 slide. The centrifuge is then operated at moderate speed and under slightly evacuated conditions (created by connecting the centrifuge to a vacuum pump or a water aspirator) until all the solvent has evaporated, leaving a thin smooth film on the glass slide. The substrate selected for the film deposition is arbitrary as long as it has a smooth surface, can be easily cleaned, and is wettable by the solution. A different and simpler procedure has recently been employed [45,71–74]: a few drops of the 2% polymer/solvent solution is deposited onto silicon wafer surfaces, of about 2×3 cm^2, inside glass dishes overnight. A thin and smooth layer of the polymer forms on the surface after the volatile solvent has evaporated. It was found that roughness of the surface is in the order of nanometers or less.

C. Dip-Coating

Substrates may be dip-coated in solutions or melts of the solid of interest. The technique has been applied, for example, to perfluorinated acids on platinum foil [175], to fluorosurfactants on glass, mica, or silicon wafer [69–72,166,172], and to the production of elastomeric films of butyl rubber [176]. The properties of the resulting film are dependent on the concentration of the solution, which is usually low, and on the speed at which the substrate is withdrawn from the liquid. The latter is best controlled by attaching the substrate to an electric motor with a withdrawal speed of between 0.2 and 10 mm/s.

D. Langmuir–Blodgett Film Deposition

Langmuir–Blodgett film deposition is basically a variation of the dip-coating process [177–179]. The technique is used to create monomolecular layers of amphiphilic molecules on high-energy substrates such as quartz, glass, mica, and metals. Such long-chain organic molecules have a polar head group.

Typically, these films are deposited using the following procedure. A flat, shallow container, such as a Langmuir–Adam surface balance, is filled with water (or other suitable liquid) and the substrate to be coated is immersed. Then a solution of the amphiphilic material, in a solvent that is insoluble in water, is deposited dropwise onto the water, thereby forming an oriented monomolecular surface film upon evaporation of the solvent. This film can then be compacted by reducing the surface area until the surface pressure reaches a maximum. The solid substrate is then withdrawn from the water through this insoluble surface film at a very low speed (approximately 2 mm/min) while the surface pressure is kept constant. This creates an oriented monolayer of the amphiphilic material on the substrate.

The nature of the adsorbed monolayer depends on the interaction of the polar head group and the substrate surface. For example, if a glass microscope slide is raised up through a barium stearate monolayer spread on distilled water, the molecules in the film will be oriented with the hydrocarbon chains outwards, and hence the adsorbed film is hydrophobic [180]. When a previously coated plate is dipped back into the surfactant-coated water, a second oriented layer will be deposited and the coated surface again becomes hydrophilic as the head groups point outward.

Langmuir–Blodgett films are not very stable and are thus difficult to use for many contact angle studies. If the adsorbed film is altered by the sessile drop, successive advancing contact angles measured at the same location will differ.

E. Self-Assembled Monolayers (SAMs)

Monolayers of certain organic molecules can be produced on metal oxide or gold surfaces by means of self-assembly [181,182]. Alkanethiols on gold have proved to be a particularly useful system because the strong specific interaction between the sulfur atom and the gold surface results in the formation of relatively robust self-assembled monolayers upon immersion in dilute solutions. Moreover, by varying the tail group of the thiol molecules, a wide range of wettabilities can be achieved; water contact angles between 118° to less than 10° have been reported on treated gold surfaces [181,182]. However, it is expected that penetration of liquids into the SAMs is inevitable. Given that two solids have the same solid surface tension and that a specific liquid may penetrate into one solid surface but not the other, the contact angles on the two chemically identical surfaces can be different: one would be due to purely energetic effects and the other to the effects of the changed energetics and penetration. In the latter case, attempts to interpret surface energetics from contact angle approaches naively could be misleading.

F. Vapor and Molecular Deposition Techniques

Vapor deposition has been found to yield very smooth surfaces for low-molecular-weight nonpolar and polar materials such as hexatriacontane and cholesterol acetate [5,7], organic pigments [98], and silicon compounds [28,183]. In all cases, the deposited material must not decompose under the elevated temperatures required to cause evaporation.

Contact Angle Techniques and Measurements 77

In general, vapor deposition is carried out in a vacuum chamber. The clean, smooth substrate is placed horizontally above the shuttered aperture of a Knudsen furnace containing the coating material in a ceramic dish. The temperature of the furnace is then raised and held constant until equilibrium is established, at which point the shutter is opened and a stream of the evaporating coating material rises to impinge and condense on the substrate. The smoothness of the resulting surface is dependent on the rate of evaporation and on the temperature of the substrate. The latter may be controlled separately with heating elements in the fixture that is used to hold the substrate above the Knudsen furnace.

Rather than depositing films by evaporation, radio-frequency sputtering has been used to form polytetrafluoroethylene (PTFE) films on quartz and stainless steel [184]. In this case, the cathode in a high-vacuum chamber is bulk PTFE, which is positioned before the target substrate along the discharge axis. The resulting sputtered films are between 0.05 to 0.06 μm thick and have properties (e.g., fluorine-to-carbon ratio, cross-linking, branching) that can be varied over a wide range.

Glow-discharge polymerization is a third film-deposition technique [185–187]. A plasma is created, e.g., by an inductively coupled (electrodeless) radio-frequency coil surrounding a glass vacuum chamber, within which is placed a clean glass substrate. A non-polymer-forming plasma of a gas such as argon is generated at a very low pressure and then a gaseous monomer is introduced into the chamber where it fragments. The molecular segments deposit on the substrate and the chamber walls, where they cross-link and form a continuous film. The properties and uniformity of the resulting film are a complex function of many parameters such as type of monomer and inert gas, pressure, flow rates, electrical power input, and chamber and substrate geometry.

G. Siliconization

There are a wide variety of silicone-based compounds that absorb strongly to clean glass, producing high-quality hydrophobic surfaces. Such surfaces have been manufactured by exposing glass to silicon oil at elevated temperature [28] or to silane solutions [166] or vapor [82]. For the latter, a typical procedure is to place clean glass slides in a vacuum desiccator containing a small amount of e.g., dimethyldichlorosilane. The desiccator is then evacuated and left at room temperature for 48 h. After removal, the slides are thoroughly washed twice in toluene, then given a final rinse in acetone, and then air-dried.

Alternatively, clean glass surfaces can be placed in a solution of 20 vol% dimethyldichlorosilane in high-purity hexane [166]. After 1 h, the glass is removed and baked for a further hour at 100°C and then given two rinses in toluene and one in acetone before drying. The resulting surfaces have an advancing water contact angle of 105° with very little hysteresis.

H. Surface Polishing

For certain materials, such as rock and coal, polishing is the typical method for the preparation of surfaces for contact angle measurements [188,189]. Polymers have also been prepared this way [190]; however, it is important to note that polished surfaces are mechanically deformed and may contain embedded contaminants from the polishing agents. The details of the procedure depend on the material being polished. For example, some polymer surfaces and natural crystals of fluoro- and hydroxyapatite might be polished on a cloth-covered rotating wheel, first with diamond paste (1 μm particle size) and then with a slurry of aluminum oxide (0.05 μm). The polished surfaces then require cleaning in an ultrasonic bath of distilled water in order to eliminate traces of the polishing media. For rocks and coal, small pieces are produced from the bulk sample and then wet-ground with a sequence of silicon carbide papers (e.g., numbers 220, 460, and 600) that result in progressively smoother surfaces. Finer surfaces can then be obtained by polishing, as above, on a cloth-covered wheel saturated with a slurry of water and grit. To minimize the risk of surface contamination, commercial polishing compounds should be avoided, if possible, since they may contain unknown chemicals. Unfortunately, when the specimen is heterogeneous, like coal, polishing exposes a new surface that may be significantly different from the natural surface.

Metals can also be electropolished to produce a smooth surface that is free of the deformed layer obtained by mechanical polishing [180]. The metal of interest is set up as the anode in a conducting liquid and undergoes a controlled corrosion reaction.

I. Preparation of Powders for Contact Angle Measurements

The process used for the preparation of powders depends on the method selected to measure the contact angle. Powdered samples are either used in their original particulate form or compressed into a cake (tablet). The former implies that the powder will be analyzed using the capillary penetration method, the sedimentation volume technique, or the solidification front technique. In all these cases, the careful cleaning and outgassing of the powders is essential to eliminate possible impurities and air bubbles that may adhere to the surface of the particles. Of course, the solvents used to clean the powder should be chemically inert and should not swell the particles. Multiple washings in an ultrasonic bath are recommended. Alternatively, crystalline powders can be purified by recrystallization, provided that the correct crystal form results. Polymorphic crystal forms can be a function of the solvent used and can have significantly different surface properties. The indirect contact angle measurement techniques mentioned above, which use the actual powder rather than a compressed tablet, usually perform best when the particle size distribution is as narrow as possible, a condition that can be achieved perhaps by sieving.

Contact angles measured directly on tablets of compressed powder are also improved if the particle size range is narrow. The surface structure (porosity and roughness) is more uniform, so that there is less variability in the contact angles. Prior to the compression, therefore, the powder should be sized, recrystallized, or otherwise cleaned by sonication in an appropriate solvent and thoroughly dried in air or vacuum. The powder should be kept in a desiccator at constant temperature and relative humidity before tablet preparation. Tablets are made by compressing a known mass of powder (usually between 0.05 and 0.5 g depending on density, size, and shape) into a cake by using an hydraulic press and a highly polished die and punch. The die should be carefully cleaned with acetone and distilled water and dried prior to tablet formation. A pressure of between 400 and 600 MPa applied for 2–5 min is found to be sufficient for most low-molecular-weight organic powders [95,97,173,191]. The exact pressure and time should be determined empirically in order to obtain the most reproducible contact angles. The use of a highly polished die provides a macroscopically smooth tablet surface; however, the surface is actually porous and, as noted above, can be altered by plastic deformation during compression [99].

It should be noted that the contact angles measured on these compressed powder cakes may not represent the thermodynamically significant angle; this procedure is meaningful only if the aim is to characterize wettability of the powder cake. However, if the attempt is to determine energetics of the powder, indirect methods, such as capillary penetration, sedimentation volume, or solidification front technique, should be employed. Alternatively, direct contact angle methods can be used when the powder structure is destroyed completely, e.g., by melting of the powder and subsequent dip-coating on another substrate, solvent casting if the material is heat-sensitive, and heat pressing if the powder is a thermoplastic.

REFERENCES

1. J. K. Spelt and D. Li, ''The Equation of State Approach to Interfacial Tensions,'' in *Applied Surface Thermodynamics* (A. W. Neumann and J. K. Spelt, eds.), Surfactant Science Series, Vol. 63, Marcel Dekker, New York, 1996, pp. 239–292.
2. R. J. Good and C. J. van Oss, in *Modern Approaches to Wettability: Theory and Applications* (M. Schrader and G. Loeb, eds.), Plenum Press, New York, 1992, pp. 1–27.
3. F. M. Fowkes. Ind. Eng. Chem. *56(12)*:40 (1964).
4. A. W. Neumann. Adv. Colloid Interface Sci. *4*:105 (1974).
5. D. Li and A. W. Neumann, ''Thermodynamics Status of Contact Angles,'' in *Applied Surface Thermodynamics* (A. W. Neumann and J. K. Spelt, eds.),

Surfactant Science Series, Vol. 63, Marcel Dekker, New York, 1996, pp. 109–168.
6. R. V. Sedev, J. G. Petrov, and A. W. Neumann. J. Colloid Interface Sci. *180*:36 (1996).
7. A. W. Neumann and R. J. Good, in *Surface and Colloid Science: Experimental Methods* (R. J. Good and R. R. Stromberg, eds.), Vol. 11, Plenum Press, New York, 1979, pp. 31–91.
8. J. K. Spelt and E. I. Vargha-Butler, "Contact Angle and Liquid Surface Tension Measurements: General Procedures and Techniques," in *Applied Surface Thermodynamics* (A. W. Neumann and J. K. Spelt, eds.), Surfactant Science Series, Vol. 63, Marcel Dekker, New York, 1996, pp. 379–411.
9. N. K. Adam and G. Jessop. J. Chem. Soc. *1152*:1863 (1925).
10. F. M. Fowkes and W. D. Harkins. J. Am. Chem. Soc. *62*:337 (1940).
11. I. Langmuir and V. J. Schaeffer. J. Am. Chem. Soc. *59*:2400 (1937).
12. T. Fort, Jr., and H. T. Patterson. J. Colloid Sci. *18*:217 (1963).
13. G. W. Longman and R. P. Palmer. J. Colloid Interface Sci. *29*:185 (1967).
14. L. R. Fisher. J. Colloid Interface Sci. *72*:200 (1977).
15. L. Wilhelmy. Ann. Physik. *119*:177 (1863).
16. A. W. Neumann, Ph.D. thesis, University of Mainz, 1962.
17. A. W. Neumann. Z. Physik. Chem. N. F. *41*:339 (1964).
18. A. W. Neumann. Z. Physik. Chem. N. F. *43*:71 (1964).
19. D. D. Jordan and J. E. Lane. Austral. J. Chem. *17*:7 (1964).
20. H. M. Princen. Austral. J. Chem. *23*:1789 (1970).
21. J. Guastalla. J. Chim. Phys. *512*:583 (1954).
22. R. E. Johnson and R. H. Dettre, in *Surface and Colloid Science* (E. Matijevic and F. R. Eirich, eds.), Vol. 2, Academic Press, New York, 1969, p. 85.
23. F. Neumann, *Vorlesungen über die Theorie der Capillarität*, B. G. Teubner, Leipzig, 1893.
24. D. Y. Kwok, D. Li, and A. W. Neumann, "Capillary Rise at a Vertical Plate as a Contact Angle Technique," in *Applied Surface Thermodynamics* (A. W. Neumann and J. K. Spelt, eds.), Surfactant Science Series, Vol. 63, Marcel Dekker, New York, 1996, pp. 413–440.
25. W. Funke, G. E. H. Hellwig, and A. W. Neumann. Angew. Macromol. Chem. *8*:185 (1969).
26. G. H. E. Hellwig and A. W. Neumann. Kolloid-Z. Z. Polym. *229*:40 (1969).
27. G. H. E. Hellwig and A. W. Neumann, 5th International Congress in Surface Activity, Barcelona, 1968, Vol. 2, Ediciónes Unidas, 1969, p. 687.
28. A. W. Neumann and D. Renzow. Z. Physik Chem. N. F. *68*:11 (1969).
29. A. W. Neumann and W. Tanner. J. Colloid Interface Sci. *34*:1 (1970).
30. J. Kloubek and A. W. Neumann. Tenside *6*:4 (1969).
31. J. B. Cain, D. W. Francis, R. D. Venter, and A. W. Neumann. J. Colloid Interface Sci. *94*:123 (1983).
32. J. B. Cain, M.A.Sc. thesis, "Dynamic Contact Angles and the Effect of Roughness," University of Toronto, 1981.
33. D. W. Francis, M.A.Sc. thesis, "Temperature Dependent Contact Angle Measurements," University of Toronto, 1978.

34. R. J. Good and M. N. Koo. J. Colloid Interface Sci. 71:283 (1979).
35. E. Kiss and C.-G. Golander. Colloids Surfaces 49:335 (1990).
36. I. Morcos. J. Electroanal. Chem. 17:7 (1968).
37. I. Morcos. J. Electroanal. Chem. 20:479 (1969).
38. I. Morcos. J. Phys. Chem. 36:2750 (1972).
39. I. Morcos. J. Electroanal. Chem. 51:211 (1974).
40. C. J. Budziak and A. W. Neumann. Colloids Surfaces 43:279 (1990).
41. C. J. Budziak, Ph.D. thesis, "Thermodynamic Status of Static and Dynamic Contact Angles," University of Toronto, 1992.
42. D. Y. Kwok, C. J. Budziak, and A. W. Neumann. J. Colloid Interface Sci. 173:143 (1995).
43. A. W. Neumann and W. Tanner. Tenside 4:220 (1967).
44. W. A. Zisman, in *Contact Angle, Wettability, and Adhesion* (R. F. Gonld, ed.), Adv. in Chemistry Series, No. 43, American Chemical Society, Washington, D.C., 1964, p. 2.
45. D. Y. Kwok, T. Gietzelt, K. Grundke, H.-J. Jacobasch, and A. W. Neumann. Langmuir 13:2880 (1997).
46. J. J. Bikerman, *Surface Chemistry*, 2d ed., Academic Press, New York, 1958, p. 343.
47. J. J. Bikerman. Ind. Eng. Chem. Anal. Ed. 13:443 (1941).
48. J. Gaydos and A. W. Neumann, "Line Tension in Multiphase Equilibrium Systems," in *Applied Surface Thermodynamics* (A. W. Neumann and J. K. Spelt, eds.), Surfactant Science Series, Vol. 63, Marcel Dekker, New York, 1996, pp. 169–238.
49. D. Duncan, D. Li, J. Gaydos, and A. W. Neumann. J. Colloid Interface Sci. 169:256 (1995).
50. P. S. de Laplace, *Méchanique Céleste*, Supplement to Book 10, J. B. M. Duprat, Paris, 1806.
51. F. Bashforth and J. C. Adams, *An Attempt to Test the Theory of Capillary Action*, Cambridge University Press and Deighton Bell, Cambridge, 1883.
52. S. Hartland and R. W. Hartley, *Axisymmetric Fluid-Liquid Interfaces*, Elsevier, Amsterdam, 1976.
53. J. F. Padday, in *Surface and Colloid Science* (E. Matijevic, ed.), Vol. 1, Wiley Interscience, New York, 1969, p. 151.
54. C. A. Smolders, Ph.D. thesis, Rijksuniversiteit, Utrecht, 1961.
55. C. A. Smolders and J. Th. G. Overbeek. Rec. Trav. Chim. 80:635 (1961).
56. R. Ehrlich. J. Colloid Interface Science 28:5 (1968).
57. D. N. Staicopolus. J. Colloid Sci. 17:439 (1962).
58. D. N. Staicopolus. J. Colloid Sci. 18:793 (1963).
59. D. N. Staicopolus. J. Colloid Sci. 23:453 (1967).
60. K. G. Pavatikar. J. Colloid Sci. 23:274 (1967).
61. C. Maze and G. Burnet. Surface Sci. 13:451 (1969).
62. C. Maze and G. Burnet. Surface Sci. 24:335 (1971).
63. Y. Rotenberg, L. Boruvka, and A. W. Neumann. J. Colloid Interface Sci. 93:169 (1983).

64. P. Cheng, D. Li, L. Boruvka, Y. Rotenberg, and A. W. Neumann. Colloids Surfaces 43:151 (1990).
65. P. Cheng, Ph.D. thesis, "Automation of Axisymmetric Drop Shape Analysis Using Digital Image Processing," University of Toronto, 1990.
66. S. Lahooti, O. I. del Río, P. Cheng, and A. W. Neumann, "Axisymmetric Drop Shape Analysis (ADSA)," in *Applied Surface Thermodynamics* (A. W. Neumann and J. K. Spelt, eds.), Surfactant Science Series, Vol. 63, Marcel Dekker, New York, 1996, pp. 441–507.
67. O. del Río and A. W. Neumann. J. Colloid Interface Sci. 196:136 (1997).
68. O. del Río, M.A.Sc. thesis, "On the Generalization of Axisymmetric Drop Shape Analysis," University of Toronto, 1993.
69. D. Li and A. W. Neumann. J. Colloid Interface Sci. 148:190 (1992).
70. D. Li, M. Xie, and A. W. Neumann. Colloid Polym. Sci. 271:573 (1993).
71. D. Y. Kwok, R. Lin, M. Mui, and A. W. Neumann. Colloids Surfaces A, 116:63 (1996).
72. O. del Río, D. Y. Kwok, R. Wu, J. Alvarez, and A. W. Neumann. Colloids Surfaces A 143:197 (1998).
73. D. Y. Kwok, C. N. C. Lam, A. Li, A. Leung, R. Wu, E. Mok, and A. W. Neumann. Colloids Surfaces A 142:219 (1998).
74. D. Y. Kwok, A. Leung, A. Li, C. N. C. Lam, R. Wu, and A. W. Neumann. Colloid Polym. Sci. 276:459 (1998).
75. R. Miller, S. Treppo, A. Voigt, W. Zingg, and A. W. Neumann. Colloids Surfaces 69:203 (1993).
76. A. M. Schwartz and F. W. Minor. J. Colloid Sci. 14:572 (1959).
77. R. J. Roe. J. Colloid Interface Sci. 50:70 (1975).
78. W. D. Bascom and J. B. Romans. Ind. Eng. Chem. Prod. Res. Dev. 7:172 (1968).
79. Y. Gu, D. Li, and P. Cheng. Colloids Surfaces A, 122:135 (1997).
80. R. J. Good and J. K. Paschek, in *Wetting, Spreading, and Adhesion* (J. F. Padday, ed.), Academic Press, New York, 1978, p. 147.
81. E. Weber, Habilitationsschrift, University of Stuttgart, 1968.
82. F. K. Skinner, Y. Rotenberg, and A. W. Neumann. J. Colloid Interface Sci. 130:25 (1989).
83. E. Moy, P. Cheng, Z. Policova, S. Treppo, D. Kwok, D. R. Mark, P. M. Sherman, and A. W. Neumann. Colloids Surfaces 58:215 (1991).
84. W. C. Duncan-Hewitt, Z. Policova, P. Cheng, E. I. Vargha-Butler, and A. W. Neumann. Colloids Surfaces 42:391 (1989).
85. B. Drumm, A. W. Neumann, Z. Policova, and P. M. Sherman. Colloids Surfaces 42:289 (1989).
86. B. Drumm, A. W. Neumann, Z. Policova, and P. M. Sherman. J. Clin. Invest. 84:1588 (1989).
87. J. I. Smith, B. Drumm, A. W. Neumann, Z. Policova, and P. M. Sherman. Infect. Immun. 58:3056 (1990).
88. D. R. Mark, A. W. Neumann, Z. Policova, and P. M. Sherman. Am. J. Physiol. 262:G171 (1992).

89. D. R. Mark, A. W. Neumann, Z. Policova, and P. M. Sherman. Pediatric Res. *35*:291 (1994).
90. B. D. Gold, P. Islur, Z. Policova, S. Czinn, A. W. Neumann, and P. M. Sherman. Clin. Invest. Med. *19*(2):92 (1996).
91. D. Y. Kwok, F. Y. H. Lin, and A. W. Neumann in *Adhesion Science and Technology*, Proceedings of the 30th International Adhesion Symposium, Yokohama, Japan, 1994 (H. Mizumachi, ed.), Gordon and Breach, 1997, pp. 25–44.
92. M. Kazayawoko, A. W. Neumann, and J. J. Balatinecz. Wood Sci. Technol. *31*:87 (1997).
93. G. Zografi and S. S. Tam. J. Pharm Sci. *65*:1145 (1976).
94. C. F. Lerk, M. Lagas, J. P. Boelstra, and P. Broersma. J. Pharm. Sci. *66*:1480 (1977).
95. W. C. Liao and J. L. Katz. J. Pharm. Sci. *68*:488 (1979).
96. V. Steiner and G. Adam. Cell Biophys. *6*:279 (1984).
97. E. I. Vargha-Butler, S. J. Sveinsson, and Z. Policova. Colloids Surfaces *58*:271 (1991).
98. A. W. Neumann, D. Renzow, H. Reumuth, and I. E. Richter. Fortschr. Kolloide u. Polymere *55*:49 (1971).
99. G. Buckton and J. M. Newton. Powder Technol. *46*:201 (1986).
100. D. T. Hansford, D. J. W. Grant, and J. M. Newton. Powder Technol. *26*:119 (1980).
101. G. Buckton and J. M. Newton. J. Pharm. Pharmacol. *37*:605 (1985).
102. A. Stamm, D. Gissinger, and C. Boymond. Drug Dev. Ind. Pharm. *10*:381 (1984).
103. N. W. F. Kossen and P. M. Heertjes. Chem. Eng. Sci. *20*:593 (1965).
104. P. M. Heertjes and N. W. F. Kossen. Powder Technol. *1*:33 (1967).
105. G. Buckton, A. E. Beezer, and J. M. Newton. J. Pharm. Pharmacol. *38*:713 (1986).
106. D. W. Fuerstenau and M. C. Williams. Colloids Surfaces *22*:87 (1987).
107. D. W. Fuerstenau and M. C. Williams. Part. Characterization *4*:7 (1987).
108. J. Diao, M.A.Sc. thesis, University of California, Berkeley, 1987.
109. D. W. Fuerstenau, M. C. Williams, and J. Diao, Paper presented at AIME Annual Meeting, New Orleans, March 1986.
110. D. W. Fuerstenau and M. C. Williams. Colloids Surfaces *22*:87 (1987).
111. M. C. Williams and D. W. Fuerstenau. Int. J. Mineral Process *20*:153 (1987).
112. D. W. Fuerstenau et al., in 1987 International Conference on Coal Science, Elsevier, Amsterdam, 1987.
113. D. W. Fuerstenau, M. C. Williams, K. S. Narayanan, J. L. Diao, and R. H. Urbina. Energy and Fuels *2*:237 (1988).
114. D. W. Fuerstenau, J. Diao, and J. Hanson, Paper presented at the American Chemical Society National Meeting, Los Angeles, September 1988.
115. D. W. Fuerstenau, J. Diao, and J. Hanson. Energy and Fuels *4*:34 (1990).
116. J. Laskowski and J. A. Kitchener. J. Colloid Interface Sci. *29*:670 (1969).
117. E. W. Washburn. Phys. Rev. *17*:273 (1921).
118. F. E. Bartell and P. M. Osterhof. Ind. Eng. Chem. *19*:1277 (1927).

119. K. Grundke, T. Bogumil, T. Gietzelt, H.-J. Jacobasch, D. Y. Kwok, and A. W. Neumann. Prog. Colloid Polym. Sci. *101*:58 (1996).
120. P. M. Costanzo, R. F. Giese, and C. J. van Oss. J. Adh. Sci. Technol. *4*:267 (1990).
121. J. Norris, M. K. Chaudhury, and R. J. Good. J. Adh. Sci. Technol. *6*:413 (1992).
122. J. Norris, R. F. Giese, C. J. van Oss and P. M. Costanzo. Clays and Clay Materials *40*(3):327 (1992).
123. Z. Li, R. F. Giese, C. J. van Oss, J. Yvon, and J. Cases. J. Colloid Interface Sci. *156*:279 (1993).
124. P. M. Costanzo, W. Wu, R. F. Giese, and C. J. van Oss, Langmuir *11*:1827 (1995).
125. G. Buckton. J. Adh. Sci. Technol. *7*:205 (1993).
126. E. Chibowski and L. Holysz. Langmuir *8*:710 (1992).
127. E. I. Vargha-Butler, T. K. Zubovits, H. A. Hamza, and A. W. Neumann. J. Disp. Sci. Technol. *6*(3):357 (1985).
128. E. I. Vargha-Butler, E. Moy, and A. W. Neumann. Colloids Surfaces *24*:315 (1987).
129. E. I. Vargha-Butler, T. K. Zubovits, D. R. Absolom, and A. W. Neumann. Chem. Eng. Commun. *33*:255 (1985).
130. A. W. Neumann, S. N. Omenyi, and C. J. van Oss. J. Phys. Chem. *86*:1267 (1982).
131. H. C. Hamaker. Physica *4*:1058 (1937).
132. S. N. Omenyi, A. W. Neumann, and C. J. van Oss. J. Appl. Phys. *52*:789 (1981).
133. A. W. Neumann, S. N. Omenyi, and C. J. van Oss. Colloid Polym. Sci. *257*:413 (1979).
134. C. J. van Oss, D. R. Absolom, A. W. Neumann, and W. Zingg. Biochim. Biophys. Acta *670*:64 (1981).
135. A. W. Neumann, R. J. Good, C. J. Hope, and M. Sejpal. J. Colloid Interface Sci. *49*:291 (1974).
136. D. Li and A. W. Neumann, "Wettability and Surface Tension of Particles," in *Applied Surface Thermodynamics* (A. W. Neumann and J. K. Spelt, eds.), Surfactant Science Series, Vol. 63, Marcel Dekker, New York, 1996, pp. 509–556.
137. E. I. Vargha-Butler, M. Foldvari, and A. W. Neumann, "Surface Characterization of Particulate Matter by Sedimentation Volume Studies," Proceedings of the 5th Conference of Colloid Chemistry, Balatonfured, Hungary, Oct. 4–7, 1988, pp. 350–353, Hungarian Chem. Soc. Pub., Budapest, Hungary, 1990.
138. E. I. Vargha-Butler, S. J. Sveinsson, and Z. Policova. Colloids Surfaces *58*:271 (1991).
139. E. I. Vargha-Butler and E. L. Hurst. Colloids Surfaces B *4*:77 (1995).
140. S. N. Omenyi, Ph.D. thesis, "Attraction and Repulsion of Particles by Solidifying Melts," University of Toronto, 1978.
141. S. N. Omenyi and A. W. Neumann. J. Appl. Phys. *47*:3956 (1976).
142. M. C. Flemings, *Solidification Processing*, McGraw-Hill, New York, 1974.
143. W. Kurz and D. J. Fisher, *Fundamental of Solidification*, Trans. Tech. Publications, Aedermannsdorf, Switzerland, 1989.

144. A. W. Neumann, C. J. van Oss, and J. Szekely. Kolloid-Z. Z. Polym. *251*:415 (1973).
145. S. Torza and S. G. Mason. Science *162*:813 (1969).
146. D. R. Uhlmann, B. Chalmers, and K. A. Jackson. J. Appl. Phys. *35*:2986 (1964).
147. V. H. S. Kuo and W. R. Wilcox. Sep. Sci. *8*:375 (1973).
148. V. H. S. Kuo and W. R. Wilcox. Ind. Eng. Chem. Process Des. Dev. *12*:376 (1973).
149. A. E. Corte. J. Geophys. Res. *67*:1085 (1962).
150. P. Hoekstra and R. D. Miller. J. Colloid Interface Sci. *25*:166 (1967).
151. J. Cissé and G. F. Bolling. J. Crystal Growth *10*:67 (1971).
152. J. Cissé and G. F. Bolling. J. Crystal Growth *11*:25 (1971).
153. A. M. Zubko, V. G. Lobonov, and V. V. Nikonova. Sov. Phys. Crystallogr. *18*:239 (1973).
154. K. H. Chen and W. R. Wilcox. J. Crystal Growth *40*:214 (1977).
155. P. F. Aubourg, Ph.D. thesis, "Interaction of Second-Phase Particles with a Crystal Growing from the Melt," Massachusetts Institute of Technology, 1978.
156. S. N. Omenyi, A. W. Neumann, and C. J. van Oss. J. Appl. Phys. *52*:789 (1981).
157. S. N. Omenyi, A. W. Neumann, W. W. Martin, G. M. Lespinard, and R. P. Smith. J. Appl. Phys. *52*:796 (1981).
158. A. W. Neumann, J. Szekely, and E. J. Rabenda. J. Colloid Interface Sci. *43*:727 (1973).
159. D. Li and A. W. Neumann, "Behavior of Particles at Solidification Fronts," in *Applied Surface Thermodynamics* (A. W. Neumann and J. K. Spelt, eds.), Surfactant Science Series, Vol. 63, Marcel Dekker, New York, 1996, pp. 557–628.
160. M. R. Soulard, E. I. Vargha-Butler, H. A. Hamza, and A. W. Neumann. Chem. Eng. Commun. *21*:329 (1983).
161. A. W. Neumann, E. I. Vargha-Butler, H. A. Hamza, and D. R. Absolom. Colloids Surfaces *17*:131 (1986).
162. A. W. Neumann, D. R. Absolom, W. Zingg, and C. J. van Oss. Cell Biophys. *1*:79 (1979).
163. S. K. Chang, O. S. Hum, M. A. Moscarello, A. W. Neumann, W. Zingg, M. J. Leutheusser, and B. Ruegsegger. Med. Progr. Technol. *5*:57 (1977).
164. W. Zingg, A. W. Neumann, A. B. Strong, O. S. Hum, and D. R. Absolom. Biomaterials *2*:156 (1981).
165. A. W. Neumann, D. R. Absolom, D. W. Francis, C. J. van Oss, and W. Zingg. Cell Biophys. *4*:285 (1982).
166. J. K. Spelt, M.A.Sc. thesis, "Surface Tension Measurements of Biological Cells Using the Freezing Front Technique," University of Toronto, 1980.
167. J. K. Spelt, D. R. Absolom, W. Zingg, C. J. van Oss, and A. W. Neumann. Cell Biophys. *4*:117 (1982).
168. S. N. Omenyi, R. S. Snyder, C. J. van Oss, D. R. Absolom, and A. W. Neumann. J. Colloid Interface Sci. *81*:402 (1981).
169. D. Li and A. W. Neumann. Colloid Polym. Sci. *270*:498 (1992).
170. J. F. Oliver, C. Huh, and S. G. Mason. Colloids Surfaces *1*:79 (1980).

171. J. K. Spelt, Y. Rotenberg, D. R. Absolom, and A. W. Neumann. Colloids Surfaces 24:127 (1987).
172. E. B. Davidson and G. Lei. J. Polym. Sci. 9:569 (1971).
173. T. R. Krishman, I. Abrahma, and E. I. Vargha-Butler. Int. J. Pharm. 80:277 (1992).
174. E. I. Vargha-Butler, T. K. Zubovits, C. J. Budziak, and A. W. Neumann. Energy Fuels 2:653 (1988).
175. E. F. Hare, E. G. Shafrin, and W. A. Zisman. J. Phys. Chem. 58:236 (1954).
176. C. J. Budziak, E. I. Vargha-Butler, and A. W. Neumann. J. Appl. Polym. Sci. 42:1959 (1991).
177. D. R. Absolom, A. W. Neumann, W. Zingg, and C. J. van Oss. Trans. Am. Soc. Artif. Int. Organs 25:152 (1979).
178. K. B. Blodgett. J. Am. Chem. Soc. 57:1007 (1935).
179. I. Langmuir. J. Franklin Inst. 218:143 (1934).
180. A. W. Adamson, *Physical Chemistry of Surfaces*, 5th ed., John Wiley, New York, 1990.
181. C. D. Bain, E. B. Troughton, Y. Tao, J. Eval, G. M. Whitesides, and R. G. Nuzzo. J. Am. Chem. Soc. 111:321 (1989).
182. C. D. Bain and G. M. Whitesides. Angew. Chem. Int. Ed. Engl. 28:506 (1989).
183. W. J. Herzberg, J. E. Marian, and T. Vermeulen. J. Colloid Interface Sci. 33:164 (1970).
184. J. J. Pireaux, J. P. Delrue, A. Herq, and J. P. Dauchot, in *Physico-Chemical Aspects of Polymer Surfaces* (K. L. Mittal, ed.), Vol. 1, Plenum Press, New York, 1983, p. 53.
185. A. Dilkes and E. Kay, in *Plasma Polymerization* (M. Shen and A. T. Bell, eds.), A.C.S. Symposium Series, No. 108, American Chemical Society, Washington, D.C., 1979.
186. A. Dilkes and E. Kay. Macromolecules 14:855 (1981).
187. H. Yasuda. J. Polymer Sci.: Macromol. Rev. 16:199 (1989).
188. E. I. Vargha-Butler, M. Kashi, H. A. Hamza, and A. W. Neumann. Coal Preparation 3:53 (1986).
189. E. I. Vargha-Butler, D. R. Absolom, A. W. Neumann, and H. A. Hamza, in *Interfacial Phenomena in Coal Technology* (G. D. Botsaris and Y. M. Glazman, eds.), Marcel Dekker, New York, 1989, pp. 33–84.
190. H. J. Busscher, A. W. J. van Pelt, H. P. deJong, and J. Arends, J. Colloid Interface Sci. 95:23 (1983).
191. G. Zografi and S. S. Tam. J. Pharm. Sci. 65:1145 (1976).

3
Measurement of Ion-Mediated and van der Waals Forces Using Atomic Force Microscopy

IAN LARSON Ian Wark Research Institute, University of South Australia, Mawson Lakes, South Australia, Australia

ANDREW J. MILLING Department of Chemistry, University of Durham, Durham, England

I.	Introduction	87
	A. Instrumentation	89
II.	Surface Force Measurement Using AFM	91
	A. Force-curve construction	91
	B. Cantilever spring-constant calibration	91
III.	Colloid-Probe Equipped AFM Measurement	93
	A. Electrical double-layer interactions	93
	B. Polyelectrolytes	100
IV.	Van der Waals Interactions	103
	A. AFM studies of attractive van der Waals interactions	104
	B. AFM studies of repulsive van der Waals interactions	106
	References	109

I. INTRODUCTION

The far-reaching importance of ion-mediated (IM) and van der Waals (vdW) interactions cannot be sufficiently stressed. Whilst a fundamental understanding of these forces is of academic interest, other fields such as mineral processing, colloidal formulation, and lubrication all rely heavily on manipulation and control of both vdW and IM forces.

Ion-mediated forces can be divided into two crude classifications, those due to simple-ion electrolytes (e.g., inorganic salts) and those due to macroionic species (e.g., polyelectrolytes and charged surfactant micelles). The former case, which is commonly referred to as the electrical-double-layer (EDL) interaction, has long been of great interest to colloid scientists. The ad hoc superposition of pairwise vdW and EDL forces forms the basis of the well known DLVO [1,2] theory of colloid stability.

When a surface is immersed in a solution it may acquire a net electrical charge through surface dissociation of ionizable groups, adsorption of charged species (e.g., ions, polymer molecules, and surfactant molecules) or through isomorphous substitution of surface atoms (analogous to the use of dopants in semiconductor fabrication). Associated with the charging of the surface is the development of co- and counterion structures in the solution adjacent to the charged surface. Together the charged surface and the counterion structure are referred to as the "electrical double layer." EDL forces commence when charged surfaces, immersed in electrolyte solution, are brought to close approach and the diffuse (distal) region of the opposing electrical double layers begin to overlap. The range of the interaction depends on the electrolyte ionic strength, and the strength of the interaction upon the magnitude of the charges on the opposing surfaces. Essentially, EDL forces are osmotically driven, the disjoining pressure being due to the relative concentration of ionic species at the mid-plane between the opposing surfaces, compared to the bulk solution.

For the case of macroions in solution, the interfacial ionic structure is somewhat more complicated, mainly because of the bulky size of these species and their (usual) polyvalency. As predicted by simple-ion theories, macroions may also be expected to form some structural features at charged interfaces due to surface–solute and solute–solute interactions. These may lead to depletion and structural ion-mediated forces; see later.

Van der Waals forces have a quite separate origin. Initially they were used without formal justification to "improve" the agreement between equations of state for gases and the observed behavior of real (nonideal) gases [3]. There are three contributions to the total vdW interaction, these being Debye [4] (dipole–dipole) forces, Keesom [5] (dipole-induced dipole) forces and London [6] (fluctional dipole–fluctional dipole) forces. In condensed matter the major contribution is the London [6] "dispersion" force, a consequence of the interaction between an instantaneous dipole on one molecule/atom and an induced dipole on a neighboring molecule/atom. In atoms, these dipoles are due to the relative instantaneous positions of the electrons around the positive nucleus. Time-averaging of this electromagnetic interaction results in a net attraction between two molecules/atoms. However, the time between the initial radiation of the electromagnetic field and the absorption of the reflected field is finite. If this amount of time is comparable to the time for internal motion, the original dipole's orientation will have changed and this will result in a decrease in the attraction between the two molecules. As the separation between the two atoms increases, the time between emission and reabsorption will increase, and the strength of the attraction will decrease. This effect is known as "retardation" [7]. Using a summation procedure, Hamaker [8] extended the London approach for intermolecular forces to the case of condensed matter. This approach has met with great success and was not superseded until the 1950s with the theory of Lifshitz [9].

It is the aim of this chapter to outline the use of the atomic force microscope (AFM) in the study of IM and vdW surface forces.

A. Instrumentation

1. Development

The AFM is a derivative of the scanning tunnelling microscope (STM, invented by Binnig and Rohrer [10] for atomic scale surface analysis). In the STM technique a tunnelling current is passed between a very fine tip and a conducting substrate. The magnitude of this current is dependent on the separation between the tip and the substrate and hence gives information about the height of features on the sample surface. As the conducting/semiconducting material is scanned under the tip in a raster pattern, the strength of the current is recorded, as is the X–Y position of the sample. A three-dimensional map of the surface can then be constructed and, with a bench-top instrument, individual atoms can be observed.

The STM is only able to image conducting or semiconducting surfaces. This limitation was overcome in 1986 by Binnig et al. [11] with the AFM. The AFM relies on "feeling" the surface of the sample with a sharp tip on a cantilever spring. Binnig et al. mounted a small diamond chip on the end of a thin gold lever that acted as a spring. As the sample was scanned underneath the tip, the force acting between the atoms at the very end of the diamond chip and the sample surface was monitored by recording the deflection of the lever with a STM. Because there was no current passing between the tip and the sample, the sample did not need to be conducting.

2. Cantilever Deflection Detection

In the first AFM, a STM was used to measure the deflection of the lever. This is a very sensitive method, but its reliability can be affected by contamination of the tunnelling tip or the back of the lever. It is also possible for the tunnelling tip to drift across the back of the lever and impair the image quality; finally, the system is prone to damage where there is spring instability. To avoid these problems a variety of other methods were developed to measure cantilever deflection:

1. Optical interferometry, both heterodyne [12,13] and homodyne [14–16] techniques.

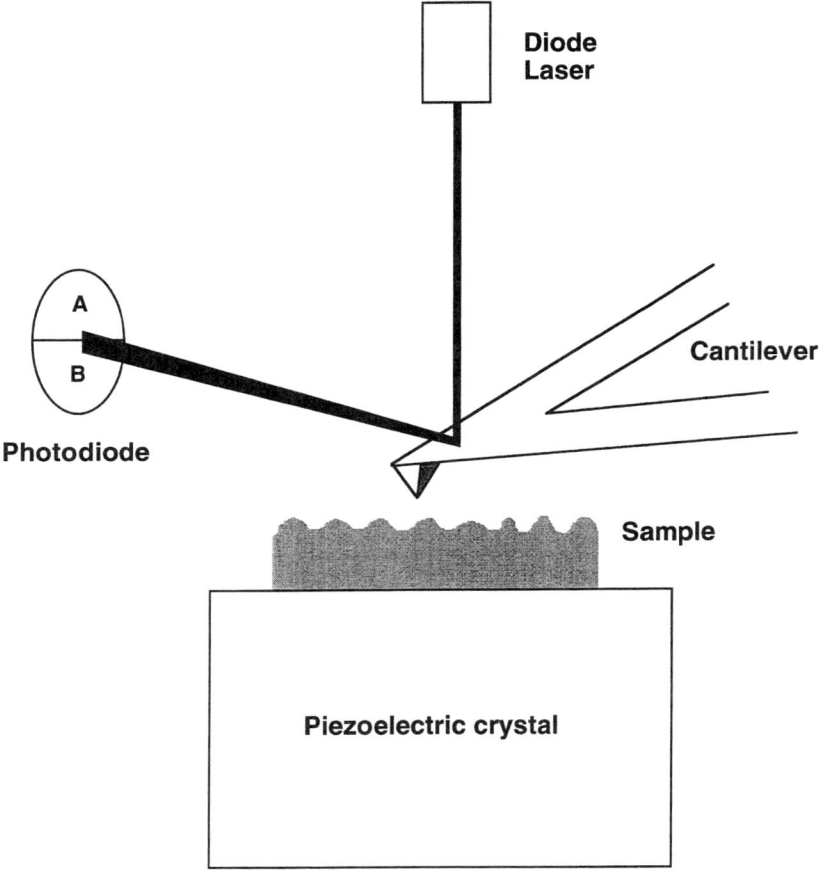

FIG. 1 Schematic illustration of a typical AFM setup.

2. Laser beam deflection [17–19], where a laser beam is reflected off the back of the cantilever onto a position-sensitive detector, e.g., a split photodiode; see Fig. 1. As the cantilever deflects, the light beam's path changes. This change results in different amounts of light hitting one half of the diode, denoted A, to the other, denoted B. By recording the relative photodiode signal A − B, the vertical movement of the cantilever is known.

3. Laser-diode feedback. This detection technique utilizes the great sensitivity of lasers to optical feedback and is used to detect movement of the cantilever. Laser light reflected off the back of the cantilever is returned back into the laser [20–22].

4. A capacitance method where the deflection of the cantilever is sensed by a counter electrode opposite to the rear side of the cantilever [23,24].

Current instruments differ little, in principle, from the first AFM developed over 10 years ago. The biggest changes have come in the methods used to detect the deflection of the cantilever spring, as mentioned above, and the development of microfabricated silicon nitride cantilevers with integrated pyramidal tips.

II. SURFACE FORCE MEASUREMENT USING AFM

A. Force-Curve Construction

Understanding the load applied to the surface via the imaging tip is of paramount importance in the interpretation of AFM images, and in many cases this load must be minimized to avoid artefacts due to surface ablation/restructuring. Subsequently it was not long before scientists started to use the AFM to measure fundamental forces [25–27]. These workers measured the forces between the AFM tips and various substrate materials. Normally in an AFM force experiment, the static deflection of the cantilever is monitored as the surface is moved to and away from the cantilever tip, and thus the raw data collected is of the form of diode voltage (A-B giving cantilever deflection) versus scanner displacement. To convert this data to more conventional force–distance form Ducker et al. [30] described a geometrical transform relying upon the identification of a "zero force" baseline (at large surface separations) and an apparent "zero separation" (where the cantilever undergoes constant compliance with piezo travel). This is schematically illustrated in Fig. 2 (software is commercially available for Nanoscope Instruments specifically for this purpose [28]). It should be noted that the "force–distance" data thus obtained should be strictly considered as force versus separation-from-compliance data, although in this chapter we shall use the abbreviation "separation."

A major advance for colloid scientists was to attach colloidal particles to the AFM tips so that surface forces relevant to colloidal dispersions could be measured [29–31]. For this purpose a "colloidal" particle (radius 2–25 μm) is glued to the cantilever tip, which enables interaction forces to be measured between this particle and a flat surface or even another particle. For ease of quantitative analysis, spherical particles are preferred.

B. Cantilever Spring-Constant Calibration

The optical deflection technique described above has the ability to detect motions of the cantilever down to the order of ca. one angstrom. By using quite weak springs (spring constants in the range of 0.01 to 0.5 N/m), force measurements with a sensitivity of ca. 10^{-12} N are realizable. As Hooke's law is

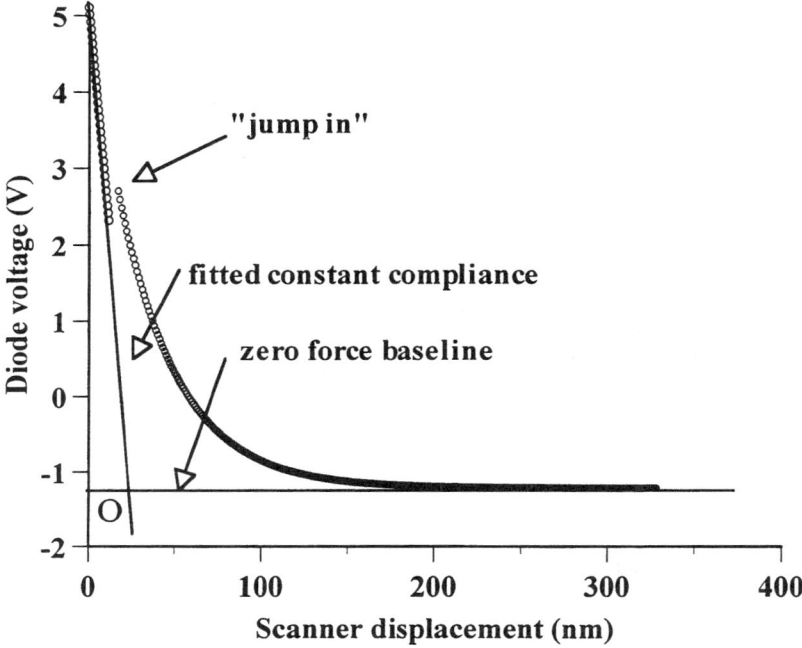

FIG. 2 Schematic illustration of the geometric constructions used to transform diode voltage versus scanner displacement data into force versus distance form.

used to convert cantilever deflection to force, accurate knowledge of the cantilever's spring constant is necessary. The calibration of the springs is usually made by attachment of small tungsten spheres to the cantilever and measuring the resonant frequency of the cantilever, as described by Cleveland et al. [32]. Several authors have used springs calibrated using a two-point method, i.e., one loaded and one free-resonant frequency measurement, or by simply by inverting the microscope head assembly and observing the displacement due to the added weight (also requiring a knowledge of the compliance gradient). Two-point methods are unsound, and measurements should be made with several differing attached masses. Other methods for cantilever spring-constant calibration have been elucidated by Sader and coworkers [33,34]. Another force-sensing technique utilizes the change in resonant frequency of the cantilever in nonlinear force fields, but unfortunately this technique is fairly insensitive in fluid media, as complex viscous damping dramatically lowers the cantilever resonant frequency Q-factor.

III. COLLOID-PROBE EQUIPPED AFM MEASUREMENT

A. Electrical Double-Layer Interactions

To date, the analysis of the EDL forces that have been measured using both AFM and the surface forces apparatus [35] have relied on DLVO [1,2] theory. This theory predicts that for like surfaces the EDL interaction is repulsive for all surface potentials. Another theory for EDL interactions, developed by Sogami and Ise [36], predicts that for high surface potentials and low ionic strengths there is a long-range attractive force between surfaces. The magnitude of this predicted attractive potential is seemingly beyond the sensitivity of present instrumentation. Its justification is based on observations of colloidal phase separation [37,38] and as such multibody interactions are implicitly involved. This topic is beyond the scope of this chapter, but the interested readers' attention is drawn to a recent paper by Schmitz [39] that discusses some of the difficulties inherent in describing multibody interactions by summation of pair potentials. In this chapter we shall follow the standard DLVO description of electrostatic interactions.

A quantitative description for the diffuse layer dates back to Gouy [40] and Chapman [41]. This model is fully described elsewhere [42], and we shall here only outline basic features of the theory. The first stage in this approach is to use the Poisson equation, Eq. (1), to describe the relationship between the electrical potential $\Psi(x)$ and the charge density ρ of ions of charge z_i at a distance x from a flat charged surface, with a regional permittivity ε:

$$\frac{d^2 \Psi_{(x)}}{dx^2} = -\frac{\rho_{(x)}}{\varepsilon} \tag{1}$$

The ions are distributed according to a Boltzmann relationship, and the well-known Poisson–Boltzmann (P–B) equation can be derived:

$$\frac{d^2 \Psi}{dx^2} = \frac{2n^0}{\varepsilon} \sum_i z_i e \sinh\left(\frac{z_i e \Psi}{kT}\right) \tag{2}$$

where n^0 is the bulk ion concentrations and e is the electronic charge.

The P–B equation can be approximately linearized using the Maclaurin expansion of the sinh function and then neglecting higher-order terms [37]. The P–B equation for ion distribution is thus simplified and yields

$$\Psi_{(x)} = \Psi_0 \exp(-\kappa x) \tag{3}$$

where Ψ_0 is the surface potential, and

$$\kappa = \left(\frac{2n^0 \sum_i z_i^2 e^2}{\varepsilon kT}\right)^{-1/2} \tag{4}$$

The quantity $1/\kappa$ has units of length and is called the Debye length; it defines the extent of the double layer, i.e., the distance in which the potential decays to $1/e$ of its initial value; κ is called the Debye–Hückel parameter. Hence within validity of this approximation (low surface potentials < 25 mV) the potential decreases exponentially away from the surface.

It should be noted that the Gouy–Chapman model is based on a number of simplifications and assumptions that ought to be remembered when applying it, namely:

1. The surface is flat and of infinite size, and the individual charges are smeared out evenly over the whole surface.
2. The electrolyte is made up of point charges distributed according to the Boltzmann distribution.
3. The dielectric constant of the solvent is assumed to be of the same value in the bulk and in the double layer, and it is the only influence that the solvent has on the double layer.

As mentioned in the introductory section, when two charged surfaces approach each other the overlap of their double layers will result in a net force between the surfaces. In the simplest case of two flat plates approaching each other, the overlap of the two double layers results in the ion concentration in the region between the plates being higher than in the bulk. This osmotic pressure produces a repulsive force between the two plates. The sign, magnitude, and behavior of this force will depend on the sign and magnitude of the surface charge and also the rate at which the two surfaces approach each other. The simplest case to deal with is that of identical flat infinite surfaces. The approach of two dissimilar flat surfaces is rather complex, and we shall return to this later. The Derjaguin approximation [43] is used to compare data collected with sphere–plate and sphere–sphere geometries to theoretical flat plate interactions.

For the case of identically charged interfaces, there are two extreme mathematical boundary conditions that can arise when identical electrical double layers overlap. The surface potential may remain constant while the surface charge density changes, or the surface may remain constant whilst the surface potential changes. At large separations, where there is little overlap, there is little difference, in terms of the interaction potential, between these two cases. Generally, though, the interaction itself will influence the surface charge and potential, so neither may remain constant but regulate themselves [44]. This "charge regulation" gives rise to an intermediate force, i.e., greater than the constant potential force (lower limit) but less than the higher constant charge limit.

The repulsive interaction between plates can be calculated by the free energy of the midplane or by using the osmotic-based argument of Hunter [42]. In the latter, the mid-plane ionic solute concentration (hence osmotic pressure) p_m is evaluated using Eq. (5), and thence the repulsive energy V_R per unit area between the plates is calculated by integration:

$$p_m = 2n^0 kT \cosh\left(\frac{\sum_i z_i e \Psi_m}{kT}\right) \tag{5}$$

$$V_R = -\int_\infty^H p_m \, dH \tag{6}$$

It is possible to calculate V_R as a function of separation exactly using numerical techniques. The method of Chan et al. [45] may be used for this purpose. Interested readers are referred to Ref. 45 for further information about these calculations. It is also possible to calculate the interaction for the charge regulation case where knowledge of the surface density of acid groups and their dissociation constants is required.

It is only recently that methods for calculating the interaction between dissimilar flat plates in a form convenient for AFM data analysis have been developed for both constant charge and constant potential interactions [46]. Previously, the approximation due to Hogg et al. [47] was widely used because of the ease of calculation. This approximation, though, is valid for constant low-potential interactions only.

When analyzing data from a dissimilar system there are two potentials involved. In Fig. 3 we show theoretical force–separation curves for different pairs of potentials that when multiplied together give the same number. For constant charge systems there is very little difference between the curves produced by the different pairs of potentials. At large separations, where theory is fitted to the experimental data to determine the diffuse layer potentials, there is little difference between the constant potential systems. Clearly, there is not a unique pair of diffuse layer potentials that fits the individual experimental force curves. Even when the constant potential interaction fits are considered, any differences between different potential pairs at small separations may be obscured if there is an extra non-DLVO short-range repulsion. For this reason it is necessary to have independently obtained values of the potentials of the materials for comparison.

The different situations that can arise between charged surfaces acting either under constant charge or constant potential conditions may be summarized thus:

1. Constant potential interactions: the force between surfaces with
   a. Unlike potentials will be attractive at all separations.

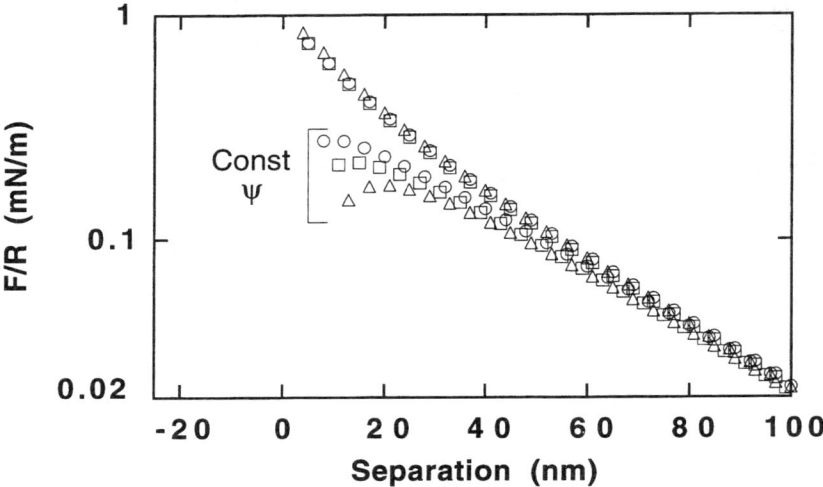

FIG. 3 Theoretical force–separation curves for different pairs of potentials that when multiplied together give the same number. The symbols refer to the different potential pairs; −35 mV and −70 mV (squares), −50 mV and −50 mV (circles), G −40 mV and −60 mV (triangles). The upper three curves are the constant charge limits, while the lower three are the constant potential boundary conditions. There is very little difference between the three constant charge curves, and at large separations there is very little difference between the constant potential curves. A Hamaker constant of $2 \times 10{-}20$ J was used in the calculations and a background electrolyte of 1×10^{-4} M NaCl.

    b. Like (but not identical) potentials will be repulsive at large separations but attractive at small separations.
    c. Identical charges will always be repulsive.
2. Constant charge interactions: the force between surfaces with
    a. Like charges will be repulsive at all separations.
    b. Unlike charges (but not equal and opposite) will be attractive at large separations but repulsive at small separations.
    c. Equal and opposite charges will always be attractive.

The majority of quantitative studies in this field have utilized the sphere–plate geometry. Of major significance has been the examination of the forces between metal oxide surfaces, for which the hydrogen ion is a potential determining ion [48].

The first reported uses of an AFM to study EDL interactions using the colloid probe technique were the studies by Ducker et al. [29,30] and by Butt [31]

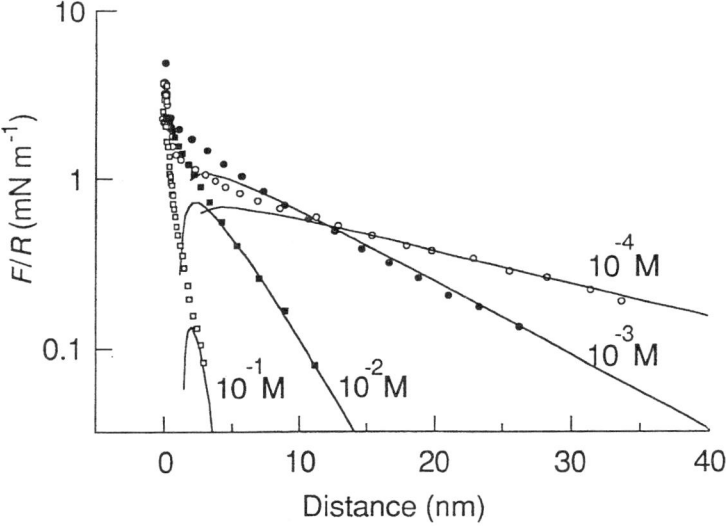

FIG. 4 Experimental force curves for a silica sphere interacting with a polished silica flat surface in the presence of simple electrolytes (concentrations as indicated) measured using an AFM. (Adapted from Ref. 29.)

for the interactions between silica surfaces. Diffuse layer potentials were estimated from the force–distance data for a variety of solution conditions, namely, varying pH and (indifferent) electrolyte concentration. Subsequently, silica has become one of the most studied model systems; smooth micron-sized spheres, possessing a chemically labile surface, and easily obtainable. Figure 4 illustrates some experimental force curves for a silica sphere interacting with a silica flat for various electrolyte concentrations and pH values, together with some theoretically generated curves. Larson et al. [49] also measured the forces between a spherical silica probe and a silica sphere glued to a flat surface (after axial alignment of the probe and the particle). Although silica has been used as a "model system" in many studies, there is still some controversy in studies made using silica from differing sources. These arguments are concerned with subtle non-DLVO effects at close approach, namely "hydration/solvation" or "gel" layers. There has yet to be a definitive study that correlates the Hofmeister series for group I metal cations with the hydration-layer hypothesis. The nonobservation of these solvation forces has been blamed on surface roughness problems and also on inadequate hydroxylation of the silica surface following some dry stage handling (for instance probe attachment, and also where some samples of spherical particles may have been lightly calcined to reduce porosity, a process that could easily be achieved via hydrothermal aging of sols). In a study most pertinent for those interested in the hydration of cementacious materials [50], Meagher [51] measured the EDL interactions between silica surfaces in the presence of calcium ions. His principal observation was that the diffuse layer potentials of the surfaces were greatly reduced, as was the Debye length (as expected for a divalent cation). However, the use of a symmetrical system entailed that the absolute sign of the surface charge could not be determined. The specific chemical interactions between the silica and the calcium ions were not addressed (the formation of an "amorphous" calcium–silicate membrane). Similarly there is a dearth of knowledge concerning forces between silica surfaces in the presence of hydrolyzable polyvalent (i.e., transition metal) cations where either precipitated hydroxide "islands" [52] or metal–silicate complexes (cf. "water gardens") [53] have been suggested. Other studies of metal oxides involve the following pairings of like materials: TiO_2–TiO_2 [54], ZrO_2–ZrO_2 [55,56], and Al_2O_3–Al_2O_3 [57]. The AFM study of interactions between titania surfaces [54] was concerned with both IM and vdW forces, and we shall discuss this study in more detail later (Section III.C).

The measurements made using zirconia probes [55,56] showed that the fitted diffuse layer potential of this material was dependent upon the thermal history of this calcined ceramic precursor material. Unlike silica surfaces where surface siloxane bridges are readily rehydrolyzed to produce hydrophilic silica [58], zircoxane surface structures are relatively inert. Thus, whilst zirconia exhibits features that detract from its use as a model system, the investigation of EDL forces is of interest per se for the handling and characterization of this material. Whilst the majority of AFM force measurements have been conducted between like surfaces, the AFM has been used to examine the forces between differing materials, e.g., SiO_2–mica [59,60] and SiO_2–Al_2O_3 [61,62]. Figure 5 illustrates some of the force curves measured by Larson et al. for the interaction between silica and alumina surfaces.

The work of Larson et al. [62] represented the first detailed study to show agreement between AFM-derived diffuse layer potentials and ζ-potentials obtained from traditional electrokinetic techniques. The AFM experimental data was satisfactorily fitted to the theory of McCormack et al. [46]. The fitting parameters used, silica and alumina zeta-potentials, were independently determined for the same surfaces used in the AFM study using electrophoretic and streaming-potential measurements, respectively. This same system was later used by another research group [63]. Hartley and coworkers [63] also compared dissimilar surface interactions with electrokinetic measurements, namely between a silica probe interacting with a polylysine coated mica flat (see Section III.B.). It is also possible to conduct measurements between a colloid probe and a metal or semiconductor surface whose electrochemical properties are controlled by the experimenter [64–66]. In Ref. 64 Raiteri et al. studied the interactions between

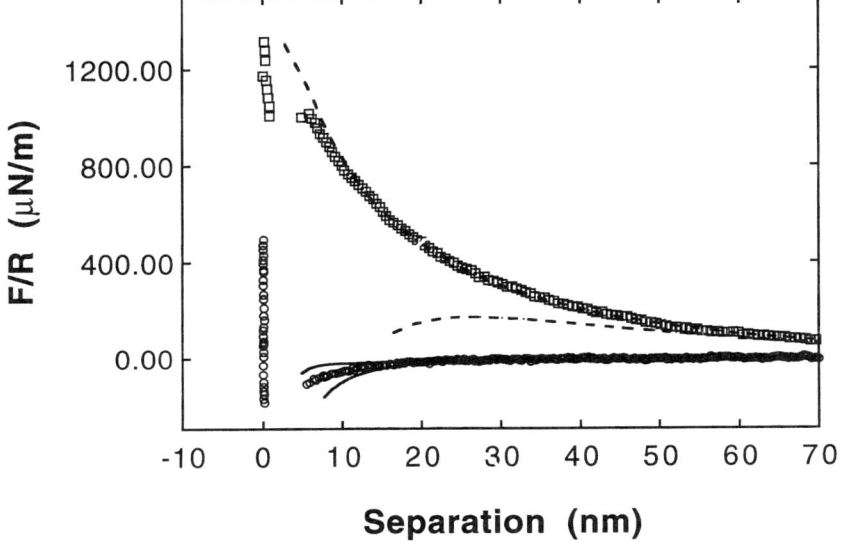

FIG. 5 Force–separation curves taken at G pH 8.2 and an electrolyte concentration of 1×10^{-4} M KNO_3 (squares), and pH 3.9 and an electrolyte concentration of 1×10^{-3} M KNO_3 (circles). The dashed lines are theoretical fits to the pH 8.2 data and the full lines to the pH 3.9 data, respectively, where the upper fitted curves are the constant charge limits (just visible in the upper left-hand corner of the pH 8.2 data) and the lower curves are the constant potential limits. The pH 8.2 experimental data lies closer to constant charge than constant potential. The fitting parameters are pH 8.2: silica potential -100 mV, alumina potential -34 mV, and Debye length $= 30.5$ nm; pH 3.9: silica potential -25 mV, alumina potential $+3$ mV, and Debye length $= 9.6$ nm. A Hamaker constant of 1.8×10^{-20} J was used in the calculations.

a silicon nitride tip and gold and platinum surfaces, whilst Hillier and coworkers [65,66] used a silica colloid probe and a gold surface.

Metal sulfides systems are of great commercial interest in metal production and optoelectronic devices. To date, the only investigations concerning EDL interactions between metal sulfide surfaces have been performed by Ralston et al. [67,68]. In the first paper the authors attempted to correlate the flotation behavior of zinc sulfide (sphalerite) with AFM force measurements. Upon comparing the zeta-potentials obtained by electrophoresis and AFM measurements, the authors concluded that a non-DLVO "hydrophobic" force was operative in the AFM force profiles. Another possible explanation is that the observed forces were due to vdW interactions (see the next section) and that the probe preparation (involving a dry handling stage) led to oxidation of the highly labile probe particle. This highlights a potential flaw in studies where correlation is attempted between colloidal dispersion behavior and AFM measurements involving oxidatively labile probe materials.

In a series of papers, Grieser and coworkers [69–71] investigated gold–gold interactions in the presence of various ions. The main impetus of these studies was to understand the fundamental mechanism of nucleation and growth mechanisms in Faraday sols, identifying (via DLVO theory) conditions where colloidal stability could be expected. The authors measured EDL forces (the surface charge being due to ion adsorption), vdW forces, and time-resolved ion displacement. Additionally, multilayer anion adsorption was observed for some polyanionic species. Hu and Bard [72] have also studied EDL interactions between gold surfaces bearing self-assembled ionogenic monolayers.

B. Polyelectrolytes

Polyelectrolytes are macromolecules that bear ionogenic functionalities. This general classification can be used to encompass a very wide spectrum of compounds, from biologically significant species such as proteins to the "simpler" synthetic polymers met with in many academic studies and industrial practices. In this section we shall only consider the effect upon surface stresses due to synthetic linear-chained polyelectrolytes, and we shall neglect polymer bridging (simultaneous adsorption of a molecule on two separate surfaces). There is presently a dearth of studies concerning AFM surface force measurements in the presence of biomolecules, the recently reported work of Bowen et al. [73] representing the only progress, to our knowledge.

In relationship to the electrical double behavior in the presence of simple electrolytes, a far more complex situation is met when considering polyelectrolyte molecules constrained between surfaces. These molecules are capable of simultaneously changing their degree of ionization, counterion distribution, and conformation (although the *inter alia* relationship between these quantities should be noted), and so polyelectrolyte and associated ion cloud distributions are extremely complex. A satisfactory generalized theory for the solution behavior of polyelectrolyte molecules is presently awaited, and in view of the given considerations a fuller understanding is most likely some way off. Scaling theories [74,75] and computer simulation [76] provide some theoretical insight into polyelectrolyte solution behavior. The recent treatment by Joanny [77] for weakly adsorbing polyelectrolytes provides the most satisfactory advancement in this area and also highlights the apparent deficiencies of mean-field-lattice-based models [78]. These factors highlight the significance of surface force measurements from both an empirical point of view and as a test for theories that are being developed in this embryonic field [76].

1. Surface Forces in the Presence of Adsorbing Polyelectrolytes

The effects (surface potential and adsorbed-layer conformation) of polyelectrolytes adsorbed at an interface have been extensively reviewed elsewhere [79]. At a basic level, the adsorption of polyelectrolyte molecules at an interface affects the electrokinetic properties of the interface. The adsorption behavior itself depends upon the solution conditions (e.g., pH and ionic strength) and shorter range polymer segment–surface interactions. Resultantly, ease of analysis is somewhat complicated, as "zero separation" is difficult to define. The SFA has been used to investigate the forces between polyelectrolyte-coated mica surfaces, for which the interested reader should see Refs. 80 and 81. AFM has been used to explore intersurface forces between zirconia surfaces in the presence of poly(acrylic acid) (PAA) [82] and silica surfaces in the presence of PAA [83] quaternized poly(vinyl pyridine) (q-PVP) [84]. Some studies [81,83] were primarily concerned with the exploration of the nonequilibrium effects of polymer bridging between the surfaces and are thus beyond the scope of this review, although it should be noted that there is apparently a fundamental difference in the form of polymer bridging forces where this force has been alleged; the force–distance profile in the latter case being more similar to the SFA-measured profile for nonaqueous bridging [85].

Hartley et al. [63] measured the forces between a mica surface bearing an adsorbed layer of poly(lysine) and a clean silica surface in simple electrolyte solution and confirmed the theoretical prediction that adsorption of a polyelectrolyte molecule at an (initially) oppositely charged interface leads to a reversal (albeit small) of the interfacial charge of the adsorbent substrate. Their experimental findings were satisfactorily explained using a fitted dissimilar surface potential model. Again, as discussed earlier for dissimilar metal oxide measurements, this analysis requires independent determination of the surface potential of at least one of the surfaces under identical solution conditions. In a related study [72], the interactions between a silica sphere and a mica flat bearing adsorbed layers of the protein bovine serum albumen were studied. The authors successfully fitted their data to a simple electrokinetic model. Milling and Vincent [82] investigated the forces between silica surfaces in the presence of adsorbing PAA. At low pH and minimal added (simple) electrolyte the authors observed a long-range EDL-type interaction between the surfaces, and upon increasing the PAA concentration the development of a secondary minimum that initially increased and then subsequently decreased in depth. The length scale of this interaction was observed to be monotonically decreasing with respect to polymer concentration. This was attributed to a depletion effect—namely, that the vacation of the polyelectrolyte molecules and their attendant ionic atmospheres from between the surfaces reduces the osmotic press relative to the bulk and thus produces

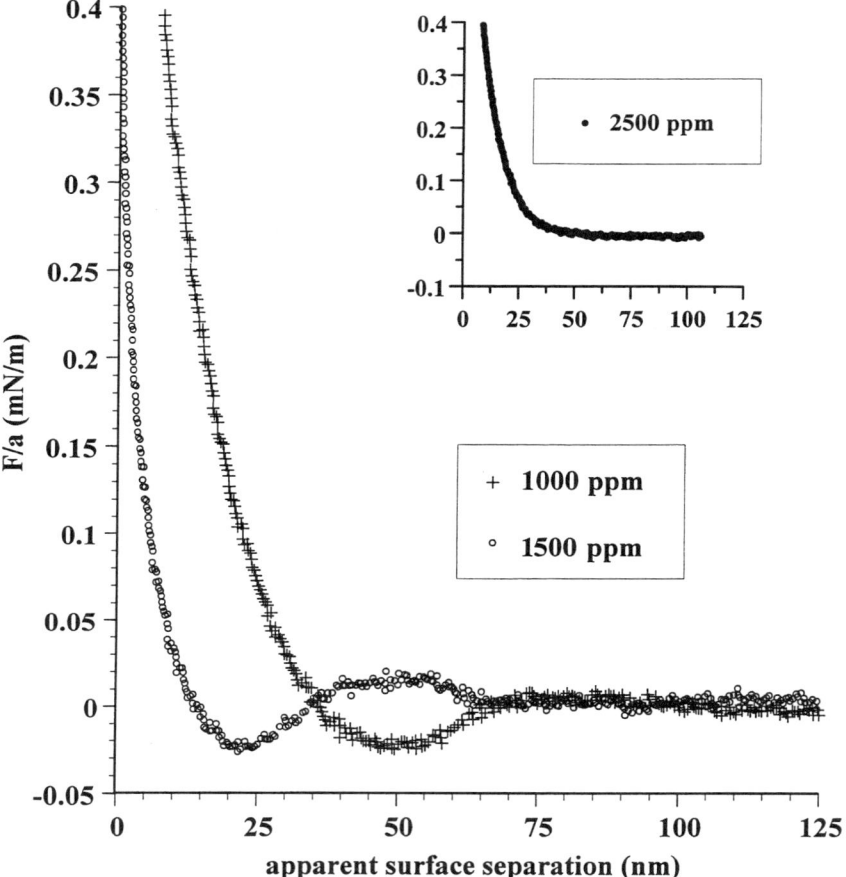

FIG. 6 Experimental force–separation behavior for a silica sphere interacting with a flat silica surface in the presence of (weakly) adsorbing poly(acrylic acid) at differing PAA concentrations. The inset shows the disappearance of the attractive force at high polymer concentrations.

an attractive force (for extensive reviews of polymer depletion phenomena see Scheutjens et al. [78], Snowden and Jenkins [86], and Milling and Vincent [87]). The force–apparent separation data correlated with an observed phase separation of a silica dispersion, under similar conditions, and was satisfactorily modelled using an ad hoc summation of a like-surface EDL interaction [46] and a simple expression for the depletion force [88]. Figure 6 illustrates some force–distance data of Milling and Vincent [86].

2. Surface Forces in the Presence of Nonadsorbing Polyelectrolytes

Studies concerning intersurface forces due to *nonadsorbing* polyelectrolyte species have until recently received scant attention. To date there has only been one published AFM study of nonadsorbing polyelectrolyte species. In this study Milling [89] used an AFM to examine the forces between silica surfaces in solutions of sodium poly(styrene sulfonate) (NaPAA). In this paper oscillatory forces between the surfaces were reported; the oscillatory length scale decreased and the magnitude increased with increasing polymer concentration. The surface separation of the secondary maximum (equivalent to interstitial depletion of the polymer molecules) increased with increasing pH, indicating that increasing the surface potential of the silica–electrolyte interface led to increased electrostatic repulsion between the (negatively charged) silica surface and the polyanion molecules. The addition of background electrolyte to the system reduced the magnitude of the observed forces, which again supports the conjecture that the intersurface forces are due to osmotic disjoining pressures (the addition of salt reduces the osmotic pressure of polyelectrolyte solutions). Several techniques other than AFM [90–92] have provided corroboratory evidence for the development of partial structure factors for polyelectrolyte molecules both in bulk solution and constrained between interfaces. These structure factors are thought to be due to electrostatic interchain self-avoidance, which in turn gives rise to a nonhomogeneous polymer segment density function at interfaces, and as discussed by Joanny [76], this leads to an oscillatory force between approaching surfaces, the final force oscillation being the depletion force as mentioned in Section III.B. Once the polyelectrolyte molecules have fully vacated the interstice (which now effectively resembles a semipermeable membrane), there remain only solvent molecules and simple ions, and the Donnan equilibrium [93] that is established between the interstice and the bulk solution entails that crucially important factors required for enumeration of the EDL force, such as local ionic strength and pH, are presently intractable quantities. Clearly in this aspect of ion-mediated surface forces, experimental observations are superseding theory.

IV. VAN DER WAALS INTERACTIONS

As mentioned in the introduction to this chapter, the first theory for condensed phase vdW interactions was due to London [6], who summated dipole interactions to describe the force between two bodies (denoted by subscripts 1 and 2) interacting via a third medium (subscript 3). Equation (7) describes this relationship for a sphere interacting with a flat surface:

$$F_{132} = \frac{aA_{132}}{6H^2} \tag{7}$$

where A_{132} is a composite Hamaker function for the materials in question, given by

$$A_{132} = (A_1^{1/2} - A_3^{1/2})(A_2^{1/2} - A_3^{1/2}) \tag{8}$$

where A_i is the Hamaker constant of material i, which is related to the polarizability of the materials. Visser [94] has extensively reviewed methods for determining this parameter. The London approach, which in many circumstances proves to be satisfactory, predicts that van der Waals forces will be both attractive (as is commonly observed for solid materials in aqueous media) and also repulsive.

A fuller theoretical analysis of vdW interactions requires recourse to Lifshitz theory [8]. Lifshitz theory requires a description of the dielectric behavior of materials as a function of frequency, and there are several reviews for the calculation of Hamaker functions using this theory. The method described by Hough and White (H–W) [95], employing the Ninham–Parsegian [96] representation of dielectric data, has proved to be most useful. The nonretarded Hamaker constant (for materials 1 and 2, separated by material 3) is given by

$$A_{132} = \frac{3kT}{2} \sum_{m=0}^{\infty} \sum_{s=1}^{\infty} \frac{(\Delta_{13} - \Delta_{23})^s}{s^3} \tag{9}$$

$$\Delta_{kl} = \frac{\varepsilon_k(i\xi_m) - \varepsilon_l(i\xi_m)}{\varepsilon_k(i\xi_m) + \varepsilon_l(i\xi_m)} \tag{10}$$

In the H–W treatment, dielectric data are used to construct the function $\varepsilon(i\xi_m)$—the dielectric constant at imaginary frequency $i\xi_m$.

The more important parameters in calculating the Hamaker constant are the effective oscillator strength C_{uv} and the relaxation frequency ω_{uv}. The best available ultraviolet optical data are often inadequate for the purpose of determining the ultraviolet part of the $\varepsilon(i\xi)$ construction [94]. Effective C_{UV} and ω_{UV} parameters can, however, be obtained from a Cauchy plot of refractive index data in the visible region. If there are no significant absorptions in the microwave region, the oscillator strength in the IR region can be worked out using the Ninham–Parsegian [95] representation of $i\xi_m$:

$$\varepsilon(i\xi) = 1 + \frac{C_{IR}}{1 + (\xi/\omega_{IR})^2} + \frac{C_{UV}}{1 + (\xi/\omega_{UV})^2} \tag{11}$$

A. AFM Studies of Attractive van der Waals Interactions

In contact mode imaging AFM, attractive vdW interactions are the principal cause of mechanical spring instabilities at close approach of the tip and sur-

TABLE 1 Accessible Force Measuring Range for vdW Interactions as a Function of Spring Constant. The Calculations Assume a Tip Radius of 5×10^{-6} m and a Composite Hamaker Constant of 1×10^{-20} J

Spring constant (N/m)	H_{max} (nm)	H_{min} (nm)	F_{max} (N $\times 10^{12}$)	$\sigma F/F_{max} \times 100$
0.05	41	6.9	173	2.9
0.10	29	5.5	275	3.6
0.20	20	4.4	437	4.6
0.50	13	3.2	805	6.2
1.00	9	2.6	1,277	7.8

face. For AFM examinations of surfaces operating in fluid media (i.e., in the absence of capillary forces), variation of the solvent environment can be used to control the tip–surface interaction (via dielectric behavior) circumventing spring instabilities. This has been theoretically discussed for various tip geometries and differing solvent environments [97–99].

There are two principal ways to estimate the Hamaker constant using AFM, direct form fitting of the experimental data to a theoretical expression (e.g., an inverse-square relationship for nonretarded systems), and (when confident that there are no other contributions to the surface force, for instance, EDL interactions), by the surface separation at jump-in (H_{jump}) according to

$$A = \frac{3k H_{jump}^3}{a} \quad (12)$$

where a is the probe radius. This expression also shows that small errors in H_{jump} will produce a large uncertainty in the estimated Hamaker coefficient. The surface separation where the (nonretarded) vdW force is theoretically detectable is given by

$$H_{max} = \left(\frac{aA}{6F_{min}}\right)^{1/2} \quad (13)$$

Choice of cantilever spring-constant to tip-radius ratio can be used to optimize the experiment. To illustrate this point, Table 1 evaluates Eqs. (12) and (13) for the theoretical maximum and minimum separations where surface forces are measurable (assuming a cantilever deflection sensitivity of 1×10^{-10} m, $a = 5 \times 10^{-6}$ m, $A_{132} = 1 \times 10^{-20}$ J) and also the fractional uncertainty associated with the sensitivity/absolute force at jump-in. Clearly the weaker the spring, the more sensitive the measurement, but weak springs may be sampling data in regions where vdW retardation may be operational. Fairly stiff springs will mainly sample nonretarded vdW forces, but over a much shorter range.

For metal oxide systems direct measurement of a system's Hamaker coefficient can be complicated by the presence of EDL forces (particularly for unlike surfaces), and to obviate this problem definitive measurements need to be taken at the isoelectric point of like materials. Such a study was undertaken by Larson et al. [56] for the interaction between titania surfaces, comprising a flat rutile surface and a polycrystalline rutile colloid probe. In their calculations of the van der Waals interaction, the authors made allowance for the uniaxial crystal structure of the rutile, and a composite function comprising weighted orthogonal terms was used. The function $\varepsilon(i\xi)$ for water was constructed using both the Gingell–Parsegian [100] and the newer Parsegian–Weiss [101] representations for water. In this instance the authors found that the difference between the results using these two representations was negligible. Figure 7 illustrates the measured forces for TiO_2 surfaces at the isoelectric point.

Biggs and Mulvaney [102] examined the vdW interactions between gold surfaces, and their results were in good agreement with Lifshitz calculations. The use of gold surfaces for this study was a particularly good test of theory, as extensive optoelectronic data is available for this metal.

B. AFM Studies of Repulsive van der Waals Interactions

The possibility of repulsive vdW interactions was implicit in the original theory of London, and it is also predicted by the Lifshitz theory. It is an intrinsic feature of repulsive vdW forces that three disparate materials (surfaces and solvent) are required. This restriction partially explains why it was not till the advent of AFM and its flexibility in choice of materials that this fundamental force could be directly examined. As mentioned earlier, theoretical studies have considered variation of fluid media to control tip–surface interactions for surface-imaging purposes; for instance, Hutter [103,104] varied the solvent media between silicon nitride AFM tips and a mica surface. In one instance a tip–surface repulsion could be attributed to a repulsive vdW force. Recently two experimental studies have reported fuller explorations of repulsive vdW forces. In the first of these studies, Milling et al. [105] studied the forces between a gold-coated tungsten sphere and a flat poly(tetrafluoroethylene) (PTFE) surface. By changing the solvent media refractive index, the authors showed that surface forces could be changed from attractive to repulsive. For low-polarity solvents, good agreement was found between experiment and calculations made using Lifshitz theory. With polar and semipolar solvents, less good agreement was found. For negative Hamaker constants, the analysis of Ducker et al. [30] had to be modified. From Eq. (12) it can be seen that the gradient of the vdW force is greater than the restoring spring constant at a finite surface separation. This has the effect that for repulsive vdW forces, compliance will be observed at a finite surface separation.

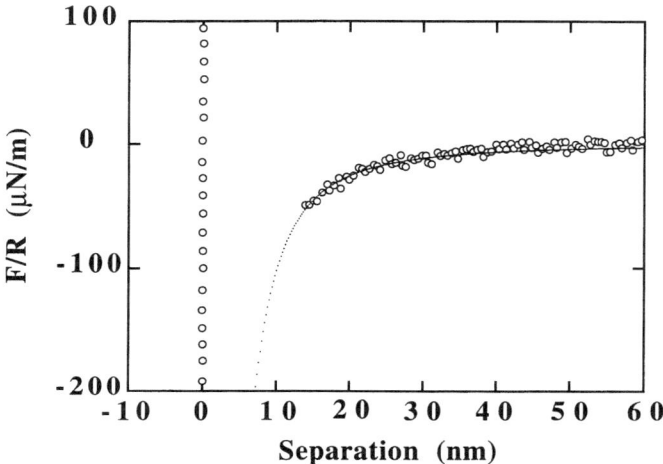

FIG. 7 Force–separation curve taken at the iep of the TiO$_2$, pH 5.6, used in this study. The force–separation points have been scaled by the effective radius of the colloid probe. The full curve is the theoretical interaction using the nonretarded Hamaker constant calculated from Lifshitz theory, $A_H = 6.1 \times 10$–20 J. The background electrolyte (KNO$_3$) concentration was 1.0×10^{-4}M dm^{-3}.

To account for this problem Milling et al. [105] modified Eq. (7) to produce Eq. (14).

$$H_{\text{apparent}} = \left(\frac{-aA}{6F}\right)^{1/2} - H_{\text{compliance}} \tag{14}$$

Hence plotting the apparent surface separation versus $(a/F)^{1/2}$ produces a straight line of gradient $(-A_{132}/6)^{1/2}$ with the (negative) intercept being the surface separation at compliance. As expected, this separation decreased with increasing k/a values. This is illustrated in Fig. 8. In the second study, Meurk et al. [106] used conventional silicon nitride AFM imaging tips and polished silica and silicon nitride surfaces in various solvents. Their results compared very well with theoretical calculations (nonretarded Lifshitz theory). For the dissimilar surfaces, both repulsive and attractive interactions could be produced, whilst for the like surface interactions, only attractive vdW forces were observed. The use of standard AFM cantilevers (with indeterminate interaction radius) entailed that quantitative agreement was not possible. If it is assumed that the tips were unaltered by the experiments, the method of Drummond [107] (for estimating the effective radius of "spherical capped" imaging tips) could have been used to estimate this factor. In this study, the smaller probe radius drastically reduces

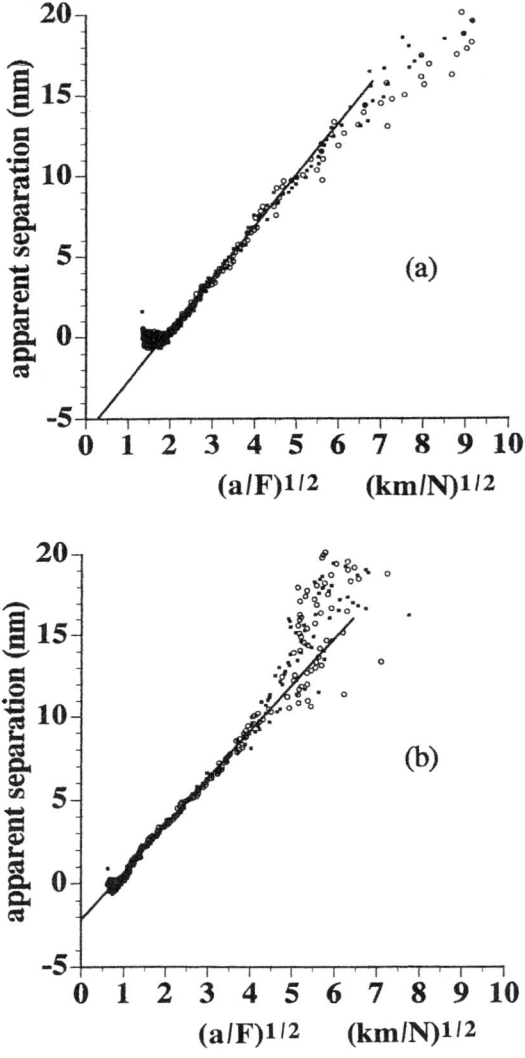

FIG. 8 Experimentally measured repulsive van der Waals interactions between a gold sphere and a smooth poly(tetrafluoroethylene) surface for two different solvents, (a) p-xylene A and (b) bromobenzene. The data is represented in the form suggested in the text using Eq. (14). The measured Hamaker constants for these systems was very similar (ca. 6×10^{-20} J). In the former diagram the ratio of k/a is ca. 9 kN m^{-2}, and in the latter this ratio is 77 kN m^{-2}. The effect of this ratio is reflected in the ordinate intercept of the graphs presented.

the compliance at finite separation problem (this finite separation crudely scaling as radius$^{1/3}$), and so this study is more directly applicable to AFM imaging applications, although considerations about surface heterogeneity are subsequently important when topologically imaging in this "noncontact" mode.

REFERENCES

1. B. V. Derjaguin and L. Landau. Acta Physiochem., 1941, *14*:633.
2. E. G. W. Verwey and T. J. G. Overbeek, *The Theory of the Stability of Lyophobic Colloids*, Elsvier, Amsterdam, 1948.
3. J. H. van der Waals, Ph.D. thesis, University of Lydon, 1973.
4. P. Debye. Z. Physik, 1920, *21*:178.
5. W. H. Keesom. Z. Physik, 1921, *22*:126, 643.
6. F. London. Z. Physik, 1930, *63*:245.
7. Casmir, H. B. G. and Polder, D. Phys. Rev., 1948, *73*:360.
8. H. C. Hamaker. Physica, 1937, *4*:1058.
9. E. M. Lifshitz. Sov. Phys., JETP, 1956, *2*:73.
10. G. Binnig and H. Rohrer. Helv. Phys. Acta, 1982, *55*:726.
11. G. Binnig, C. Quate and G. Gerber. Phys. Rev. Letts., 1986, *56*:930.
12. Y. Martin, C. C. Williams, and H. K. Wickramasinghe. J. Appl. Phys., 1987, *61*:473.
13. (i) Y. Martin and H. K. Wickramasinghe. Appl. Phys. Letts., 1987, *50*:1455. (ii) Y. Martin, D. Rugar, and H. K. Wickramasinghe. Appl. Phys. Letts., 1988, *52*:244. (iii) Y. Martin, D. W. Abraham, and H. K. Wickramasinghe. Appl. Phys. Letts., 1988, *52*:1103.
14. G. M. McClelland, R. Erlandsson, and S. Chiang, *Review of Progress of Non-Destructive Evaluation* (D. O. Thompson and D. E. Chimenti, eds.), Plenum Press, New York, 1987, *6*:307.
15. R. Erlandsson, G. M. McClelland, C. M. Mate, and S. Chiang. J. Vac. Sci. Technol. A, 1988b, *6*:266.
16. A. J. den Boef. Appl. Phys. Letts., 1989, *55*:439.
17. G. Meyer and N. M. Amer. Appl. Phys. Letts., 1988, *53*:1045.
18. S. Alexander, L. Hellemans, O. Marti, J. Schneir, V. Elings, P. K. Hansma, M. Longmire, and J. J. Gurley. Appl. Phys., 1989, *6*:164.
19. B. Drake, C. B. Prater, A. L. Weisenhorn, S. A. C. Gould, T. R. Albrecht, C. F. Quate, D. D. Cannell, H. G. Hansma, and P. K. Hansma. Science, 1989, *243*:1586.
20. D. Sarid, D. Iams, V. Weisenberger, and L. S. Bell. Opt. Lett., 1988, *13*:1057.
21. D. Sarid and B. J. Elings. Vac. Sci. Technol. B., 1991, *9*:431.
22. W. Denk and D. W. Pohl. Appl. Phys. Lett., 1991, *59*:2171.
23. T. Göddenhenrich. H. Lanke, U. Hartmann, and C. Heiden. J. Vac. Sci., Technol, A., 1990a, *8*:383.
24. G. Neubauer, S. R. Cohen, G. M. McClennand, D. Horne, and C. M. Mate. Rev. Sci. Instrum., 1990, *61*:2296.

25. E. Meyer, H. Heinzelmann, P. Grutter, T. Jung, H. R. Hidber, H. Rudin, H. J. Guntherodt. Thin solid films, 1989, *181*:527–544.
26. N. A. Burnham and R. J. Colton. J. Vac. Sci. Technol., (A), 1989a, *7*:2906.
27. Y. N. Moiseev, V. M. Mostepanenko, V. I. Panov, and I. Y Sokolov. Zh. Tekh. Fiz. (USSR), 1989, *15*:5; 1990, *60*:141.
28. D. Y. C. Chan (Dept. of Mathematics, University of Melbourne, Victoria, Australia).
29. W. A. Ducker, T. J. Senden, and R. M. Pashley. Nature, 1991, *353*:239.
30. W. A. Ducker, T. J. Senden, and R. M. Pashley. Langmuir, 1992, *8*:1831.
31. H.-J. Butt. Biophys., 1991, *60*:1438.
32. J. P. Cleveland, S. Manne, D. Bocek, and P. K. Hansma. Rev. Sci. Instrum., 1993, *64*:403.
33. J. E. Sader. Rev. Sci. Instrum, 1995, *66*:4583.
34. J. E. Sader, I. Larson, P. Mulvaney, and L. R. White. Rev. Sci. Instrum, 1995, *66*:37891.
35. D. Tabor and R. H. S. Winterton. Proc. Roy. Soc. A., 1969, *312*:435.
36. I. Sogami and N. Ise. J. Chem. Phys., 1984, *81*:6320.
37. N. Ise, in *Proc. 19th Yamada Conference on Ordering and Organisation in Ionic Solutions* (N. Ise and I. Sogami, eds.), World Scientific, Singapore, 1988.
38. K. Ito, H. Yoshida, and N. Ise. Science, 1994, *263*:66.
39. K. S. Schmitz. Langmuir, 1997, *13*:5849.
40. G. J. Gouy. J. Phys. Radium, 1910, *9*:457.
41. D. L. Chapman. Phil. Mag., 1913, *25*:475.
42. R. J. Hunter, *Foundations of Colloid Science I*, Oxford University Press, Oxford, 1989.
43. B. V. Derjaguin. Kolloid-Z., 1934, *69*:155.
44. B. W. Ninham and V. A. Parsegian. J. Theor. Biol., 1971, *31*:405–428.
45. D. Y. C. Chan, R. M. Pashley, and L. R. White. J. Colloid Interface Sci., 1980, *77*:283.
46. D. McCormack, S. L. Carnie, and D. Y. C. Chan. J. Colloid Interface Soc., 1995, *169*:177.
47. R. Hogg, T. W. Healy, and D. W. Furstenau. Trans. Far. Soc., 1966, *62*:1638.
48. T. W. Healy and L. R. White. Adv. Colloid Interface Sci., 1978, *9*:303.
49. P. G. Hartley, I. Larson, and P. J. Scales. Langmuir, 1997, *13*:2207.
50. I. Odler, in *Lea's Chemistry of Cement and Concrete*, 4th ed. (P. C. Hewlett, ed.), Arnold Press, London, 1998.
51. L. Meagher. J. Colloid Interface Sci., 1992, *152*:293.
52. R. O. James and T. W. Healey. J. Colloid Interface Sci., 1972, *40*:42; 1972, *40*:65.
53. J. D. Birchall, A. J. Howard, and D. D. Double. Cement and Concrete Research, 1980, *10*:145.
54. I. Larson, C. J. Drummond, D. Y. C. Chan, and F. Grieser. J. Am. Chem. Soc., 1993, *115*:11885.
55. M. Prica, S. Biggs, F. Grieser, and T. W. Healey. Colloid Surf. A., 1996, *119*:205.
56. S. Biggs, P. J. Scales, Y. K. Leong, and T. W. Healey. J. Chem. Soc. Faraday Trans., 1995, *91*:2921.

57. H. G. Pederson, J. W. Hoy, and J. Engell, *Proceedings of the Fourth European Ceramics Society Conference*, 1995, 2:31.
58. R. K. Iler, *The Chemistry of Silica*, Wiley-Interscience, New York, 1979.
59. P. G. Hartley, I. Larson, and P. J. Scales. Langmuir, 1997, *13*:2207.
60. G. Toikka and R. A. Hayes. J. Colloid Interface Sci., 1997, *191*:102.
61. I. Larson, C. J. Drummond, D. Y. C. Chan, and F. Grieser. Langmuir, 1997, *13*:2109.
62. S. Veeramasoneni, M. R. Yalamanchili, and J. D. Miller. J. Colloid Interface Sci., 1996, *184*:594.
63. P. G. Hartley and P. J. Scales. Langmuir, 1998, *14*:6948.
64. R. Raiteri, M. Grattarola, and H.-J. Butt. J. Phys. Chem., 1996, *100*(41):16700.
65. A. C. Hillier, S. Kim, and A. J. Bard. J. Phys. Chem., 1996, *100*(48):18808.
66. K. Hu, F.-R. F. Fan, A. J. Bard, and A. C. Hillier. J. Phys. Chem. B, 1997, *101*:8298.
67. T. H. Muster, G. Toikka, R. A. Hayes, C. A. Prestidge, and J. Ralston. Colloids Surf. A, 1996, *106*:203–211.
68. G. Toikka, R. A. Hayes, and J. Ralston. Langmuir, 1996, *12*:3783–3788.
69. S. Biggs, M. K. Chow, C. F. Zukoski, and F. Grieser. J. Colloid Interface Sci., 1993, *160*:511.
70. S. Biggs, P. Mulvaney, C. F. Zukoski, and F. Grieser. J. Am. Chem. Soc., 1994, *116*:9150.
71. I. Larson, D. Y. C. Chan, C. J. Drummond, and F. Grieser. Langmuir, 1997, *13*:2429.
72. K. Hu and A. J. Bard. Langmuir, 1997, *13*:5114.
73. W. R. Bowen, N. Hilal, R. W. Lovitt, and C. J. Wright. J. Colloid Interface Sci., 1998, *197*:348.
74. P. G. de Gennes, P. Pincus, R. M. Velasco, and F. Brochard. J. Phys. (Paris), 1976, *37*:1461.
75. A. V. Dobrynin, R. H. Colby, and M. Rubinstein. Macromolecules, 1995, *28*:1859.
76. G. A. Christie and S. L. Carnie. J. Chem. Phys., 1990, *92*:7661.
77. X. Chatellier and J. F. Joanny. J. Phys. France II, 1996, *6*:1669.
78. M. A. G. Dahlgren and F. A. M. Leermakers. Langmuir, 1995, *11*:2996.
79. G. J. Fleer, M. A. Cohen Stuart, J. M. H. M. Scheutjens, T. Cosgrove, and B. Vincent, *Polymers at Interfaces*, Chapman and Hall, London, 1993.
80. P. M. Claesson, M. A. G. Gahlgren, and L. Eriksson. Colloids Surf. (A), 1994, *93*:293.
81. P. M. Claesson, O. E. H. Paulson, E. Blomberg, and N. L. Burns. Colloids Surf. (A), 1997, *123*:341.
82. S. Biggs. Langmuir, 1995, *11*:156.
83. A. J. Milling and B. Vincent. J. Chem. Soc. Faraday Trans., 1997, *93*:3179.
84. S. Biggs and A. D. Proud. Langmuir, 1997, *13*:7202.
85. J. Klein, and P. F. Luckham. Nature, 1984, *308*:836.
86. P. D. Jenkins and M. J. Snowden. Adv. Poly Interface Sci., 1996, *68*:57.
87. A. Milling and B. Vincent, in *Colloid-Polymer Interaction: From Fundamentals to Practise*, John Wiley, London, to appear.

88. G. J. Fleer, J. M. H. M. Scheutjens, and B. Vincent. *Polymer Adsorption and Dispersion Stability*, ACS Symp. Ser., 1984, *240*:245.
89. A. J. Milling. J. Phys. Chem., 1996, *100*:8986.
90. A. Sharma, S. N. Tan, and J. Y. Walz. J. Colloid Interface Sci., 1997, *191*:236.
91. A. Asnacios, A. Espert, A. Colin, and D. Langevin. Phys. Rev. Letts., 1992, *78*:4974.
92. N. Ise, T. Okubo, Y. Hiragi, H. Kawai, T. Hashimoto, M. Fujimura, A. Nakajima, and H. Hayashi. J. Am. Chem. Soc., 1979, *101*:5836.
93. J. G. Donnan and E. A. Guggenheim. Z. Physik Chem., 1932, *162*:346.
94. J. Visser. Adv. Colloid Interface Sci., 1981, *15*:157.
95. D. B. Hough and L. R. White. Adv. Colloid Interface Sci., 1980, *14*:3.
96. P. W. Ninham and V. A. Parsegian. J. Chem. Phys., 1970, *52*:4578.
97. U. Hartmann. Phys. Rev. B, 1991, *43*:2404.
98. F. O. Goodman and N. Garcia. Phys. Rev. B, 1991, *43*:4728.
99. N. Garcia and V. T. Binh. Phys. Rev. B, 1992, *46*:7946.
100. D. Gingell and V. A. Parsegian. J. Theor. Biol., 1972, *36*:41.
101. V. A. Parsegian and G. H. Weiss. J. Colloid Interface Sci., 1981, *81*:285.
102. S. Biggs and P. Mulvaney. J. Chem. Phys., 1994, *100*:8501.
103. J. L. Hutter and J. Bechhoefer. J. Appl. Phys., 1993, *73*:4123.
104. J. L. Hutter and J. Bechhoefer. J. Vac. Sci. Technol. B, 1994, *12*:2251.
105. A. Milling, P. Mulvaney, and I. Larson. J. Colloid Interface Sci., 1996, *180*:460.
106. A. Meurk, P. F. Luckham, and L. Bergström. Langmuir, 1997, *13*:3896.
107. C. J. Drummond and T. J. Senden. Colloid Surf. A, 1994, *87*:217.

4
Measurement of Electro-osmosis as a Method for Electrokinetic Surface Analysis

NORMAN L. BURNS Institute for Surface Chemistry, Stockholm, Sweden

I.	Characterization of the Aqueous Solution–Solid Interface	114
II.	Electro-osmosis: Theory and Modeling	115
	A. The electrical double layer	115
	B. Origin of electro-osmosis	117
	C. Origin of surface charge	118
	D. Model considerations	119
III.	Measurement of Electro-osmosis	119
	A. The cylindrical cell	120
	B. The rectangular cell	123
	C. Experimental considerations	125
IV.	Electro-osmosis and Surface Analysis	126
	A. Quartz	126
	B. Silane modified surfaces	127
	C. Poly(ethylene glycol) modified surfaces	129
	D. Plasma polymer modified surfaces	130
V.	Future Applications	135
	Symbols	137
	References	139

I. CHARACTERIZATION OF THE AQUEOUS SOLUTION–SOLID INTERFACE

In the field of surface and colloid science, characterization of surfaces has become an important utility in accounting for the observed behavior of any two-phase system. Many factors can affect the state of an interfacial region including van der Waals, electrostatic, acid–base, and covalent interactions between phase components. A complete account of the forces operating at an interface does much to predict and clarify the behavior of a system. In this context surface characterization is essential.

In the colloidal realm, given the large surface-to-volume ratio and the relatively small range of force that can sway the disposition of a colloidal particle, it is easy to appreciate the importance of controlling surface properties. Research literature abounds with the characteristics of colloid systems and model systems that mimic colloid surfaces. Applications permeate the fields of materials processing, adhesion, coatings, food science, and medicine.

For well-dispersed colloid systems, particle electrophoresis has been the classic method of characterization with respect to electrostatic interactions. However, outside the colloidal realm, i.e., in the rest of the known world, the measurement of other electrokinetic phenomena must be used to characterize surfaces in this respect. The term ''electrokinetic'' refers to a number of effects induced by externally applied forces at a charged interface. These effects include electrophoresis, streaming potential, and electro-osmosis.

Classically, electrokinetic phenomena are described by the equation of Navier–Stokes along with the continuity equation [1–3]. The Navier–Stokes equation accounts for the balance of forces in the electrokinetic problem. For steady laminar fluid flow the Navier–Stokes equation takes the following form in electrokinetics:

$$\eta \nabla^2 \mathbf{v} - \nabla p + \rho_e \mathbf{E} = \mathbf{0} \tag{1}$$

with the continuity equation

$$\nabla \cdot \mathbf{v} = \mathbf{0} \tag{2}$$

accounting for the viscous forces $\eta \nabla^2 \mathbf{v}$, the pressure gradients ∇p, and the electrical body forces $\rho_e \mathbf{E}$ acting on a parcel of fluid, where η is the viscosity of the fluid medium, $\mathbf{v}$ the fluid velocity, p the pressure, ρ_e the local charge distribution, and $\mathbf{E}$ the electric field.

Examination of Eq. 1 provides insight into the source and the type of information obtained from electrokinetic phenomena. With this equation it can be shown that the application of an electric field upon a fluid region possessing a net charge would result in translational motion and/or a pressure gradient. Applied to a solid–liquid system, if the solid is mobile, the result would be migration of the solid (electrophoresis). The region in solution bearing countercharges would migrate as well, resulting in fluid flow. Such an electrically induced fluid flow is known as electro-osmosis. Conversely, a flow forced about the solid region would result in an electric field (streaming potential).

To characterize a surface electrokinetically involves the measurement of one of the above electrokinetic effects. With disperse colloidal systems it is practical to measure the particle electrophoretic mobility (induced particle velocity per unit applied electric field strength). However, for a nondisperse system one must measure either an induced streaming potential or an electro-osmosis fluid flow about the surface.

In the following sections an account of the origin and measurement of electro-osmosis is elicited. Furthermore, it is shown how to employ its measurement as a characterization technique. The discussion will focus on the measurement of electro-osmosis in cylindrical chambers and in a novel rectangular chamber whereby electro-osmosis can be measured at small sample plates. Examples of using the measurement of electro-osmosis as a surface characterization technique are discussed in terms of interpretation of the source of electro-osmosis according to classical electrokinetic theory.

II. ELECTRO-OSMOSIS: THEORY AND MODELING

Much research has focused on determining surface charging properties and interpreting changes in properties upon surface chemical modification. To this end it has been useful to measure the pH dependence of electro-osmosis. To extract information from electro-osmosis measurements, a mechanism of pH-dependent charge formation at the surface must be forwarded, and an appropriate model of charge distribution in the interfacial region is necessary. For this purpose a model for the pH dependence of electro-osmosis is adopted from classical electrokinetic theory [4]. Such a description is afforded by a Gouy–Chapman type description of the double layer and solutions to the Poisson and Boltzmann equations. Surface charge is represented in terms of acid–base site dissociation of ionizable surface groups after Healy and White [5].

A. The Electrical Double Layer

Figure 1 shows a simple schematic representation of the electrical double layer according to Gouy and Chapman. Net bound charge at the solid surface (σ_0) is countered by diffuse charge in solution of net value σ_d such that

$$\sigma_0 + \sigma_d = 0 \tag{3}$$

The charge density in the volume adjacent to the surface (ρ_e) is the sum of the N ionic species of concentration c^i carrying respective charges $z^i e$, where e is Coulombic charge.

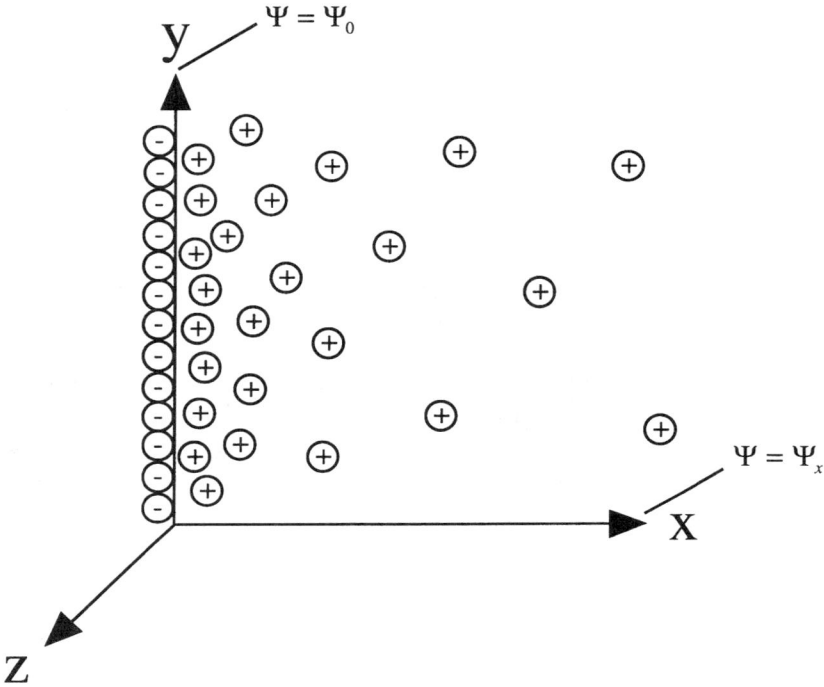

FIG. 1 The Gouy–Chapman electrical double layer. Net bound charge at a solid surface ($x = 0$) is countered by diffuse charge in solution. Coions in the diffuse double layer are omitted for clarity.

$$\rho_e = -e \sum_{i=1}^{N} z^i c^i \qquad (4)$$

The electric field flux within a fluid volume of charge density ρ_e in a continuous dielectric medium, such as water, can be described in terms of the permittivity ε of the medium and electric potential Ψ by Poisson's equation

$$\rho_e = -\varepsilon \nabla^2 \Psi \qquad (5)$$

Assuming an equilibrium balance of electrical and diffusional forces, the charge distribution in the interfacial region can be described according to

$$kT \nabla \ln c^i + z^i e \nabla \Psi = 0 \qquad (6)$$

Electro-osmosis in Electrokinetic Surface Analysis

which in the scheme of Fig. 1 yields the Boltzmann equation

$$c_0^i = c_\infty^i \exp\left(\frac{-z^i e \Psi}{kT}\right) \tag{7}$$

This expression can be used to relate ion concentrations at the surface (c_0^i) to those in the bulk (c_∞^i). Ψ_0 is the potential at the surface ($x = 0$) relative to that in bulk solution.

The diffuse double layer charge over a surface of area S can be taken by integration of the charge density over the surface (defined by Ω).

$$\sigma_d S = \iint_\Omega \rho_e \, dS \tag{8}$$

For the flat double layer and a 1 : 1 electrolyte in the scheme of Fig. 1,

$$\sigma_d = \int_0^\infty \rho_e \, dx = -(8C\varepsilon kT)^{1/2} \sinh\left(\frac{e\Psi_0}{2kT}\right) \tag{9}$$

Note that the potential (Ψ) decays from its surface value to a value of zero at some point in the bulk. Inserting Eqs. (4) and (7) into Poisson's equation and applying the appropriate boundary conditions ($\Psi = \Psi_0$ at $x = 0$ and $\Psi = 0$ at $x = \infty$), the resulting equation can be integrated twice to yield

$$\tanh\left(\frac{e\Psi_x}{4kT}\right) = \tanh\left(\frac{e\Psi_0}{4kT}\right) \exp\left[-\left(\frac{2ce^2 x^2}{\varepsilon kT}\right)^{1/2}\right] \tag{10}$$

This expression relates the potential at the surface to that at some distance x (Ψ_x).

B. Origin of Electro-osmosis

With a model of the distribution of charges near the solid–liquid interface, it is now possible to discuss the phenomenon of electro-osmosis. Referring back to Eq. 1, the Navier–Stokes equation can be solved for electro-osmosis given the appropriate boundary conditions. In the scheme of Fig. 1, consider application of an electric field of strength E_∞ tangential to the surface (in the z-direction). Assuming constant viscosity and permittivity, and negligible pressure gradients as a result of the fluid flow, integration of Eq. (1) from the hydrodynamic plane of shear where the fluid velocity is zero ($v = 0$) to a point in the bulk where the potential is zero and the fluid velocity is constant (v_{eo}) gives

$$v_{eo} = -\frac{\varepsilon E_\infty}{\eta}\zeta \tag{11}$$

where ζ is the potential at the plane of shear.

This equation, known as the Helmholtz–Smoluchowski equation, relates the potential at a planar bound surface region to an induced electro-osmosis fluid velocity [6]. Recall that in the previous section surface charge was related to a potential in solution. In the following section surface charge will be related to the chemistry of the surface. A model for the development of surface charge in terms of acid–base dissociation of ionizable surface groups is introduced.

C. Origin of Surface Charge

In the absence of specific adsorption of electrolyte ions, surface charge is considered to originate from acid–base dissociation of ionizable groups. In terms of acid groups (AH) and basic groups (B), the respective pH-dependent equilibrium between surface sites and solution at the interface can be represented as

$$AH \rightleftharpoons A^- + H^+ \tag{12}$$

$$B + H^+ \rightleftharpoons BH^+ \tag{13}$$

The respective equilibria can be described by

$$K_a = \frac{[A^-][H^+]_s}{[AH]} \tag{14}$$

$$K_b = \frac{[BH^+]}{[B][H^+]_s} \tag{15}$$

where K_a is an acid dissociation constant, K_b is the base dissociation constant, and $[H^+]_s$ is the hydronium ion concentration at the surface. The total number of respective acid and basic sites, N_a and N_b, can be represented as the sum of charged and uncharged species:

$$N_a = [A^-] + [AH] \tag{16}$$

$$N_b = [BH^+] + [B] \tag{17}$$

Given i distinct acid types and j basic types on the surface, each with its own equilibrium, the total surface charge (σ_s) can be represented by

$$\begin{aligned}\sigma_s &= e \sum_j [BH^+]_j - e \sum_i [A^-]_i \\ &= \sum_j \left[\frac{eN_b}{1 + 10^{(\text{pH}_s - \text{p}K_b^j)}} \right] - \sum_i \left[\frac{eN_a}{1 + 10^{(\text{p}K_a^i - \text{pH}_s)}} \right]\end{aligned} \tag{18}$$

D. Model Considerations

If in the above equations ζ can be identified with the surface potential Ψ_0 at the experimental surface (or some potential Ψ_x), and σ_s identified with σ_0, then the equations in the previous sections provide us with a means to relate electro-osmosis to pH in terms of the number density of ionizable surface sites, pK of the surface site, position of the hydrodynamic plane of shear, ionic strength, and temperature.

The model adopted considers only a homogenous planar region of charge bearing groups at a surface. Electrolyte species are assumed to behave as point charges in a continuous dielectric medium distributed in the mean potential created by the surface and by the ions located in their mean position. Thus correlations between the instantaneous positions of the ions in solution are neglected. The charges are assumed to obey the Poisson–Boltzmann equation up to the planar region. Furthermore, in using the Helmholtz–Smoluchowski equation, viscosity and permittivity are assumed to be constant in the interfacial region with a well-defined plane of shear.

Many more-sophisticated models have been put forth to describe electrokinetic phenomena at surfaces. Considerations have included distance of closest approach of counterions, conduction behind the shear plane, specific adsorption of electrolyte ions, variability of permittivity and viscosity in the electrical double layer, discreteness of charge on the surface, surface roughness, surface porosity, and surface-bound water [7]. Perhaps the most commonly used model has been the Gouy–Chapman–Stern–Grahame model [8]. This model separates the counterion region into a compact, surface-bound "Stern" layer, wherein potential decays linearly, and a diffuse region that obeys the Poisson–Boltzmann relation.

A problem with using the "more-sophisticated" models is the concomitant increase in the number of parameters needed to describe electrokinetic phenomena, in particular electrokinetically inaccessible parameters. Model parameters must be assumed, or acquired from independent experimental techniques. Another problem is that it is difficult to incorporate more than one or two considerations into the same model.

In light of these considerations experimentally derived parameters from the model presented above should be viewed as effective. Absolute quantitative values are not their essence. The experimentally derived parameters have proven useful as a qualitative basis for discussion of the properties of specific systems.

III. MEASUREMENT OF ELECTRO-OSMOSIS

The basic experimental approach when using electro-osmosis to characterize surfaces is to measure fluid mobility at a surface when electro-osmosis is induced by an applied electric field. Several methods, direct and indirect, can be employed including measurement of bulk fluid flow through a porous sample plug or capillary [4]. In this chapter discussion will be limited to the use of closed, free-fluid microparticle electrophoresis cells to measure electro-osmosis [9,10]. In this method particles suspended in a fluid medium are electrophoresed in an electrode chamber. Particle mobility is determined optically through the chamber wall. However, if the chamber surface is charged, electro-osmosis is induced at the chamber walls by the applied electric field. A hydrodynamic circulatory flow in the chamber results. If particle mobility is of interest, this hydrodynamic flow complicates mobility determination, since the flow imposes upon the particle's mobility. Hydrodynamic flow must be taken into account. If the chamber surface is of interest, however, the hydrodynamic flow reflects electro-osmotic fluid flow at the surface, and the particle can be used as a tracer to visualize fluid flow. Thus a chamber surface may be characterized electrokinetically, according to electro-osmotic fluid flow, with an accurate description of the hydrodynamics of the fluid flow in the chamber.

The hydrodynamics of fluid flow in microparticle electrophoresis chambers are described by solution to the Navier–Stokes equation for steady laminar fluid flow, Eq. (1), with boundary values defined by the chamber geometry. Considering the dimensions of an experimental chamber compared to the thickness of the double layer (mm to nm), fluid flow at the surface would appear to move at a constant velocity. In other words, the region of varying velocity, viscosity, charge density, potential, etc., in the electrical double layer is not observable. Thus viscosity is constant in the bulk hydrodynamic problem. Accordingly the electrical force term $\rho_e \mathbf{E}$ in Eq. (1) can be ignored, since the net charge density in the observable region is zero. However, it should be noted that such considerations would appear in any attempt to interpret the source and surface characteristics responsible for the measured electro-osmotic velocity. This includes interpretation in terms of the Helmholtz–Smoluchowski equation; hence the elaborate discussion in the previous section of a model of the origin of electro-osmotic fluid flow. Though it is beyond the intent of this summary to present the derivation of the hydrodynamic equations, it is useful to discuss relevant consequences of the hydrodynamics and chamber geometry with respect to experimental practice. Following is a discussion of measurement of electro-osmosis in the two most common chamber geometries: cylindrical and rectangular.

A. The Cylindrical Cell

In one experimental design, microparticle electrophoresis can be observed in cylindrical quartz capillaries, such as with the modified Rank Brothers Mark I apparatus (Cambridge, U.K.) schematically illustrated in Fig. 2 [11–13]. In this design, the capillary (typically quartz, 2 mm internal diameter) serves as an

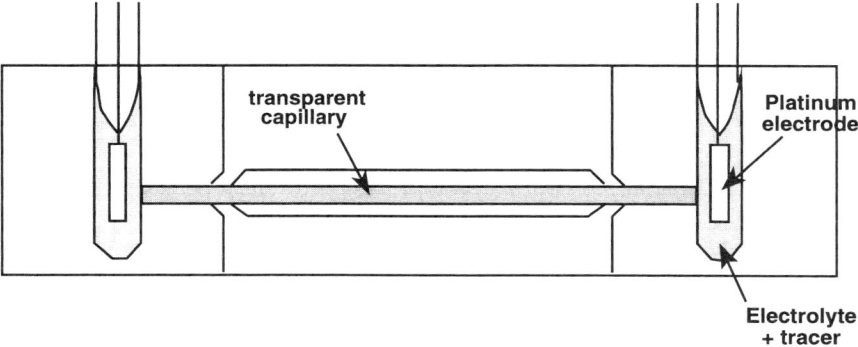

FIG. 2 Schematic of the modified Rank Mark II electrophoresis chamber.

electrophoresis chamber and experimental substrate. The chamber is filled with a suspension of tracer particles, on the order of 1 μm, and fitted with electrodes (blank platinum, platinum black, palladium, or silver/silver chloride, depending upon application) at the ends of the chamber. The chamber is then immersed in a thermostatted water bath, and electrophoresis of particles is observed with a microscope at various locations across the diameter of the capillary. A DC power supply provides electric field strengths in the range 1–150 V/cm. Particle velocity is taken across an ocular graticule in the microscope assembly where location of the particle is determined by two-axis distance micrometers mounted on the Rank assembly.

Defining mobility as velocity per unit electric field strength, the apparent particle mobility U_p, the particle velocity v_p per unit electric field strength E_∞, at a given location in the cell is calculated according to

$$U_p \equiv \frac{v_p}{E_\infty} = \frac{d\kappa\pi R^2}{tI} \tag{19}$$

where v_p is the observed particle velocity, E_∞ the applied field strength, d the distance the particle traveled over time t, R the radius of the capillary, κ the conductivity of the fluid medium, and I the electric current. Current is monitored with a voltmeter during electrophoresis, and conductivity of the fluid medium is also measured.

Figure 3 shows typical pH-dependent particle mobility profiles across the diameter of a closed cylindrical chamber for a negatively charged particle and chamber surface. In the closed cell, fluid flow at the surface is compensated by a return flow down the center of the cell, resulting in a parabolic distribution of particle velocities. Particle velocity at a given location is the sum of the intrinsic electrophoretically induced particle velocity and fluid velocity.

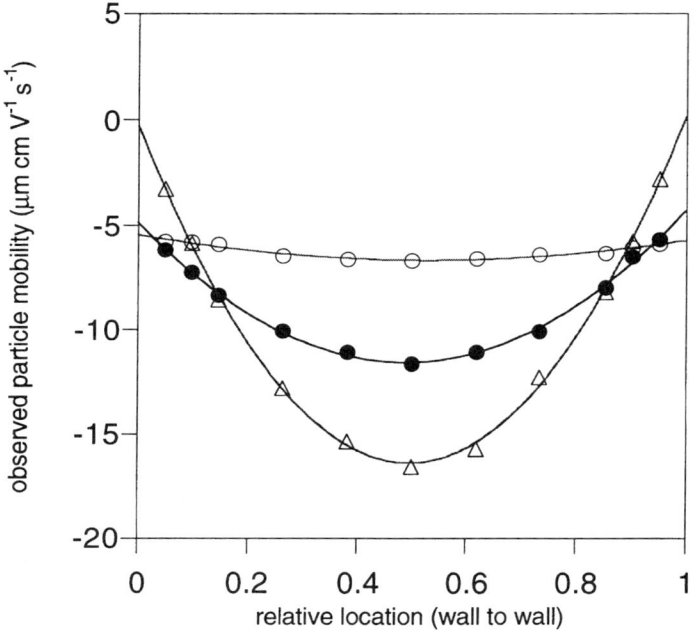

FIG. 3 Apparent particle mobility in 7.5 mM NaCl of a sulfated polystyrene latex particle across the diameter of a closed quartz cylindrical chamber. Profiles are associated with varying degrees of electro-osmosis at pH 2 (○), pH 6 (●), and pH 11 (△).

Description of the hydrodynamics in the cylindrical capillary experimental design is fairly simple. Considering only electrostatic and fluid frictional forces acting upon the suspended particles, apparent particle mobility at a given location r across the diameter of the capillary may be represented by a solution to the Navier–Stokes equation in the scheme of a coordinate system with the origin in the center of the capillary by

$$U_p = 2U_{eo}\left(\frac{r}{R}\right)^2 - U_{eo} + U_{el} \tag{20}$$

where U_{eo} is wall electro-osmotic fluid mobility, R is the capillary radius, and U_{el} is the intrinsic electrophoresis mobility of the particle. Given Eq. (20), electro-osmotic fluid mobility is taken from a linear least-square fit of the experimental apparent particle mobility versus relative location (r/R) data.

However straightforward the hydrodynamics of the cylindrical chamber, the experimental substrate is limited to transparent materials that can be drawn into

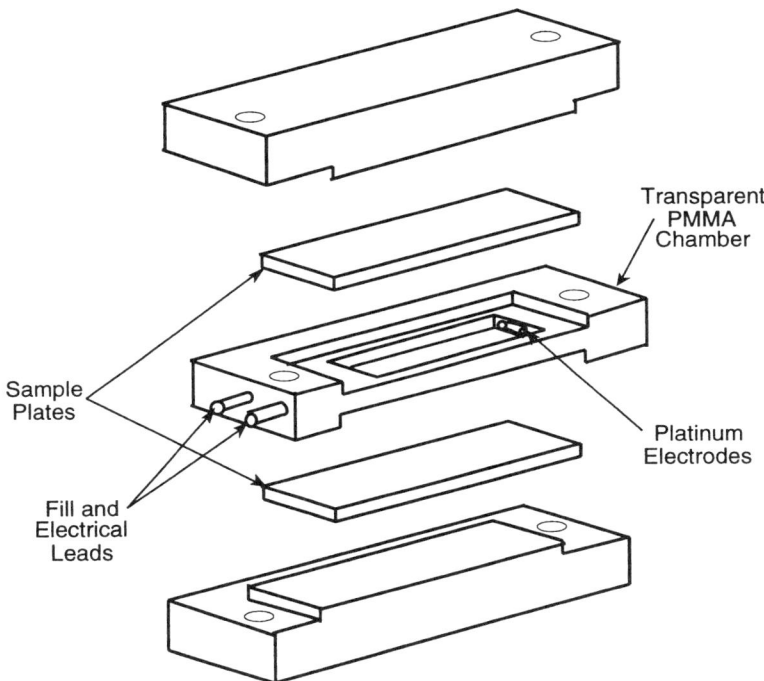

FIG. 4 The rectangular electrophoresis chamber for electrokinetic characterization of macroscopic surfaces.

a capillary and retain the optical properties necessary to view the reference particles. To characterize nontransparent surfaces, other cell geometries must be used.

B. The Rectangular Cell

A successful geometry for the study of nontransparent surfaces in this work has been that of the rectangular cell [14]. The rectangular cell developed in the context of this work is shown in Fig. 4. This configuration allows for characterization of a wide variety of surfaces as long as they are not highly conducting. In this design 1×2 cm sample plates are clamped onto either side of a hollow spacer to form the upper and lower walls of the electrophoresis chamber. The spacer, made of optically transparent poly(methyl methacrylate), forms a chamber of dimensions $2 \times 5 \times 15$ mm. Blank platinum electrode wires mounted inside the spacer at the lengths induce electro-osmotic fluid flow at the plates, as

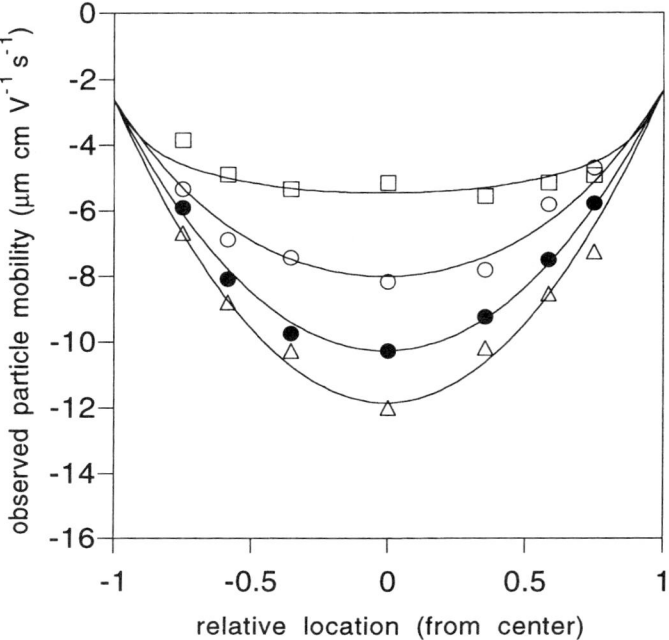

FIG. 5 Particle mobility profile (from plate to plate) between two thermally oxidized silicon wafers in 1 mM NaCl at various locations from the sidewall: 0.1 mm (□), 0.4 mm (○), 0.902 mm (●), and 2.5 mm (△). Drawn lines correspond to a linear least squares curve fit to the hydrodynamic equations describing fluid flow between the plates.

well as along the sidewalls. Tracer particles are observed through the side of the spacer at various locations between the sample plates using the same standard microparticle electrophoresis instrumentation as for the cylindrical chamber.

A drawback to the rectangular cell design is that the hydrodynamic description of fluid flow is much more complicated than in the cylindrical cell. Though it is beyond the intent of this discussion to go into full detail, analytical solutions to the Navier–Stokes equation for steady laminar fluid flow have been derived that can be used for calculation of electro-osmosis at flat plates in the rectangular cell configuration where electro-osmosis may differ at the upper, lower, and side chamber walls [15].

Figure 5 shows a typical particle mobility profile with a negatively charged particle and identical negatively charged sample plates. As with the cylindrical cell, electro-osmotic fluid mobilities at the walls (sample plates) can be taken from linear least squares fits to the equations describing fluid flow in the chamber [15]. Alternatively, if only a single sample plate is of interest, the electro-osmotic fluid mobility can be obtained from a particle mobility measurement at a single location in the cell given a particle with no intrinsic electrophoretic mobility, e.g., polymer-coated latex particles [11,12]. Thus a single 1 × 2 cm sample plate can be characterized from a single particle mobility determination in the rectangular cell. This has clear advantages over the electrokinetically equivalent flat plate streaming potential technique, which requires determination of potential versus pressure curves along a rectangular chamber formed between two identical sample plates [16,17].

C. Experimental Considerations

In theory, there are no limits to the accessibility of electrokinetic data. However, there are a number of physical situations that limit the range of electrokinetic data that can be obtained in the above-described experimental setups. The experimental techniques described above require visual determination of particle velocities and are limited to the range 3–100 μm/s. Additionally, according to Eq. (19), current and solution conductivity affect particle velocity. In experimental practice, the current limit in the cell is around 300 μA owing to the use of blank platinum electrodes. Under higher current, electrolytic reactions at the electrodes result in electrode polarization, heating, and subsequent formation of gas bubbles and thermal convection in the cell. Furthermore, solution conductivity measurements are not reliable below about 100 μS/cm. The above limitations restrict the typical range of solution ionic strength at which one can work to 1–10 mM.

Another factor affecting particle velocity is the relative contribution of electro-osmotic fluid flow and electrophoresis on the particle. Since the observed velocity is the sum of the intrinsic electrophoretically induced particle velocity and fluid velocity, there exist situations where the two combine to render the particle velocity unobservable within the range discussed above. For similar reasons, there are situations where particle velocity may be observable outside the range discussed above. Choice of particle is important in being able to extract the electrokinetic information.

To increase the working range it has been useful to use a video camera in combination with high magnification (2000×) to increase the observable velocity range to 1–100 μm/s. This has allowed the use of particles having no intrinsic electrophoretic mobility in measurement, and simplification of the determination of electro-osmosis (the need to account for intrinsic electrophoretic particle velocity is eliminated). Such particles, having no intrinsic mobility in the working range discussed above, have been produced in this work by grafting poly(ethylene glycol) to an amine modified polystyrene latex particle [15].

Electro-osmosis at a sample plate wall can then be taken from a single particle mobility measurement.

One way to increase the range of measurement may be in the use of light scattering (Doppler electrophoresis) to determine particle velocities. Such methods are used in contemporary commercially available analytical particle electrophoresis apparatuses. However, presently available equipment is not designed for ready exchange (replacement) of chamber surfaces for electro-osmosis studies.

IV. ELECTRO-OSMOSIS AND SURFACE ANALYSIS

Having accounted for the origin of electro-osmosis and for a means to measure the phenomenon at a wide variety of surfaces, and armed with the equations of the previous sections, we now can interpret the pH dependence of electro-osmosis of a solid surface. Such an analysis, though it is not an entirely complete account of the interface, enables us to discuss charging phenomena at an interface and gives a semiquantitative account of ionizable surface groups. In the following specific examples, the use of electro-osmosis as a complementary characterization technique is discussed, taking advantage of what the electrokinetic data can reveal.

A. Quartz

Quartz is a crystalline form of SiO_2 where silicon is tetrahedrally bound to four oxygen atoms. At the surface of quartz and other siliceous substances, acidic silanol groups formed from hydrolysis are present. Surface silanols are titratable in aqueous solution and are a source of pH-dependent surface charge. Fully hydroxylated silicas are considered to have a silanol population of $\sim 5/nm^2$ with an intrinsic pK_a from 5.8 to 7.2 [18].

Figure 6 shows pH-dependent electrokinetic profiles in 7.5 mM NaCl for a clean fully hydroxylated quartz surface and the same surface heated to 190°C. A trend of increasing electro-osmotic mobility with increasing pH is observed, indicating a negative surface at all pH values (note that positive electro-osmotic mobility indicates negative surface charge). In model terms, the quartz surface is best described by two distinct acid sites at the surface. The clean surface is consistent with a fully hydroxylated surface with the majority of silanols having a pK_a of 6.9 and a small number of additional acid groups of pK_a 3.6. Upon heating, a tenfold reduction in the number of weak acid groups was observed along with a threefold increase in the number of stronger acid groups.

Köhler and Kirkland noted that for chromatographic column packings analyzed by diffuse reflectance infrared Fourier transform (DRIFT) spectroscopy and chromatographic behavior, silica appears to possess "associated" and "isolated" surface silanol groups [19]. The associated silanol (possessing weak-acid characteristics) is identified with a silanol hydrogen-bonded to its nearest neigh-

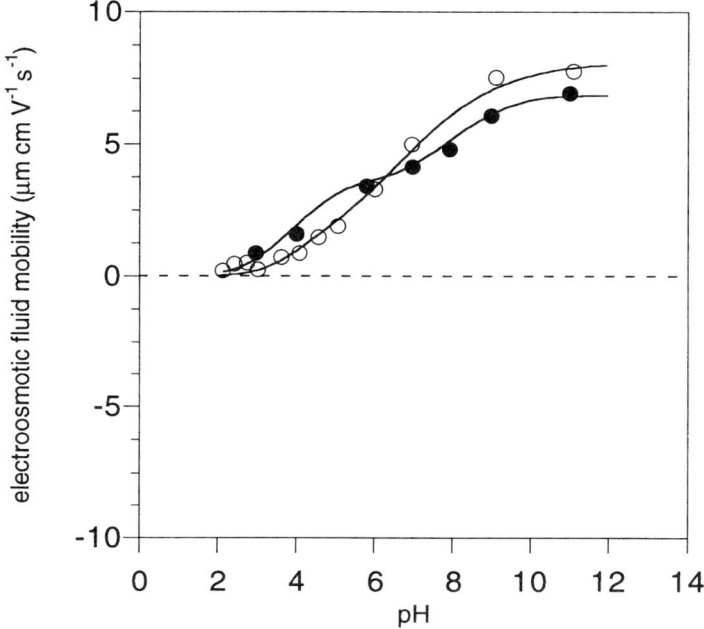

FIG. 6 pH-dependent electrokinetic profile for clean (○) and heat treated (●) quartz in 7.5 mM NaCl.

bor. Isolated silanols do not participate in hydrogen bonding and possess strong acid characteristics. Heat treatment of silica reduced the total number of silanols and increased the relative amount of acidic silanols.

Recent surface force measurements revealed a similar trend [20]. Comparing steam-treated to flame-treated silica sheets using site-dissociation/site-binding model, a decrease in silanol surface sites and apparent decrease in average pK_a was observed upon heat treatment. Furthermore, a repulsive force other than double-layer and van der Waals forces was observed 15 Å from the surface. This repulsion was attributed to hydration of the surface and was found to be independent of surface treatment and electrolyte concentration. In Burns' treatment, an arbitrary plane of shear was introduced to provide a best model fit [13]. A value of 9 Å from the surface for the plane of shear was determined from electro-osmosis measurements.

B. Silane Modified Surfaces

In the last 25 years, 3-aminopropyltriethoxysilane (APS) has become a universally applied coupling agent for covalently linking inorganic surfaces to or-

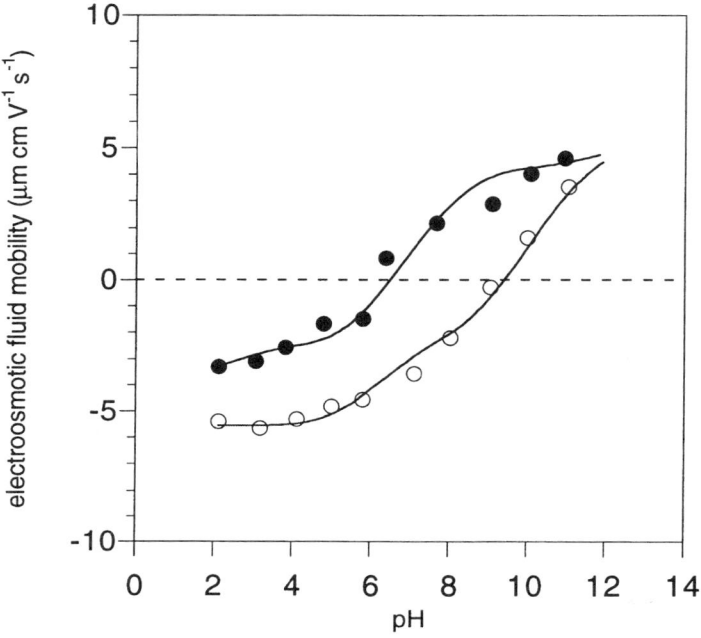

FIG. 7 pH-dependent electro-osmosis profiles for APS coated capillaries, APS 1 (○) and APS 2 (●) in 7.5 mM NaCl.

ganic materials. Applicable fields include composite and adhesive materials and bonded phase chromatography [21,22]. Despite widespread use, its chemistry at oxide surfaces has only recently been understood so as to take advantage of its full potential as a coupling agent. Recent studies employing DRIFT spectroscopy, solid-state NMR, and ellipsometry have focused on the effects of deposition conditions on the chemical nature of APS films [23]. Factors such as surface moisture, solvent, and curing affect the surface state of adsorbed APS. In this case APS deposition was performed on highly hydrated quartz surfaces from anhydrous toluene with postreaction curing in a vacuum oven (190°C). Curing serves to condense silane silanols both intermolecularly and with quartz surface silanols. These conditions are considered optimal for a maximum number of silanes covalently bound to the substrate.

Figure 7 shows pH dependent electrokinetic profiles in 7.5 mM NaCl for APS-coated quartz. The APS surface is best described by a two acid–single base model of the surface. The acid sites appear to be silanols of the same pK_a as for clean glass, and the basic site an APS amine of pK_b 10, which is consistent with a value obtained from thermometric enthalpy titration [24].

The two coatings of Fig. 7 differ by precuring conditions. The APS 1 coating was not rinsed with toluene prior to curing, but coating 2 was rinsed repeatedly with toluene prior to curing. The lower density of basic sites on coating 2 suggests that rinsing removed some physisorbed silane prior to curing. Amine surface density in both cases is low compared to the 1–3 nm^{-2} density for thick physisorbed layers of APS observed on silica and controlled-pore glass. However, this low density is consistent with a chemisorbed hydrothermally stable monolayer.

The initial measurement with APS-treated capillaries produced distorted particle-mobility parabolas consistent with desorption onto particles near the wall. Subsequent measurements did not produce such an effect and thus indicated that a stable APS layer remained. It should also be noted that a hydrothermally stable sublayer of density 0.48 nm^{-2} was observed on controlled-pore glass by Fang [25].

Culler et al. were able to distinguish between chemisorbed and physisorbed APS in DRIFT spectra of APS-coated silicon powder [22]. Removal of physisorbed silane by exposure to warm water reveals a hydrolytically stable layer of APS with less than monolayer coverage, covalently bound to the surface. They concluded that intermolecular silane bonds are easily hydrolyzed and that the overall degree of condensation was unimportant to hydrothermal stability. Hydrothermal stability was only enhanced by increasing the number of direct surface bonds. The electro-osmosis study supports this view of APS surface chemistry.

C. Poly(ethylene Glycol) Modified Surfaces

Surface-bound, neutral, hydrophilic polymers such as polyethers and polysaccharides dramatically reduce protein adsorption [26–28]. The passivity of these surfaces has been attributed to steric repulsion, bound water, high polymer mobility, and excluded volume effects, all of which render adsorption unfavorable. Consequently, these polymer modified surfaces have proven useful as biomaterials. Specific applications include artificial implants, intraocular and contact lenses, and catheters. Additionally, the inherent nondenaturing properties of these compounds has led to their use as effective tethers for affinity ligands, surface-bound biochemical assays, and biosensors.

While the advantages and applications of immobilizing hydrophilic polymers on surfaces are well documented, full characterization of surface-localized polymers in terms of density and film thickness is rare. Many authors have reported optimal thicknesses, molecular weights, or immobilization techniques without concern for grafting density. As a result it is not clear what concentration is required to produce the desired effects [29,30]. According to the theory of Jeon for protein interactions with terminally grafted poly(ethylene glycol) (PEG), grafting density and polymer chain length are the critical parameters [31]. Recent experiments with polysaccharide grafts suggest that grafting density is more critical than chain extension [32].

Previous studies indicated correlations between the ability of various polymer coatings to reduce electro-osmosis (or particle electrophoretic mobility) and to control both protein adsorption and phase wall wetting [33,34]. Other studies have shown that adsorbed neutral, hydrophilic polymers attenuate electrokinetic effects, presumably owing to displacement of the hydrodynamic plane of shear [35–39]. In this example, PEGs were grafted onto quartz capillary surfaces and characterized with respect to the pH dependence of induced electro-osmotic fluid flow [13]. PEG derivatives were tethered to quartz surfaces activated *a priori* with a sublayer containing a reactive functional group. Sublayers employed include 3-aminopropyltriethoxysilane (APS), 3-mercaptopropyltriethoxysilane (MPS), and poly(ethylene imine) (PEI). In addition, PEG-PEI conjugates were adsorbed directly to an unmodified quartz surface.

Figure 8 shows the effect of molecular weight on the pH dependence of electro-osmosis for a succinimidyl carbonate derivatized linear PEG grafted to APS modified quartz. As compared to the APS control, the PEG graft resulted in a shift in the point of zero electro-osmosis and a molecular-weight-dependent shift in the plateaus attained at the high and low pH values. From model fits it was possible to derive graft density and effective hydrodynamic thickness for each of the grafts. For this specific system, graft density was independent of molecular weight, while the effective hydrodynamic thickness increased monotonically with molecular weight. Also noteworthy was a shift in the acid pK associated with the silanol group increasing by half a pK unit. This can be explained in terms of the PEG hydrogen bonding with the silanol [40].

In an extension of this work, Emoto et al. have used electro-osmosis measurement and the model above to follow trends in graft density and effective hydrodynamic thickness with grafts of the glycidyl ether of PEG to aminopropylsilane modified quartz [41]. The effects of salt concentration and temperature on the grafting reaction were investigated. Results suggested that PEG chains increasingly extend normal to the surface as graft density increases and that the PEG conformation in solution influences grafting density.

D. Plasma Polymer Modified Surfaces

The deposition of thin polymeric films from a cold plasma in a radio-frequency glow discharge apparatus has become an important means of modifying surfaces in materials applications [42]. Applications receiving much attention recently have been the use of plasma polymerization to obtain biocompatible materials, and to produce functional surfaces for attachment of biologically active substances [43–45]. In this respect, many studies of protein adsorption have been

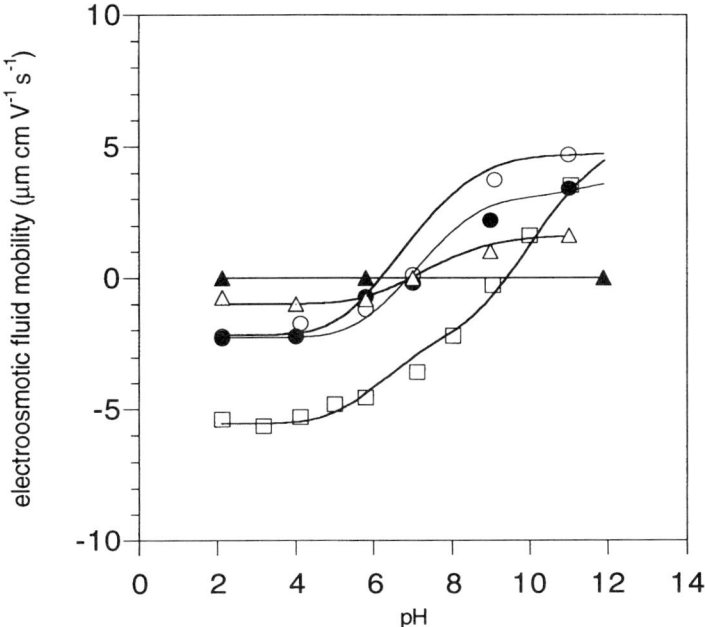

FIG. 8 The effect of molecular weight on the pH dependence of electro-osmosis for PEG grafted to an APS modified quartz surface (□) for PEG MW 500 (○), 3400 (●), 8000 (△), and 35000 (▲) in 7.5 mM NaCl.

performed on model plasma polymer surfaces. Among these studies, relevant surface properties include wetting, charge, composition, and "conformational mobility." Thus in characterizing these surfaces, determination of charging properties and functional group density is desired. Such information can be obtained from the pH dependence of electro-osmosis.

Figure 9 shows the pH dependence of electro-osmosis in 1 mM NaCl for three plasma polymer surfaces having different functional characteristics. The respective surfaces of Fig. 9 are plasma polymerized acrylic acid, hexamethyldisiloxane (HMDSO), and 1,2-diaminocyclohexane (DACH). It is evident from the figure that these surfaces have very different electrokinetic surface properties. This surface titration clearly distinguishes the acid–base properties of the respective surfaces.

The plasma polymerized acrylic acid surface is a hydrophilic surface containing carboxylic acid groups in the surface layer. The film deposited can be described as a cross-linked polyacrylic acid [46]. From the pH dependence of

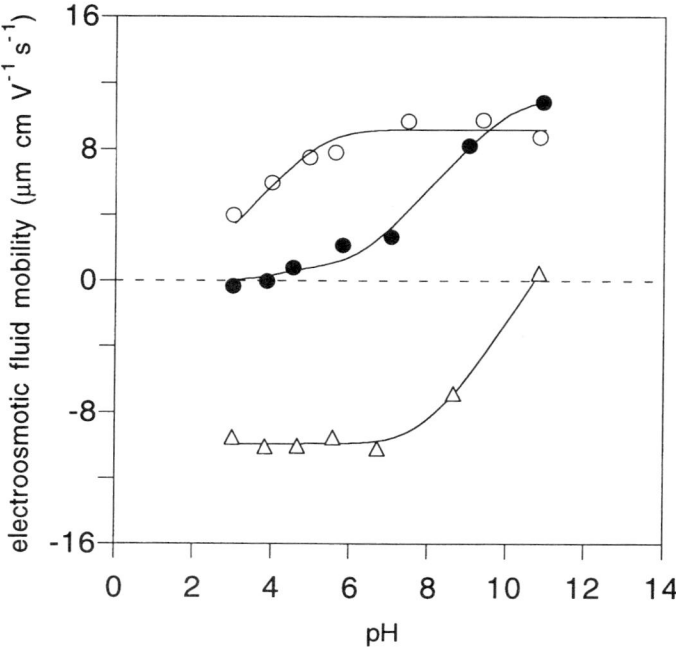

FIG. 9 The pH dependence of electro-osmotic fluid flow for three plasma polymer surfaces in 1 mM NaCl: acrylic acid (○), HMDSO (●), and DACH (△).

electro-osmotic fluid flow the acrylic acid surface of Fig. 9 behaves as having a single-acid ionizable surface group of pK_a 2.9 and a density of 8.8 nm^2 per acid group; these values are consistent with surface carboxylic groups.

Plasma polymerized HMDSO is a hydrophobic cross-linked polysiloxane network. The HMDSO surface of Fig. 9 behaves as having two distinct acid surface groups, one of pK_a 7.3 and a density of 5.6 nm^2 per group, and one of pK_a 4.2 and a density of 200 nm^2 per group. These values are consistent with a surface containing a considerable number of silanol groups (of pK_a 7.3) and a small number of organic acid groups (of pK_a 4.2), perhaps resulting from oxidation of the surface induced by residual polymer free radicals [47].

The DACH surface of Fig. 9 is a freshly prepared surface, less than one day old. It can be modeled as having a single basic ionizable group of pK_b 10.4 and a density of 5.0 nm^2 per group, and an acid group of $pK_a < 2$ and a density of 15 nm^2 per group. The pK of the acid site in this case is somewhat ambiguous, since the highly positively charged surface would bring the surface pH outside the range of the acid group (i.e., the bulk pH of 3 would correspond

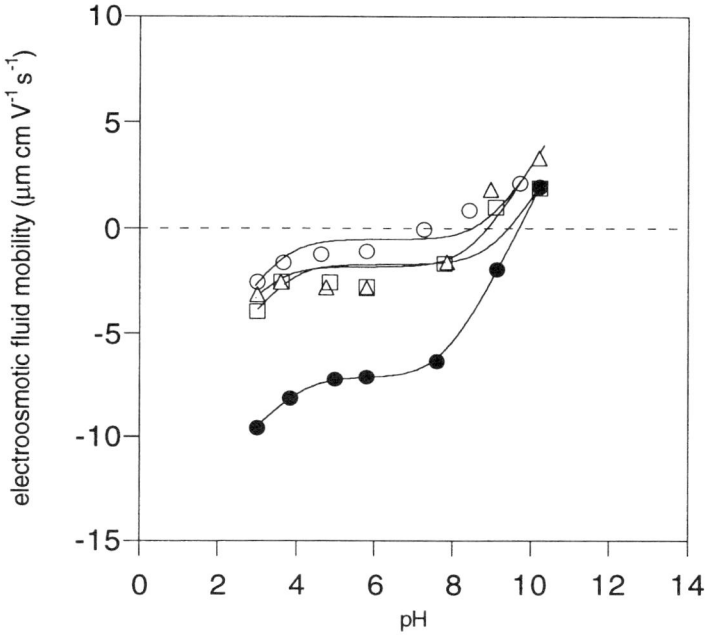

FIG. 10 The pH dependence of electro-osmosis for the DACH surface in 1 mM NaCl after exposure to atmosphere for 1 day (●), 8 days (□), 15 days (△), and 64 days (○).

to a surface pH of 5; in effect the titration begins at pH 5). It has been observed that plasma polymerized DACH changes its surface composition upon standing under ambient conditions. Previously, this aging effect was presumed to be due to oxidation initiated by free radical processes [48]. However, recent evidence suggests that the change in surface composition is due to the formation of alkyl ammonium carbamates from reaction of amine groups with atmospheric carbon dioxide [49]. This appears to be a phenomenon common to amino-functional surfaces [50,51]. In Fig. 10 the effect on charging properties upon storing the polymerized DACH surface under ambient conditions is shown. Upon aging there is a reduction in basic group density, with a corresponding appearance of an acid group ($pK_a \approx 3-5$) to a point where the acid-to-base ratio in one. This behavior is consistent with the formation of alkyl ammonium carbamates at the surface.

Lassen and Malmsten have performed ellipsometrically determined *in situ* protein adsorption measurements on the above plasma polymer surfaces with human serum albumin (HSA), human immunoglobulin (IgG), and human fibrinogen (Fgn) at pH 7.2 [52]. In this study correlations between protein adsorption and surface charge, and origin of charge, could be made. All three proteins adsorbed to the greatest extent at the DACH surface, slightly less on the polymerized HMDSO surface. With the acrylic surface, much less protein adsorbed, and adsorption kinetics were slower. These observations can be explained in terms of the respective charges at the surface and protein. At pH 7.2 all three of these proteins carry a net negative charge. The positively charged DACH surface would attract the protein and result in pronounced protein adsorption due to electrostatic attraction. For the acrylic acid surface the negative charge would repel the protein and slow adsorption kinetics. The acrylic acid and DACH surfaces have similar wetting properties, both being hydrophilic, suggesting that surface charge and/or specificity of the functional group are the governing properties with respect to the differences in protein adsorption. As for polymerized HMDSO, the surface has a very low charge at pH 7.2 and is quite hydrophobic; however, there are a high number of polar "ionizable" groups at the surface, namely silanol. In light of this it is possible that, in addition to hydrophobicity, specific interactions with these functional groups could play a role in protein adsorption at this surface.

Since the polymerized DACH surface spontaneously changes its acid-to-base ratio and charging characteristics with time, it served as an interesting model substrate for protein adsorption. As such, protein adsorption experiments were performed with HSA and Lys using an ELISA technique [49]. It was found that lysozyme adsorbs in much higher amounts that HSA irrespective of the charge at the surface. Whereas adsorption of HSA was not dependent on the change in surface charge, adsorption of lysozyme decreased with aging, i.e., as the ratio of acid to base sites at the surface approached unity. An interesting finding was that the more lysozyme was adsorbed the more positively charged was the surface, even at pH values where the protein carried a strong net positive charge.

Another plasma polymer system studied has been that of *n*-butane. Malmsten et al. prepared plasma polymer surfaces of *n*-butane at different energy inputs to determine the effect of cross-link density on protein adsorption [53]. Furthermore, novel copolymer surfaces of *n*-butane and nitrogen at different ratios were prepared and investigated. The surfaces were characterized according to the pH dependence of electro-osmosis and by ESCA. The adsorption of human serum albumin (HSA) and fibrinogen at *n*-butane plasma polymer surfaces was investigated with *in situ* ellipsometry. Within experimental uncertainty, the adsorption of both fibrinogen and HSA was constant over the power range used for the preparation of the *n*-butane surfaces and corresponds to that found for other hydrophobic surfaces under similar conditions. Addition of nitrogen in the plasma gas mixture during deposition resulted in the appearance of amine groups at the surface and an increased density of these groups with increasing nitrogen content, as evidenced by electro-osmosis and ESCA. This, in turn, was found to result in an increased fibrinogen adsorption but in a weak decrease in HSA adsorption.

V. FUTURE APPLICATIONS

In the design of the novel rectangular cell described in this work, the range of surface that can be characterized electrokinetically is greatly expanded. In addition to the examples discussed there are an even larger number of systems for which electrokinetic characterization would be useful. The technique has the advantage that it requires a relatively small area compared to other techniques and is not limited to disperse systems. The only limitation to the material characterized is that it not be highly conducting. The electro-osmosis technique for characterizing macroscopic surfaces with respect to zeta potential and its pH dependence has many potential applications. Following are examples that have been chosen to illustrate the broad range of potential applications of the technique. Relevant systems include biological surfaces, nondispersible particles, paper surfaces, and ceramic mineral powders.

Electrokinetic characterization of biological surfaces has been useful in the past. Systems under study have ranged from electrophoresis of human cells to streaming potential across the urinary bladder of the goat [10,54]. Biological surfaces also extend into the range of materials that can now be characterized by electro-osmosis measurement at the flat plate. A solid surface brought in contact with a body fluid is almost immediately covered by a thin organic film, the main constituents of which are proteins. Adherence of cells or bacteria is usually the next event in the *in vivo* situation. Pertinent examples are biomaterials on which a relatively thick biofilm is gradually formed after implantation and dental plaque, which forms almost instantaneously on all bare surfaces in the oral cavity.

Whereas the biofilm formed on implants is normally rich in platelets and other cells, the main constituent of dental plaque is oral streptococci. The common denominator of the two cases is that cell and bacterial adherence takes place on top of the proteinaceous layer rather than at the bare surface. In turn, the properties of the bare surface influence the structure and composition of the protein layer [55]. It is known that both cells and bacteria may adhere by a specific mechanism, mediated by surface-absorbed proteins [56]. However, nonspecific cell and bacterial binding also occurs in many instances [57]. Regardless of the binding mechanism, characterization of the proteinaceous film and the bare surface in terms of charge density and acid and base strength of charged or chargeable sites would be of great value in improving the understanding of the binding processes [58].

Another potential use of electro-osmosis is in the surface characterization of solid particles. Characterizations of solid particles are normally made by determining the zeta potential, which is usually calculated fro electrophoretic mobility data. A prerequisite of electrophoretic mobility measurements is that the particles can be dispersed in water. Disintegration of aggregates and lumps into primary particles must be smooth and irreversible under the conditions used. Aggregate disintegration can be a problem with hydrophobic, low-charged particles such as carbon black and certain organic pigments. With such particles, determination of electrophoretic mobility is not a straightforward matter. In preliminary experiments we have found that by immobilizing such particles on a macroscopic surface, characterization by electro-osmosis on flat surfaces is possible. Though the effects of surface roughness on the quantitative aspects of this type of measurement have not been fully explored, the technique is promising, and good qualitative results have already been obtained.

Surface characterization of paper is also important in order to understand the surface chemistry behind printing, such as the mechanism responsible for adherence of printing ink pigments to the paper surface as well as interactions between fountain solution and paper. Surface characterizations are equally important for coated and for uncoated paper, since both are used as substrates for printing; magazines and newspapers are common examples. The common characterization methods employed today are ESCA (electron spectroscopy for chemical analysis) and contact angle determinations. ESCA gives the chemical composition of the paper surface, and contact angle measurements, using series of liquids of varying polarity, give information about surface free energy and acid–base properties of the paper surface [59]. Information about densities of charged groups at the surface should obviously be useful, since it would contribute to the understanding of the importance of electrostatic interactions in the printing process. However, there is no straightforward analytical method in use today for determining surface charge density. Electro-osmosis could be a useful tool in this respect.

Information about surface charge density and isoelectric point (i.e.p.) is essential in all types of ceramic processing, specifically in the optimization of dispersing conditions for ceramic powders. For the majority of materials, such as Al_2O_3, MgO, and SiC, potentiometric titration is useful for this type of surface characterization, and values can be determined with a high degree of accuracy. However, for some types of particles, notably Si_3N_4, potentiometric titration is not feasible owing to the large particle surface area–to–solution volume ratio required for the potentiometric response [60,61]. Dissolution of silica at the surface above pH 8 interferes with the measurement, severely limiting the range of titration and the accuracy and reproducibility of results.

Electrophoretic mobility measurement has been useful in this context, as the measurement is made at very low surface area–to–solution volume ratios. However, particle electrophoretic mobility is related to surface charge in complex ways depending upon particle size and morphology [2]. Quantitative analysis

Electro-osmosis in Electrokinetic Surface Analysis 137

according to theories is only valid for spherical monodisperse particles, which is not the case for milled ceramic powders. In such a case interpretation of charging properties is restricted to the position of the i.e.p. This limits the number of free model parameters in any type of analysis.

With electro-osmosis data, on the other hand, interpretation is not subject to the complexities of the electrophoretic measurement. Analysis of zeta potential is straightforward, and a wide range of pH can be employed. In this light it would be promising to characterize ceramic and mineral materials of a wide variety of compositions and forms, e.g., powders and processed plates.

In the preceding pages a need for the characterization of a number of macroscopic surfaces in terms of effective surface charge was expressed. For this purpose the measurement of the pH dependence at quartz capillaries and at flat plates was employed. To extract information about surfaces from the electro-osmosis measurements a simple model of the pH dependence of electro-osmosis was put forward. A number of surfaces of biomaterial interest were characterized, serving to elucidate the surface behavior in a number of systems. The general utility of the measurement of electro-osmosis at flat plates was indicated. In addition, a general description of the hydrodynamics of the rectangular electrophoresis chamber, where electro-osmosis may vary at the chamber walls, was derived. It has also been shown that neutral, hydrophilic polymer coatings may be used to reduce probable error in analytical particle electrophoresis, as a result of both focusing errors and asymmetries in the chamber with respect to electro-osmosis.

Though the model presented and used does not give a complete account of the interface and the origin of measured electro-osmotic fluid mobility, it was proven useful in interpretation of surface properties. The range of electrolyte concentration that can be used in the manual particle electrophoresis chamber developed in this work is limited, and this limits the model of the origin of electro-osmosis that can be tested, such as inclusion of a Stern layer.

The promise of the technique, both in terms of developing electrokinetic theory and in characterizing surfaces, would be greatly expanded with an increase in the range of measurement and with automation. This could perhaps be accomplished through design of a rectangular chamber adaptable to a commercially available light-scattering analytical particle electrophoresis apparatus where flat plates could be readily exchanged for the chamber walls. The author looks forward to the greater utilization of the measurement of electro-osmosis to characterize surfaces.

SYMBOLS

E electric field, V m^{-1}
E_∞ applied field strength, V m^{-1}
p pressure, N m^{-2}

η	Newtonian shear viscosity, kg m^{-1} s^{-1}
ρ_e	local charge density, C m^{-3}
v	Fluid velocity vector, m s^{-1}
σ_0	surface charge over area S, C m^{-2}
σ_s	surface charge due to acid–base dissociation over area S, C m^{-2}
σ_d	diffuse double layer charge over area S, C m^{-2}
c^i	number density of ith species, m^{-3}
c^i_0	number density for ith species immediately adjacent to the interface, m^{-3}
c^i_∞	number density for ith species in free solution far from the interface, m^{-3}
z^i	valence of ith species
e	elementary charge, C
ε	permittivity of fluid medium, $D\varepsilon_0$, C V^{-1} m^{-1}
D	relative permittivity, dimensionless
ε_0	permittivity of free space, C V^{-1} m^{-1}
Ψ	electric potential, V
Ψ_0	equilibrium double layer potential, V
Ψ_x	potential at some distance x out from a surface, V
ζ	zeta potential, V
S	surface area, m^{-2}
Ω	function defining spatial boundaries of a surface, $\Omega(x, y, z)$
AH	undissociated Brønsted acid group
A^-	dissociated Brønsted acid group
BH^+	undissociated Brønsted basic group
B	dissociated Brønsted basic group
H^+	hydronium ion
$[AH]$	number density of undissociated acid group, m^{-2}
$[A^-]$	number density of dissociated acid group, m^{-2}
$[BH^+]$	number density of undissociated basic group, m^{-2}
$[B]$	number density of dissociated basic group, m^{-2}
$[H^+]_s$	number density of hydronium ions immediately adjacent to the interface, m^{-3}
pH$_s$	$-\log[H^+]_s$
K_a	acid dissociation constant, m^{-3}
pK_a	$-\log K_a$
K_b	base dissociation constant, m^3
pK_b	$-\log K_b$
N_a	sum of dissociated and undissociated acid group number densities, m^{-2}
N_b	sum of dissociated and undissociated basic group number densities, m^{-2}

v_{eo}	observed wall electro-osmotic fluid velocity, m s^{-1}
v_p	observed particle velocity, m s^{-1}
U_{eo}	wall electro-osmotic fluid mobility, m^2 V^{-1} s^{-1}
U_{el}	intrinsic electrophoretic particle mobility, m^2 V^{-1} s^{-1}
U_p	observed particle mobility, m^2 V^{-1} s^{-1}
d	distance increment, m
t	time increment, s
κ	bulk solution conductivity, S m^{-1}
R	radius of a cylindrical capillary, m
r	radial distance from the center of a cylindrical capillary, m
I	electric current, C s^{-1}
APS	3-aminopropyltriethoxysilane
DACH	1,2-diaminocyclohexane
DRIFT	diffuse reflectance infrared Fourier transform spectroscopy
ELISA	enzyme linked immunosorbent assay
ESCA	electron spectroscopy for chemical analysis
HMDSO	hexamethyldisiloxane
HSA	human serum albumin
Lys	human lysozyme
MPS	3-mercaptopropyltriethoxysilane
NMR	nuclear magnetic resonance spectroscopy
PEG	poly(ethylene glycol)
PEI	poly(ethylene imine)

REFERENCES

1. W. B. Russel, D. A. Saville, and W. R. Schowalter, *Colloidal Dispersion*, Cambridge University Press, Cambridge, 1989.
2. S. S. Dukhin and B. V. Deraguin, in *Surface and Colloid Science*, Vol. 7 (E. Matijevic, ed.), John Wiley, New York, 1974, Chaps. 2 and 3.
3. J. Th. G. Overbeek, in *Colloid Science* (H. R. Kruyt, ed.), Elsevier, Amsterdam, 1952, Chap. V.
4. R. J. Hunter, *Zeta Potential in Colloid Science*, Academic Press, London, 1981.
5. T. W. Healy and L. R. White. Adv. Colloid Interface Sci. 9:303 (1978).
6. S. S. Dukhin, in *Surface and Colloid Science*, Vol. 7 (E. Matijevic, ed.), John Wiley, New York, 1974, Chap. 1.
7. J. Lyklema, *Fundamentals of Interface and Colloid Science*, Vol. II, Academic Press, London, 1995, Chaps. 3 and 4.
8. R. O. James and G. A. Parks, in *Surface and Colloid Science*, Vol. 12 (E. Matijevic, ed.), John Wiley, New York, 1982.
9. A. M. James, in *Surface and Colloid Science*, Vol. 11 (R. J. Good and R. R. Stromberg, eds.), Plenum Press, New York, 1979.

10. G. V. F. Seaman, in *The Red Blood Cell*, 2d ed. (D. M. Surgenor, ed.), Academic Press, New York, 1975.
11. B. J. Herren, S. G. Shafer, J. M. Van Alstine, J. M. Harris, and R. S. Snyder. J. Colloid Interf. Sci. *115*:46 (1987).
12. J. M. Van Alstine, N. L. Burns, J. A. Riggs, K. Holmberg, and J. M. Harris. Colloids Surf. A *77*:149 (1993).
13. N. L. Burns, J. M. Van Alstine, and J. M. Harris. Langmuir *11*:2768 (1995).
14. N. L. Burns. J. Colloid Interf. Sci. *183*:249 (1996).
15. N. L. Burns. J. Colloid Interf. Sci. *183*:249 (1996).
16. R. A. Van Wagenen and J. D. Andrade. J. Colloid Interface Sci. *76*:305 (1980).
17. B. D. Bowen. J. Colloid Interf. Sci. *106*:367 (1985).
18. R. K. Iler, *The Chemistry of Silica*, John Wiley, New York, 1979.
19. J. Köhler and J. J. Kirkland. J. Chromatogr. *385*:125 (1987).
20. A. Grabbe and R. G. Horn. J. Colloid Interface Sci. *157*:75 (1993).
21. K. L. Mittal, *Silanes and Other Coupling Agents*, VSP, Utrecht, The Netherlands, 1992.
22. S. R. Culler, H. Ishida, and J. L. Koenig. J. Colloid Interface Sci. *106*:334 (1985).
23. D. E. Leyden, *Silanes, Surfaces and Interfaces*, Gordon and Breach, New York, 1986.
24. D. J. Kelly and D. E. Leyden. J. Colloid Interface Sci. *147*:213 (1991).
25. Z.-H. Fang, M.S. thesis, University of Alabama in Huntsville, Huntsville, AL (1991).
26. J. M. Harris, *Poly(ethylene glycol) Chemistry*, Plenum Press, New York, 1992.
27. W. R. Gombotz, W. Guanhui, T. A. Horbett, and A. S. Hoffman. Biomed. Mater. Res. *25*:1547 (1991).
28. E. W. Merrill and E. W. Salzmann. ASAIO *6*:60 (1983).
29. D. Gingell and N. J. Owens. J. Biomed. Mater. Res. *28*:491 (1994).
30. Z. Zhao, A. Malik, and M. L. Lee. Anal. Chem. *65*:2747 (1993).
31. S. I. Jeon, J. H. Lee, J. D. Andrade, and P. G de Gennes. J. Colloid Interface Sci. *142*:149 (1991).
32. E. Österberg, K. Bergström, K. Holmberg, T. P. Schuman, J. A. Riggs, N. L. Burns, J. M. Van Alstine, and J. M. Harris. J. Biomed. Mat. Res. *29*:741 (1995).
33. E. Österberg, K. Bergström, K. Holmberg, J. A. Riggs, J. M. Van Alstine, T. P. Schuman, N. L. Burns, and J. M. Harris. Colloids Surf. *77*:159 (1993).
34. J. F. Boyce, B. A. Hovanes, J. M. Harris, J. M. Van Alstine, and D. E. Brooks. J. Colloid Interface Sci. *149*:153 (1992).
35. N. V. Churaev, I. P. Sergeeva, and V. D. Sobolev. J. Colloid Interface Sci. *169*:300 (1995).
36. L. K. Koopal, V. Hlady, and J. Lyklema. J. Colloid Interface Sci. *121*:49 (1988).
37. M. A. Cohen Stuart, F. H. W. H. Waajen, and S. S. Dukhin. Colloid Polymer Sci. *262*:423 (1984).
38. F. J. Nordt, R. J. Knox, and G. V. F. Seaman, in *Hydrogels for Medical and Related Applications* (J. D. Andrade, ed.), ACS Symposium Series 31, American Chemical Society, Washington, DC, 1976.
39. D. E. Brooks. J. Colloid Interface Sci. *43*:687 (1973).

40. B. V. Erenmenko, B. E. Platonov, I. A. Uskov, and I. N. Lyubchenko. Kolloidnyi Zh. *36*:240 (1974).
41. K. Emoto, J. M. Harris, and J. M. Van Alstine. Anal. Chem. (in press).
42. H. Yasuda, *Plasma Polymerization*, Academic Press, Orlando, FL, 1985.
43. B. Lassen, C.-G. Gölander, A. Johansson, and H. Elwing. Clinical Mat. *11*:99 (1992).
44. M. Malmsten, B. Lassen, K. Holmberg, V. Thomas, and G. Quash. J. Colloid Interface Sci. *177*:70 (1996).
45. C.-G. Gölander, B. Lassen, K. Nilsson-Ekdahl, and U. R. Nilsson. J. Biomater. Sci. Polymer Edn. *4*:25 (1992).
46. D. L. Cho, P. M. Claesson, C.-G. Gölander, and K. Johansson. J. Appl. Polym. Sci. *41*:1373 (1990).
47. J. L. Parker, P. M. Claesson, J.-H. Wang, and H. K. Yasuda. Langmuir *10*:2766 (1994).
48. C.-G. Gölander, M. W. Rutland, D. L. Cho, A. Johansson, H. Ringblom, S. Jönsson, and H. K. Yasuda. J. Applied Polymer Sci. *49*:39 (1993).
49. N. L. Burns, C. Brink, and K. Holmberg. Colloids Surf. B *5*:161 (1995).
50. K. P. Battjes, A. M. Barolo, and P. Dreyfuss, in *Silanes and Other Coupling Agents* (K. L. Mittal, ed.), VSP, Zeist, The Netherlands, 1992, p. 199.
51. R. Zhou, A. Hierleman, U. Weimar, D. Schmeisser, and W. Göpel, in *Digest of Technical Papers: The 8th International Conference on Solid-State Sensors and Actuators*, Vol. 1, 225-PD6, 1995, p. 886.
52. B. Lassen and M. Malmsten. *J. Colloid Interface Sci.* (in press).
53. M. Malmsten, J.-Å. Johansson, N. L. Burns, and H. K. Yasuda. Colloids Surf. B *6*:191 (1996).
54. P. C. Shukla and G. Misra. Langmuir *8*:1149 (1992).
55. C. A. Haynes and W. Norde. Colloids Surf. B *2*:517 (1994).
56. R. J. Gibbons ad D. I. Hay, in *Molecular Mechanisms of Bacterial Adhesion* (L. Switalski, M. Höök, and E. Beachey, eds.), Springer-Verlag, New York, 1989.
57. I. H. Pratt-Terpsta, A. H. Weerkamp, and H. J. Busscher. J. Dent. Res. *68*:463 (1989).
58. W. Norde. Adv. Colloid Interface Sci. *25*:267 (1986).
59. J. E. Elftonson and G. Ström, in *Proceedings of the TAPPI Advanced Coating Fundamentals Symposium*, Dallas, TX, 19–20 May, 1995.
60. L. Bergström, Ph.D. thesis, Royal Institute of Technology, Stockholm, Sweden, 1992.
61. J. Sonnefeld. Colloids Surf. A *208*:27 (1996).

5
X-Ray Photoelectron Spectroscopy (XPS) and Static Secondary Ion Mass Spectrometry (SSIMS) of Biomedical Polymers and Surfactants

KEVIN M. SHAKESHEFF and MARTYN C. DAVIES School of Pharmaceutical Sciences, University of Nottingham, Nottingham, England

ROBERT LANGER Department of Chemical Engineering, Massachusetts Institute of Technology, Cambridge, Massachusetts

I.	Introduction	143
II.	Description of the Techniques	145
	A. X-ray photoelectron spectroscopy (XPS)	145
	B. Static secondary ion mass spectrometry (SSIMS)	148
	C. Complementary techniques in surface analysis	150
III.	Examples of XPS and SSIMS Analysis of Surfactant Polymers	151
	A. Surface enrichment of copolymer components	151
	B. Microparticle surface chemistry	155
	C. Blend surface segregation	161
	D. End-group contributions to surface chemistry	163
	E. Desorption	170
IV.	Conclusions	170
	References	171

I. INTRODUCTION

Interactions between a solid material and its environment are determined by the properties of the surface of the material. In many modern technologies, controlling these interactions, via the engineering of surfaces with specific properties, is a powerful method of improving materials' performance [1].

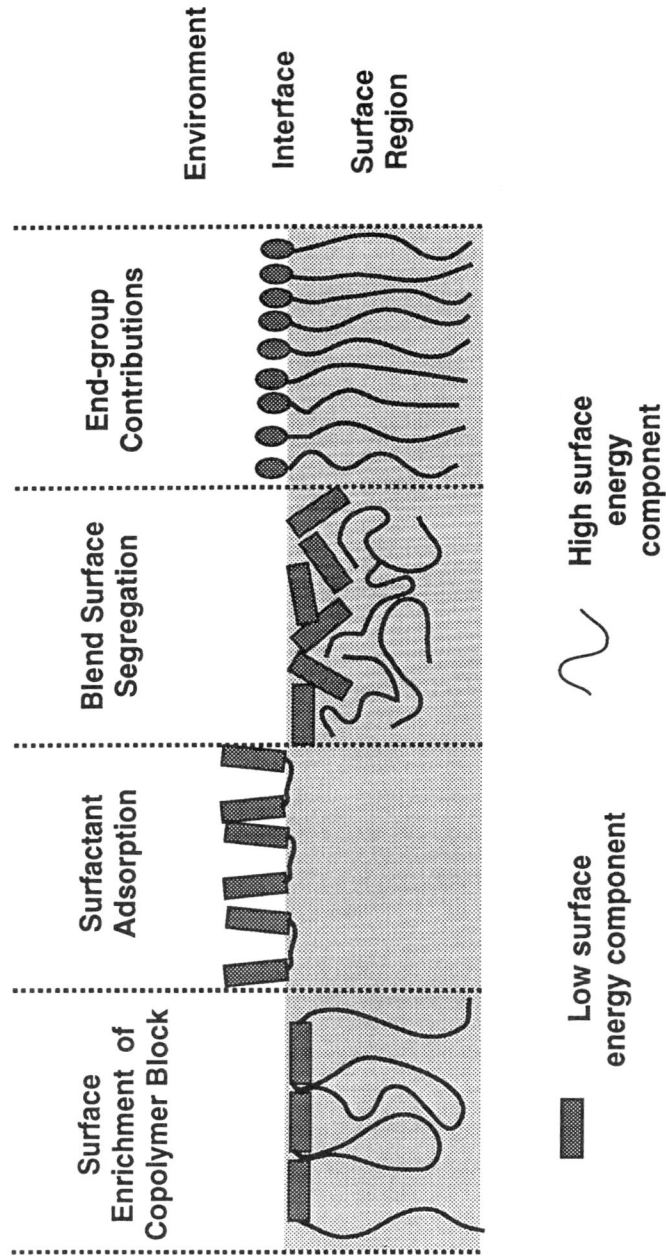

FIG. 1 Surface engineering strategies based on surfactant behavior at surfaces.

XPS and SSIMS of Biomedical Polymers

The importance of surface engineering is highlighted by developments in our central area of research, the design and analysis of polymeric biomaterials [2–4]. Biomaterials are materials that interface with the living environment of the patient and interact with this environment to improve the health of the patient. Polymeric biomaterials are currently employed as prosthetic implants [5], vehicles for the controlled delivery of drugs [4,6,7], and templates for tissue regeneration [8]. A key factor in determining the success of a biomaterial is its surface properties. For example, a polymeric material that possesses a hydrophobic surface will initiate a series of protein adsorption and cell recognition events that can cause the body to reject the material [9]. In contrast, polymers with engineered surfaces can initiate controlled interactions with biomolecules and cells allowing the biomaterial to perform such functions as controlled cell spreading and growth [10], tissue-specific localization of drug delivery devices [11], and long-term implant biocompatibility [12].

Surfactants have played a major role in polymer surface engineering because their inherent ability to enrich surfaces with specific chemical groups provides a simple method of controlling surface chemistry. Figure 1 shows four surface engineering strategies that are dependent on the material's inherent surface activity.

Successfully developing a surface engineering strategy based on surfactant behavior at interfaces requires surface characterization techniques that can validate and quantify surface chemistry changes. This review describes the role of two surface chemistry analysis techniques that have proven highly successful in surfactant analysis: x-ray photoelectron spectroscopy (XPS) and static secondary ion mass spectrometry (SSIMS). In Section II, the methods by which these techniques analyze surface chemistry are described. In Section III, recent examples of their application in surfactant-based surface engineering are described.

II. DESCRIPTION OF THE TECHNIQUES

The description of XPS and SSIMS will concentrate on the underlying physical processes that generate surface chemical data. Information on the design of instruments and specific details of operating procedures and conditions have been provided by a number of excellent books and review articles [13–18].

A. X-Ray Photoelectron Spectroscopy (XPS)

Core electrons are held in their orbitals by characteristic binding energies. These binding energies are dependent on orbital type, atom type (due to nuclear charge and size) and the nature of neighboring atoms. Surface chemistry analysis by XPS is based on the measurement of the binding energies of electrons ejected from sample surfaces.

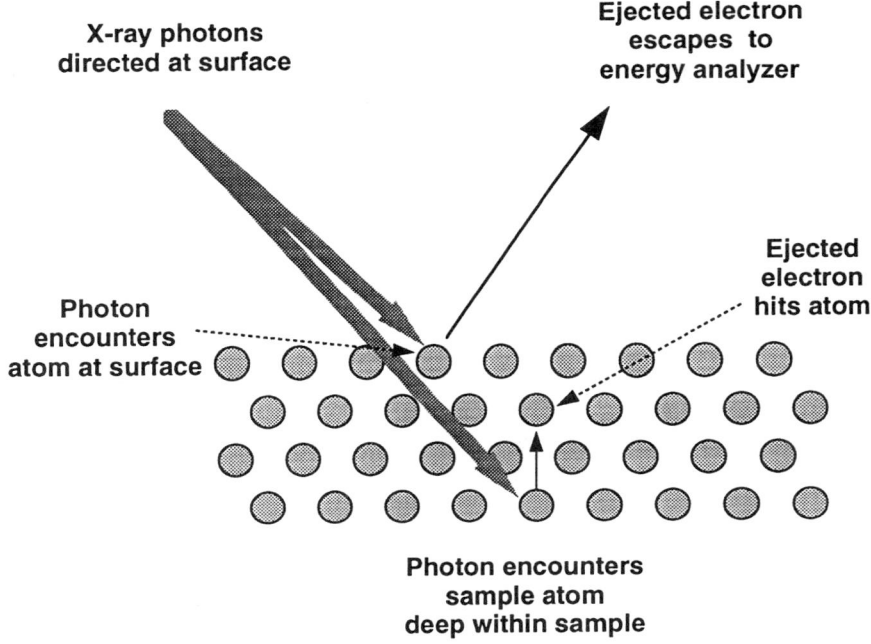

FIG. 2 Schematic representation of the XPS technique.

Photoionization is used as the mechanism of ejecting electrons from surfaces. This process is shown schematically in Fig. 2. A low-energy monochromated x-ray beam of photons is shone onto the sample surface. The photons penetrate the surface to a depth of approximately 1 μm. If a photon interacts with a sample atom, the energy from the photon will ionize core electrons and cause their ejection from the atom. Within the bulk of the sample the mean free path of ejected electrons is very small, owing to the high probability of interaction with surrounding atoms. However, electrons ejected from atoms within the top 5 nm layer of the sample have a high probability of escaping from the sample without further interactions. These electrons will pass into the ultrahigh vacuum (UHV) conditions within the spectrometer. UHV conditions are required to ensure that electrons are not scattered by gas molecules and to limit surface contamination during analysis.

The kinetic energy E_K of the ejected electrons is measured by an electron energy analyzer. The original binding energy E_B experienced by the electron in the sample can be calculated from E_K using the simple relationship

$$E_K = h\nu - E_B - \Phi$$

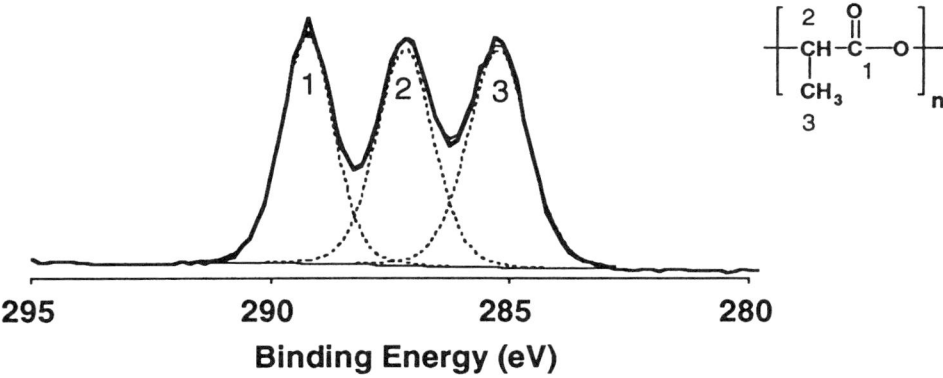

Peak	Binding Energy (eV)	Experimental %	Theoretical %
1	289.20	31.6	33.3
2	287.05	32.9	33.3
3	285.00	35.4	33.3

FIG. 3 Carbon 1s (C1s) data obtained from a sample of poly(lactic acid) (PLA).

where $h\nu$ is the incident photon energy and Φ is the spectrometer work function, and measured by calibration from standard samples. Calculation of E_B allows the derivation of detailed surface chemistry information. Wide scans over a binding energy range of 0 to 1000 eV reveal information on the type of elements present in the surface region, and the atomic percentages of these elements can be calculated.

High-resolution scans of individual elemental peaks can reveal information on the type of chemical groups and bonds present in molecules at the surfaces of samples. An example of this type of data is shown in Fig. 3, in which the binding energies of the carbon 1s orbitals for a poly(lactic acid) (PLA) sample are shown. Three peaks are evident from this data, and using deconvolution algorithms the precise binding energy and area of these peaks can be measured. The presence of three peaks is explained by the presence of three types of carbon groups with the PLA structure. The methyl group carbon generates a peak at a binding energy of 285.0 eV. In contrast, the peak from the carboxyl carbon is shifted by +4.2 eV to a value of 289.2 eV. This shift is caused by the electron withdrawing effect of the two oxygen atoms bonded to this carbon. This displaces valence electrons closer to the oxygen nucleus, resulting in the positive charge of the carbon atom nucleus exerting a stronger force on the core electrons. The middle peak of the PLA spectrum is shifted by +2.0 eV compared to the methyl group. This peak is generated by the carbon α to the carboxyl group. This carbon experiences electron withdrawing from the ether oxygen that is directly bonded to. In addition, this carbon experiences a secondary shift caused by the induced electron withdrawing nature of the neighboring carbon of the carboxyl group. This is termed a secondary shift because the electron withdrawing nature of the neighboring is not inherent to the atom type but is induced by its own neighboring atoms. Peak shifts in the C1s region are highly predictable based on knowledge of the types of atoms present in the sample. For polymeric samples an excellent database of standard shifts has been published [15]. The data in Fig. 3 also demonstrates that accurate quantification of chemical species present at sample surfaces is possible by measurement of relative peak areas within the C1s region.

Information available from XPS data includes valence band structure and the detection of π orbitals. Procedures for obtaining these types of data are not discussed further in this review because they are not commonly encountered in surfactant materials analysis. Reviews of these procedures have been published [14,15].

A major strength of XPS analysis is the ability to quantify changes in chemical composition with increasing sampling depth. The sampling depth of XPS analysis is dependent on the probability of electrons escaping the sample surface without interacting with atoms. If an electron is ejected from an atom that is 10 nm beneath the sample surface, then the minimum distance traveled by the electron to escape the sample is 10 nm. In this case the electron would have to travel perpendicular to the sample surface. If an electron from a sample travels towards the surface at an angle less then 90° then the distance it will travel before escaping the surface will be greater than 10 nm. The greater the angle of deviation from perpendicular the greater this distance will be. As a consequence, electrons traveling at nonperpendicular angles have an increased probability of hitting atoms in the solid and therefore not becoming available for energy analysis. By rotating the sample to different angles it is possible to change the "takeoff" angle, the angle between the sample surface and the detector. This allows XPS analysis to measure the energy of electrons leaving the surface at different specified angles. At low takeoff angles the electrons must have originated from thin surface layers close to the sample surfaces. As takeoff angle is increased, the thickness of this sampling layer increases. 95% of electrons reaching the analyzer are derived from a surface layer of thickness $3\lambda \sin \alpha$ (where α is the takeoff angle and λ is a sample-dependent constant). Values of λ can be estimated for many polymers.

B. Static Secondary Ion Mass Spectrometry (SSIMS)

SSIMS analysis is dependent on a phenomenon in which kinetic energy from the collision of a particle with a surface is dissipated by a subsurface collision

XPS and SSIMS of Biomedical Polymers

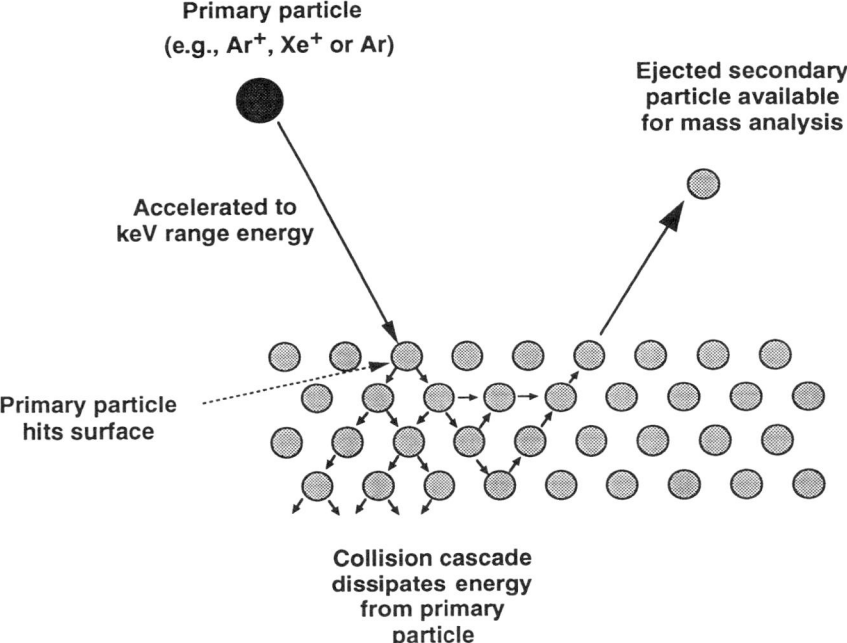

FIG. 4 Schematic representation of the SSIMS technique.

cascade that can result in the expulsion of surface chemical fragments. This process is shown schematically in Fig. 4. The formation of the collision cascade is vital to this whole process because it ensures that fragments are removed from the surface at a remote point away from the site of particle collision. The surface chemical fragments are ejected from the outer 1 nm of surface, ensuring that SSIMS is one of the most surface-specific techniques available. As with XPS analysis, the SSIMS techniques requires UHV conditions.

The fragments from the surface are a mixture of neutral species, cations, and anions. A mass spectrum of the charged species of this mixture is taken using either a quadrupole or a time-of-flight (ToF) mass spectrometer. Conventional rules of interpreting mass spectra can then be employed to identify the chemical structure of the fragments [13,18]. When quadrople detectors are employed, fragments with m/z (mass/charge) values of between 0 and 800 can be detected. For polymeric samples, fragments in this range can provide a detailed "fingerprint" of the complete polymer surface chemistry. Examples of the utility of "fingerprint" spectra in the analysis of surfactants are provided in Section III. A drawback with the use of quadrople detectors is the low efficiency of the detector (approximately 1% of charged fragments are captured for mass analysis). In addition, the detection of fragments is dependent on the mass of the fragment, with inefficient capture of higher mass fragments. This imposes limitations on the use of quadrople SSIMS as a quantitative surface analysis technique. However, carefully validated studies with quadrople systems can reveal interesting quantitative surface data. Indeed, the early definitive studies on SSIMS quantification used quadrople systems.

ToF–SIMS analysis overcomes some of the limitations of the quadrople systems because the detector is sensitive to lower fragment concentrations and detects all fragment masses simultaneously. In polymer science, the ToF–SIMS technique possesses the major advantage of allowing high molecular weight to be detected. For polymers with average molecular weights of less than 10,000 it has been demonstrated that complete polymer chains can be detected [19,20]. In addition, ToF–SIMS retains the ability to obtain "fingerprint" spectra at low m/z values.

The term "static" is used for analysis in which the total dose of primary particles hitting the surface is below 10^{13} particle/cm^2. Below this value surface damage is low enough to ensure that the distribution of fragments ejected from the surface is not influenced by the analysis process (i.e., surface chemistry is not changed by the analysis to a great enough extent to shift the probability of specific fragments being generated).

Quantification of SSIMS data is complicated by uncertainty and variations in the efficiency of fragment ejection (sputtering) and ionization processes. These variations are collectively termed "matrix effects." Matrix effects can result in the intensity of peaks from a specific fragment varying in a nonlinear manner as the true surface chemistry changes (e.g., due to a change in monomer ratios within a copolymer). Therefore quantification of surface chemistry variations cannot be performed by measuring peak area. Approaches to overcome this problem have been proposed that involve comparison of intensities derived from a number of fragments [21,22]. All authors using such approaches are careful to highlight the uncertainties inherent in SSIMS quantification, and in general all such data is treated as "semiquantitative."

C. Complementary Techniques in Surface Analysis

In the surface analysis of polymeric materials it is rare that any one technique can completely characterize a surface. To some extent all surface analysis techniques have limitations that impose restrictions on the type of surface chemistry data they can generate. Therefore it is often necessary to combine information for different instruments to build a consistent characterization of surfaces. The complementary use of XPS and SSIMS provides an example of this approach. XPS data is often excellent at quantifying simple surface chemistry where either different elements or characteristic peak shifts are diagnostic of one surface component. For many polymeric biomaterial surfaces the surface chemistry of the material is too complex to characterize fully with this XPS-based approach. For these samples, SSIMS is often an excellent technique because complex surface structures can generate diagnostic peaks. When used in combination, the quantitative XPS data can augment the complex peak distribution obtained by SSIMS. The use of complementary techniques is not limited to XPS and SSIMS. Techniques such as the use of the scanning probe microscope [23], surface plasmon resonance analysis [24], ellipsometry [25], and dynamic contact angle analysis [26] can all generate synergistic data to that produced by XPS and SSIMS.

III. EXAMPLES OF XPS AND SSIMS ANALYSIS OF SURFACTANT POLYMERS

A. Surface Enrichment of Copolymer Components

The design of copolymer systems is well established in biomaterials science as a method of generating new materials with tailored properties [4]. The general approach to copolymer design is to mix two monomers that produce differing properties in the polymeric material. Ideally, by changing the proportions of each monomer type, it is possible to vary the material's characteristics between the two extremes of the homopolymers of each monomer. Properties that have been controlled in this way include biodegradation kinetics, glass transition temperature, crystallinity, and biocompatibility.

As discussed in Section I, the surface chemistry of any biomaterial has a central role in determining *in vivo* responses. If a copolymer is designed with two monomers with differing surface energies, surface chemistry can deviate significantly from bulk chemistry. Therefore a large number of investigations with XPS and SSIMS have been devoted to the analysis of surface enrichment of copolymer constituents.

The importance of surface enrichment in the determination of copolymer surface properties is demonstrated by a number of surface analysis studies of polyurethanes [22,27–29]. The versatility of polyurethanes as biomaterials is derived from the ability to control physicochemical and biological properties of the material by altering the proportions of hard and soft segments.

A recent example of XPS analysis of polyurethane surfaces has been provided by Yoon et al. [30]. In these studies polyurethanes were prepared with 4,4′-methylenebis(phenyl isocyanate) (MDI) and hexafluoro-1,5-pentanediol (FP) extenders as the hard segment, and poly(tetramethylene glycol) (PTMO) as the soft segment. The authors introduced hydrophobic soft segments, either poly(dimethylsiloxan) (PDMS) or polyisobutylene (PIB) into this polyurethane structure. The concept underlying this polymer design was the desire to modify surface properties with the new soft segments whilst maintaining the excellent bulk properties of the original polyurethane.

Elemental analysis of the PDMS-containing polymers provided conclusive evidence of significant enrichment of the PDMS in the outer 2 nm of solvent-case films. The elemental analysis was performed using takeoff angles of 55° and 80°. These two angles result in approximate sampling depths of 8 and 2 nm respectively. The data obtained at 80° takeoff from a polymer that contained 20% w/v of PDMS in the bulk polymer (PDMS20–PEU) revealed that 25% of the atoms in the outer 2 nm of the sample were Si. This is the maximum atomic percentage of Si obtainable indicating a PDMS-only outer 2 nm.

An example of the high-resolution C1s spectra from the PDMS-containing polyurethanes is shown in Fig. 5. The top spectrum shows the curve deconvolution proposed for the 55° angle takeoff of PDMS20–PEU polymer. The peaks at 291.2 and 287.8 eV result from the presence of the fluorine-substituted carbons and the ether carbon of the FP unit, respectively. The peak at 286.4 results from the ether carbon of the PTMO units. The peak at 285.0 eV indicates the presence of carbons experiencing no chemical shifts; such carbons result from unsubstituted carbons and the methyl groups of PDMS. When the takeoff angle was changed to 80° it was observed that the FP and PTMO related peaks disappeared from the C1s spectrum and the 285.0 eV was the only remaining peak. This data supports the theory that the PDMS component dominates the surface chemistry of this polyurethane.

Polyurethane materials have been used to investigate the ability of SSIMS analysis to generate quantitative data on surface enrichment processes. Hearn et al. performed foundation work for this field of SSIMS analysis when they investigated the enrichment of poly(propylene glycol) (PPG) soft segments over MDI/ethylene diamine hard segments [22]. Their approach to quantifying SSIMS surface data was to analyze model homopolymers composed of only hard or soft segment components. This data was analyzed to find unique peaks that could be assigned to one component but not the other. For example, analysis of PPG generated an intense cationic peak at m/z 59 whilst the hard segment generated a unique cationic peak at m/z 106.

$$\left[\begin{array}{c} CH_3 \\ | \\ CHCH_2-OH \end{array} \right]^+ \qquad CH_2=\!\!\!\!\bigcirc\!\!\!\!=NH_2^+$$

m/z 59 $\qquad\qquad\qquad$ m/z 106

The authors proposed that the extent of soft segment surface enrichment in a range of polyurethanes could be quantified by dividing the intensity of a soft segment peak (I_{ss}) by the intensity of a hard segment peak (I_{HS}). This theory was tested by analyzing a series of polyurethanes with increasing amounts of the PPG

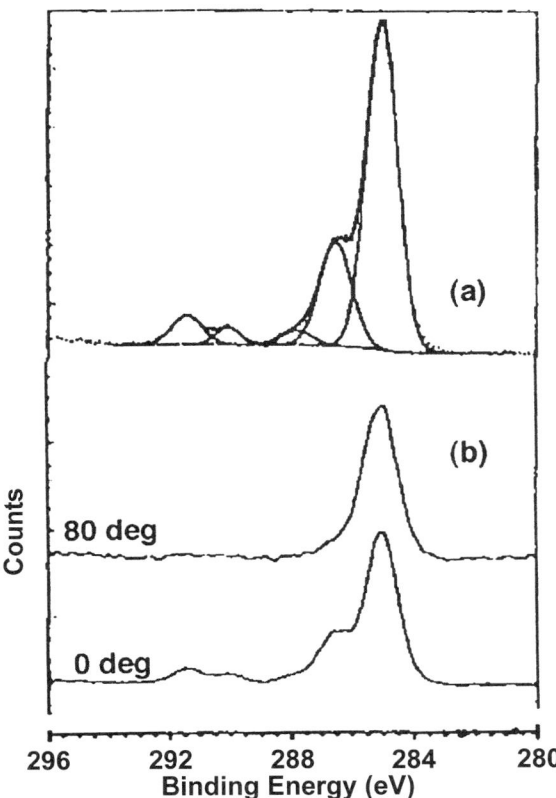

FIG. 5 C1s XPS spectra obtained from surfaces of PDMS-containing polyurethanes. (From Ref. 30.)

segment in the bulk polymer structure. Increases in the value of I_{SS}/I_{HS} were found to follow increases in the PPG segment contribution to the polyurethane, suggesting that quantification of the SSIMS data from series of structurally related copolymers was possible.

In recent years many new biodegradable block and random copolymers have been designed for biomaterials applications, especially for use as drug delivery devices and tissue engineering scaffolds [4,31,32]. Increasingly in the field of biomaterials design there has been an appreciation of the need to understand how bulk chemical changes in biodegradable polymer structures translate into changes in surface chemistry and hence into changes in *in vivo* surface properties. A number of combined XPS and SSIMS studies of biodegradable polymer structure have been described. For example, Davies et al. investigated the surface chemistry of linear poly(ortho esters) composed of alternating sequences of 3,9-diethylidene-2,4,8,10-tetraoxaspiro[5.5]undecane (DETOSU) and either a flexible diol 1,6-hexanediol (HD) or a rigid diol *trans*-cyclohexanedimethanol (*t*-CDM) [21].

CH_3CH_2
OCH_2
CH_2O
CH_2CH_3
C
C
C
O
OCH_2
CH_2O
O
R
n

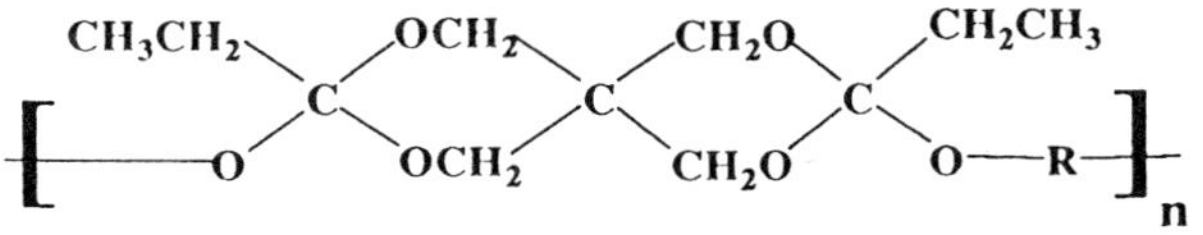

DETOSU

where *R* is either:

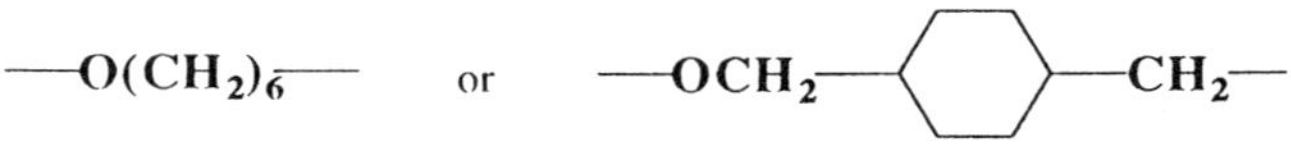

$O(CH_2)_6$
or
OCH_2
CH_2

HD **t-CDM**

This poly(ortho ester) system was originated by Heller as a material whose mechanical properties could be precisely controlled by varying the ratios of HD to *t*-CDM. When a series of poly(ortho esters) were analyzed by XPS, the expected peaks within the C1s data were observed. However, the similarly in structure between the HD and *t*-CDM components in terms of XPS recorded chemical shifts prohibited quantification of their relative surface concentration. Using ToF–SIMS, and a similar interpretation strategy to that employed by Yoon et al. for the polyurethanes, Davies et al. were able to distinguish the presence of HD and *t*-CDM at the polymer surfaces. Cationic peaks diagnostic of DETOSU, HD, and *t*-CDM included the peaks at *m/z* 127, 101, and 109, respectively. These peaks were assigned to the fragment structures here.

CH_2
O
O
CH_2CH_3
+
$(CH_2)_6OH$ +
+CH_2
CH_2

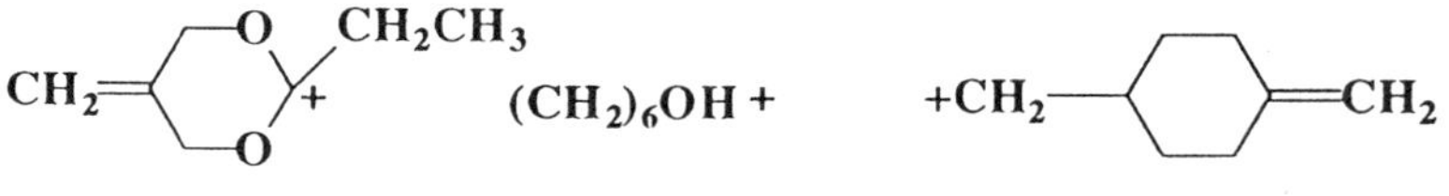

***m/z* 127** ***m/z* 101** ***m/z* 109**

By comparing the peak areas of these diagnostic peaks for a series of polymers with varying HD : *t*-CDM ratios is was possible to semiquantify polymer surface chemistry. The graph in Fig. 6 shows a plot of the ratios of the *t*-CDM diagnostic peak 109 and the HD diagnostic peak 101 compared to the DETOSU diagnostic peak 127. It is evident that the quantities of both HD and *t*-CDM appear to increase linearly with increasing bulk composition.

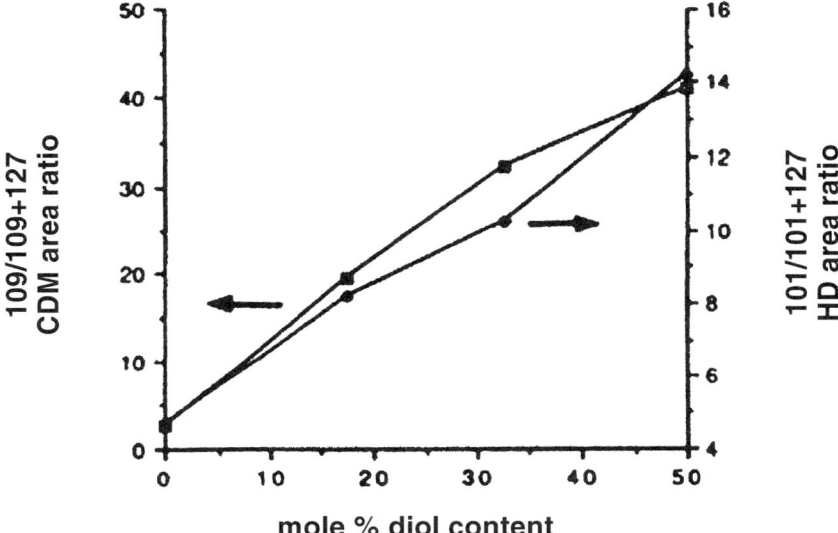

FIG. 6 Semiquantification of SSIMS data from poly(ortho esters) using peaks diagnostic of the presence of each monomer species. (From Ref. 21.)

An important development in ToF–SIMS analysis of copolymer structures has been described by Briggs and Davies in their analysis of methyl methacrylate-poly(ethylene glycol) methacrylate copolymers with average PEG molecular weights of 1000 [33]. High-mass ToF–SIMS analysis of this copolymer is shown in Fig. 7. The remarkable feature of this data is the presence of two series of peaks assigned to $[CH_2(OCH_2CH_2)_nNa]^+$ and $[CH_2(OCH_2CH_2)_nK]^+$ with the cationization process of the fragments involving a well-known interaction between ethylene glycol and either Na^+ or K^+ ions. The author proposed that the intensity of peaks at m/z values in excess of 900 suggested that oligomeric intensity distribution of fragments centered around m/z 1000, i.e., consistent with the known average molecular weight of the PEG units. This represents a first example of the measurement of polymer molecular mass distributions of tethered surfactant polymer brushes at surfaces and therefore has important consequences in the analysis of polymeric biomaterials.

B. Microparticle Surface Chemistry

Methods of preparing polymeric microparticles with tailored surface chemistries are of technological importance in a wide range of applications including drug delivery, biochemical diagnostics, and paint/coating products [34]. Microparticle

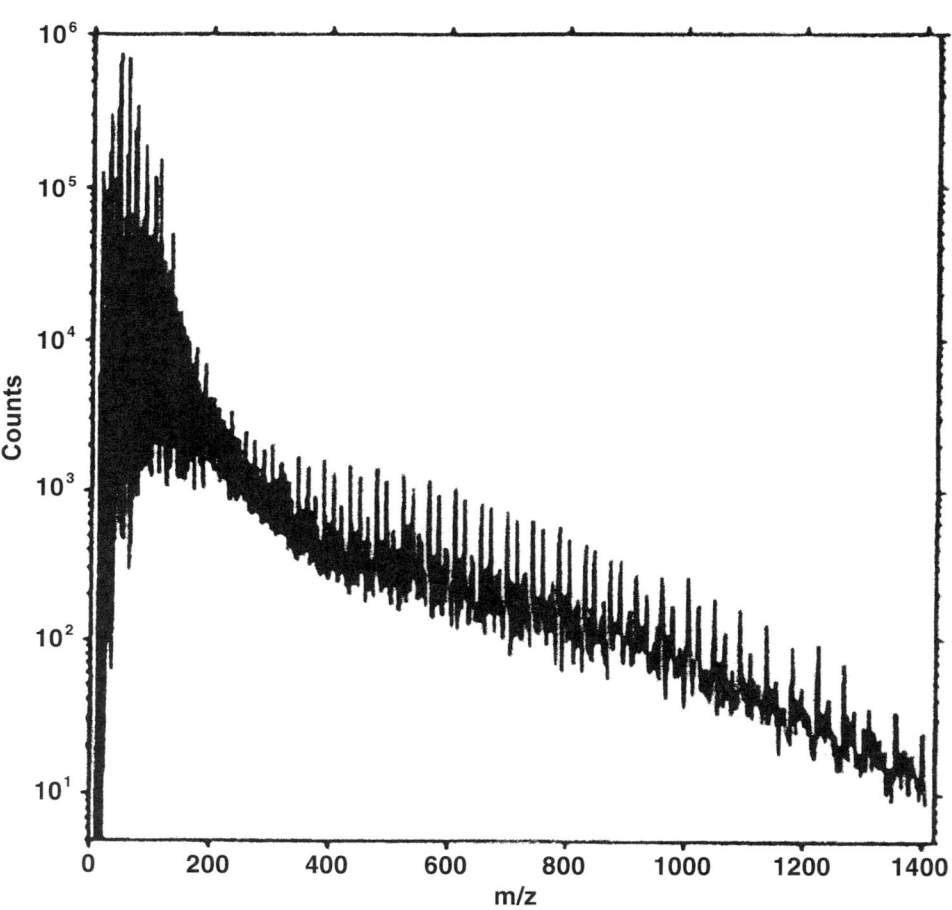

FIG. 7 Positive ion ToF–SIMS spectra of methyl methacrylate-poly(ethylene glycol) methacrylate copolymers with average PEG molecular weights of 1000. (From Ref. 33.)

surface chemistry is vital both in the mechanism of colloid stabilization and in the functioning of the final material. In general, the formation of a large surface area within microparticle formulations promotes the attachment to or enrichment of low-surface-energy molecules/components at the forming microparticle surface. Here we consider two types of surface-energy driven microparticle surface chemical changes; the adsorption of a polymeric surfactant and the covalent attachment of hydrophilic polymeric molecules.

XPS and SSIMS of Biomedical Polymers

1. Polymeric Surfactant Adsorption

In the field of site-specific drug delivery there is considerable research effort invested in the development of methods of changing microparticle surface chemistry with the aim of altering the distribution of the particles within the body [35]. For example, Illum et al. demonstrated that if microparticles were coated with surfactant block copolymers (e.g., Pluronic A-B-A polymers, where A is PEG and B is poly(propylene oxide) (PPO)), then the microparticles avoided uptake by the liver and therefore stayed in the blood circulation for long periods [36]. The ability to change the behavior of biological systems to microparticles via surfactant attachment is based on the alteration of the adsorption of proteins to the microparticles. This alteration often involves inhibiting nonspecific adsorptions that normally trigger liver cells to remove the particle from the circulation.

Surfactant adsorption is an excellent method of tailoring microparticle surface chemistry because the surfactant can be designed to stabilize the forming particle surface during an emulsion-based fabrication procedure. Surface chemistry analysis is established as a vital component of the microparticle design process because it allows the amount of surfactant adsorbing to a surface to be quantified.

Shakesheff et al. provided an example of this type of analysis in their XPS study of the adsorption of poly(vinyl alcohol) (PVA) to biodegradable microparticles [37]. The microparticles were composed of either poly(DL-lactic acid) (PLA), poly(lactic acid-*co*-glycolic acid) (PLGA), or block copolymers of either PLA or PLGA with PEG. The PVA was 88% mol hydrolyzed, meaning that 88% of the monomers were vinyl alcohol and 12% of the monomers were vinyl acetate. Figure 8 shows the XPS recorded change in polymer surface chemistry between a PLA film and PLA microparticles formed in presence of PVA. In Fig. 8a the deconvoluted C1s data for PLA films shows the expected three peaks generated by the ester group (peak **1**, 289.1 eV), the methine group (peak **2**, 287.1 eV), and the methyl group (peak **3**, 285.0 eV). The difference in surface chemistry produced by microparticle fabrication is evident from Fig. 8b, where the C1s region for the PLA microparticles is shown with the C1s region of the PLA films. Clearly, the microparticle surface chemistry generated significantly increase intensities between binding energies of 285.0 and 286.5 eV. Careful deconvolution of the C1s data for the microparticles demonstrated that this increased intensity could be explained by the adsorption of PVA. To generate a physically meaningful deconvolution it was necessary to constrain the curve fitting procedure so that the algorithm varied the relative proportion of PVA to PLA but did not change the relative size of peaks within either component (i.e., PVA peaks maintained the same area ratios to each other). The deconvoluted data is shown in Fig. 8c with the PVA and PLA components dissected out of the

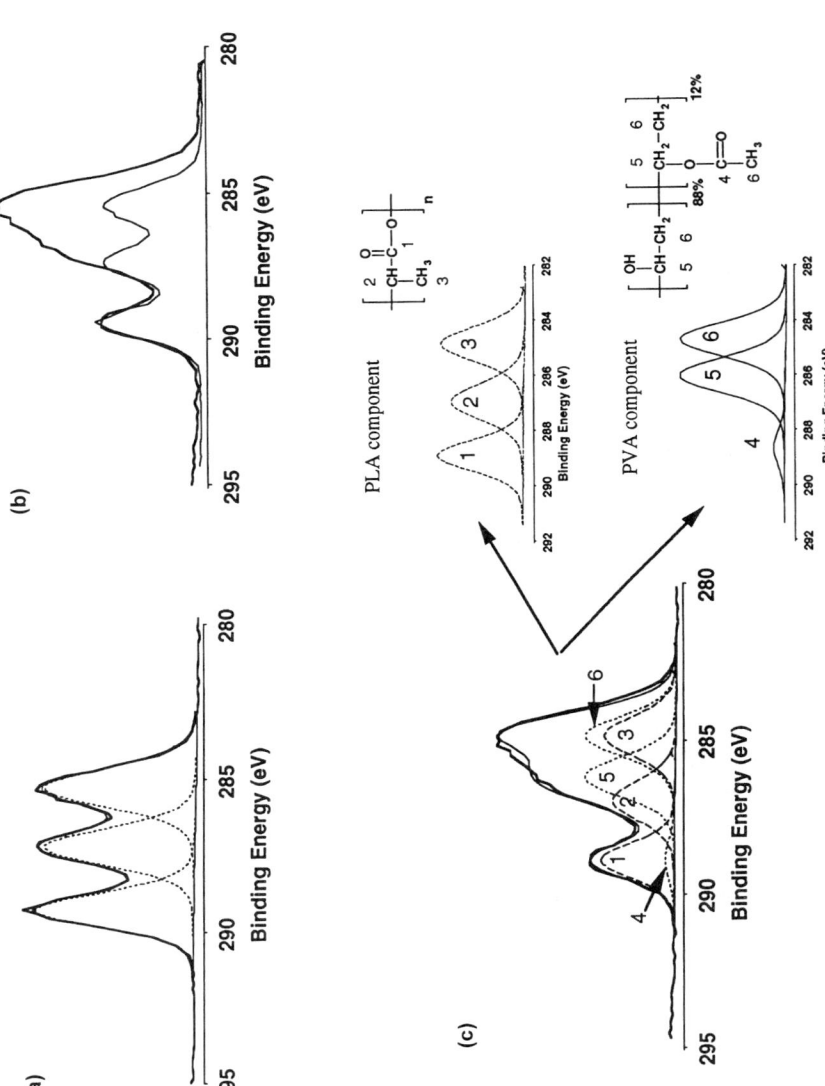

FIG. 8 XPS analysis of PVA adsorption to PLA microparticle surfaces. (a) C1s data from a PLA film surface. (b) Comparison between PLA film and PLA microparticle surfaces. (c) C1s data from PLA microparticles. Deconvolution indicated the presence of PLA and PVA components. (From Ref. 37.)

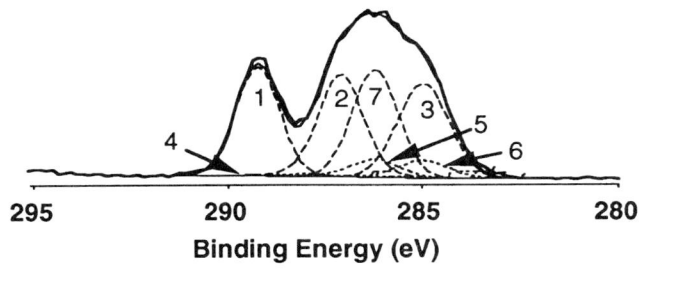

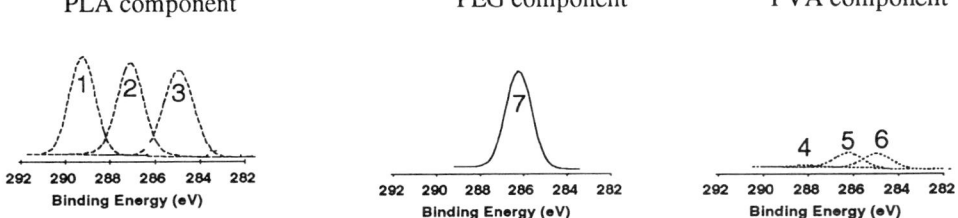

FIG. 9 XPS analysis of PLA-PEG microparticle surfaces showing the contribution to surface chemistry of the lactic acid, ethylene glycol, and vinyl alcohol groups. (From Ref. 37.)

deconvolution and shown individually. Quantification of the C1s data indicated that the surface chemistry of the PLA microparticles, within the sampling depth of the XPS technique, was composed of approximately 45% PVA and 55% PLA.

This study went on to highlight an interesting contrast in the mechanism of surface stabilization of PLA and PLA-PEG containing microparticles. As described above, the PLA microparticles adsorbed relatively large quantities of PVA to stabilize the new surfaces formed within an aqueous environment. For the PLA-PEG microparticles, the hydrophilic PEG component decreased the surface energy of the polymer surfaces within the aqueous environment, compared to the PLA homopolymer material. Therefore significantly less PVA adsorption occurred during the emulsion fabrication. The data in Fig. 9 shows the full deconvolution of the C1s data of the microparticles. By comparing the data in Fig. 9 and Fig. 8c it is evident that much less PVA has adsorbed to the PLA-PEG microparticles. Measurement of the peak areas showed that the surface of these particles was composed of only 12% PVA.

Another example of surfactant adsorption quantification of XPS was described by Coombes et al. [38]. They introduced the concept of using PEG as anchoring segments as part of PEG-dextran (PEG-DEX) conjugates. The C1s XPS data in Fig. 10 shows that the incorporation of PEG-DEX could be observed

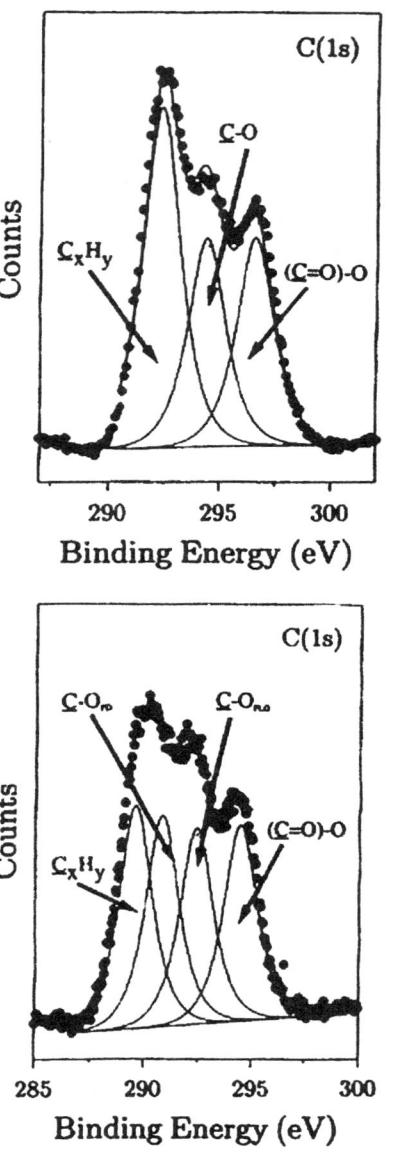

FIG. 10 XPS analysis of anchoring PEG-DEX conjugates into the surfaces of PLG microparticles. The top C1s data was recorded from a PLG control sample and displays considerable hydrocarbon contamination. The bottom C1s region proves the presence of the PEG-DEX conjugate at the microparticle surface due to the presence of a significant $\underline{C}$-O_{PD} peak. (From Ref. 38.)

by deconvolution and measurement of relative intensities of the ether carbon peaks associated with PEG-DEX (labeled $\underline{C}$-O_{PD}) and the biodegradable polymer, poly(lactic acid-*co*-glycolic acid) (labeled $\underline{C}$-O_{PLG}). Immunoassay studies were then performed to show that the dextran component could bind dextran-specific antisera. These studies showed that after thorough washing the particles displayed a strong specific binding of the antisera, indicating that dextran molecules were present at the surface.

2. Covalent Attachment

An elegant example of the analysis of colloid surfaces containing covalently attached hydrophilic species has been provided by Brindley et al., who studied the surface chemistry of polystyrene colloids with surface grafted polyethylene glycol groups [39]. These colloids were prepared by surfactant-free copolymerization of styrene with PEG using potassium persulphate as an initiator. The XPS analysis of these microparticles is shown in Fig. 11.

If we consider in more detail the data from Fig. 11f, the polymerization formula indicates that for every styrene monomer present in the bulk there were 0.061 PEG molecules with average molecular weights of 2000. Converting that formula into a ratio of monomeric units, for every styrene monomer present there were 2.7 ethylene glycol units. In the C1s region of the final colloid generated from this formula, ethylene glycol units were found to generate 41.6% of the total peak area, and styrene units generated 58.4% (correcting for the presence of potassium persulphate initiator). Now each styrene unit contains 8 carbon atoms, all ejecting electrons with binding energies of 285.0 eV during XPS analysis, and each ethylene glycol unit contains two carbon atoms ejecting electrons at 286.7 eV. Therefore the surface monomer ratios translate to 1 styrene monomer for every 2.8 ethylene glycol monomers. This result indicated that the surface composition of the microparticles was closely related to the polymerization formula.

C. Blend Surface Segregation

Closely related to the polymeric surfactant stabilization of microparticle surfaces is the stabilization of surfaces by the segregation of one component of a blend. This powerful method of surface engineering involves the inherent thermodynamically driven enrichment of surfaces by the blend component possessing the lowest surface energy.

Chen and Gardella used this surface engineering strategy to create siloxane-rich surfaces [40]. Their approach involved the blending of a homopolymer (A) with a block copolymer composed of a block with the same chemical identity as the homopolymer (A) and a block of PDMS. For all homopolymer types studied (polystyrene, poly(α-methylstyrene) and Bisphenol A polycarbonate), XPS analysis of Si:C ratios revealed a significant enrichment of the PDMS

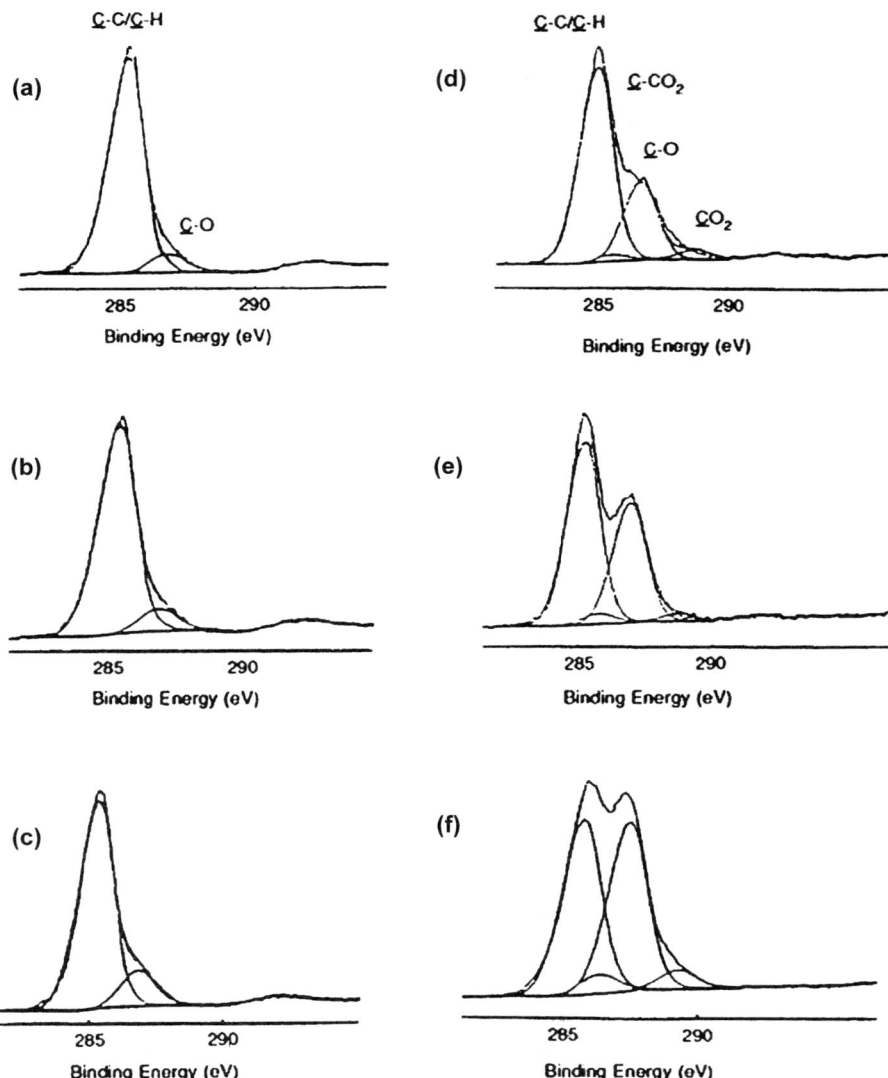

FIG. 11 Quantification of expression of covalently attached PEG at the surface of polystyrene microparticles. The amount of PEG can be calculated from the relative intensity of the C-O peak, which is diagnostic of the ether repeat unit. Increasing amounts of PEG (mw 2000) were present in the polymerization formula from a through f. Polymerization formula ratios of styrene:PEG2000 were (a) 1:0.0044, (b) 1:0.0090, (c) 1:0.0250, (d) 1:0.0380, (e) 1:0.0530, and (f) 1:0.0610. (From Ref. 39.)

block. Typically, the PDMS block needed to be only 6% (by weight) of the original blend composition for the surface region to be composed of at least 95% PDMS. The authors determined a number of relationships between the extent of surface segregation and the block molecular weight or copolymer architecture. For example, studies using polystyrene as the homopolymer were performed with the molecular weight of the homopolymer much smaller than the molecular weight of the A block of the copolymer, and it was found that A-B copolymer architecture gave higher PDMS surface concentrations than either A-B-A or B-A-B architectures.

XPS and SSIMS analysis of biodegradable polymer blends has also revealed an important role for surface segregation in determining the final surface properties of materials. Davies et al. have used these techniques to quantify surface segregation in blends of poly(DL-lactic acid) (PLA) and poly(sebacic anhydride) (PSA) [41]. For low-molecular-weight PLA (2K), bulk studies indicated that the polymer blend was miscible. However, by quantifying the area of the peak resulting from the ether carbon of PLA in a range of blends with varying PLA:PSA ratios, it was found that significant surface enrichment by the PLA component occurred. This is shown in Fig. 12, where the surface molar percentage of PLA is plotted against the bulk molar percentage of PLA. For all bulk molar percentages of PLA up to approximately 80%, the surface molar percentage was found to be higher. For higher molecular weight PLA (50K), phase separation of the PSA and PLA occurred as determined by atomic force microscopy (AFM). An interesting phenomenon in the XPS data appeared to be dependence on this blend surface morphology. Deconvolution of the C1s data for this blend was only possible if it was assumed that insufficient charge compensation occurred on the PLA domains as compared to the PSA domains. This resulted in the PLA-derived peaks being shifted + 1 to 1.4 eV compared to the PSA peaks.

D. End-Group Contributions to Surface Chemistry

The modification of polymer end-group chemistry has been utilized as a method of controlling the surface activity of polymeric materials. The synthesis of polymers with low surface energy end-groups can generate sufficient thermodynamic driving forces to dictate that polymer surface chemistry is dominated by end-groups.

This approach to polymer surface engineering has been extensively used in the synthesis of new fluorine-containing polymers. Fluoropolymers possess a number of beneficial properties including resistance to chemical attack and low coefficients of friction. However, the large-scale synthesis of the polymers is often inherently difficult and generates highly toxic by-products. Therefore interest has been stimulated in the use of surface-modifying fluorine-containing chemicals.

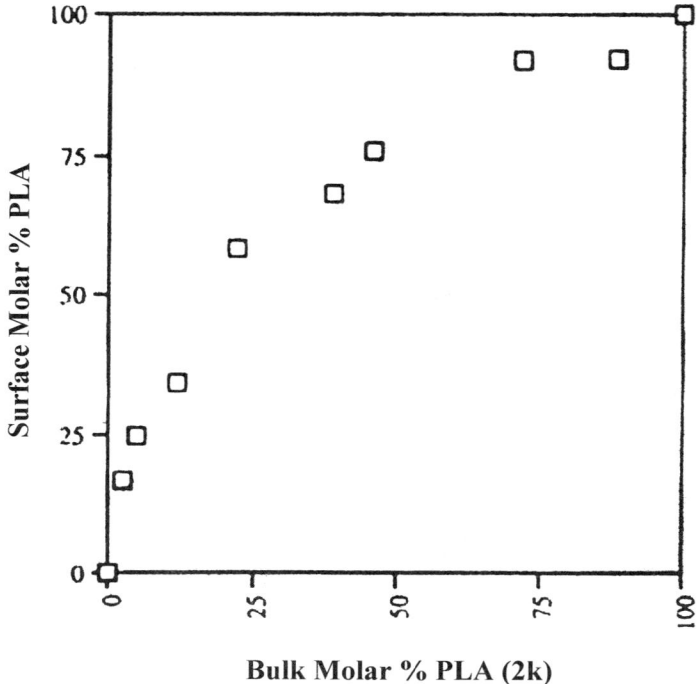

FIG. 12 Blend surface segregation in PLA/PSA blends. Surface enrichment is evident from the higher surface molar percentage of PLA compared to the bulk surface percentage of PLA. (From Ref. 41.)

Hunt et al. described the synthesis of perfluoroalkyl-terminated polystyrene via the terminal functionalization of living anionic polymerizations using chlorosilane derivatives [42]. Tof–SIMS analysis of the resulting end-capped polymers showed an intense series of peaks that corresponded to fragments with n styrene groups attached to the perfluoroalkyl end-groups and one silver atom (from the substrate) per fragment (see Fig. 13). The distribution of molecular mass of fragments, although likely influenced by a complex convolution of the relative probabilities of fragment formation processes, was in close agreement with masses of the polymers determined by gel permeation chromatography (GPC).

Su et al. have undertaken a detailed examination of the mechanism by which perfluoroalkyl end-groups enrich the surfaces of blends of homopolymers with their end-group fluorinated equivalents [43]. Their studies aimed to assess the parameters involved in determining whether the enthalpic gain from fluorine surface enrichment outweighed the entropic penalty of polymer chains having

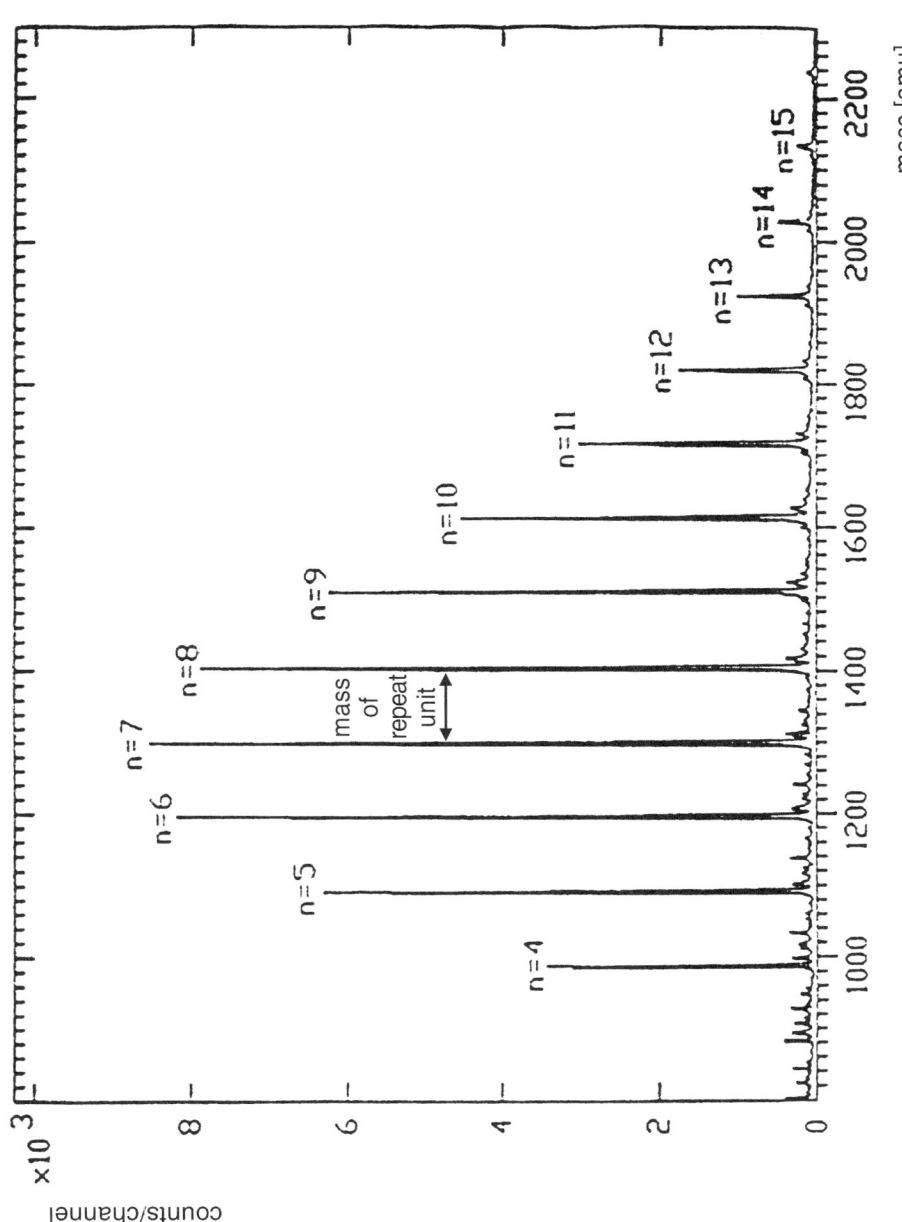

FIG. 13 ToF–SIMS mass spectrum of a perfluoroalkyl-terminated polymer. (From Ref. 42.)

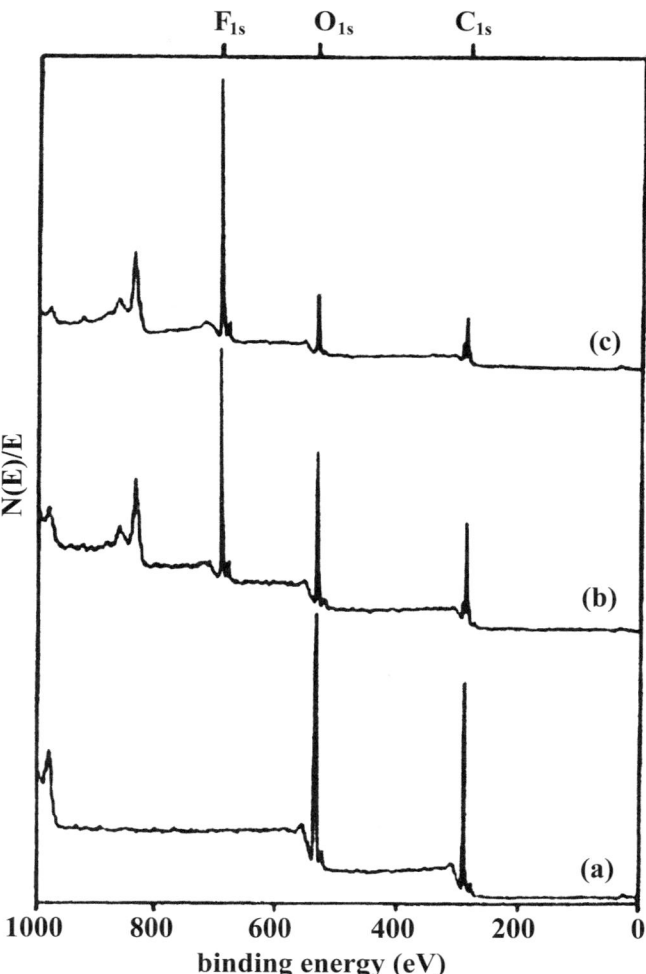

FIG. 14 Elemental XPS analysis comparing the surface chemistry of 3K mw PEO and 3K mw PEO^{2F}. (a) 3K mw PEO analyzed with a 75° takeoff angle. (b) 3K mw PEO^{2F} analyzed with a 75° takeoff angle. (c) 3K mw PEO^{2F} analyzed with a 15° takeoff angle. (From Ref. 43.)

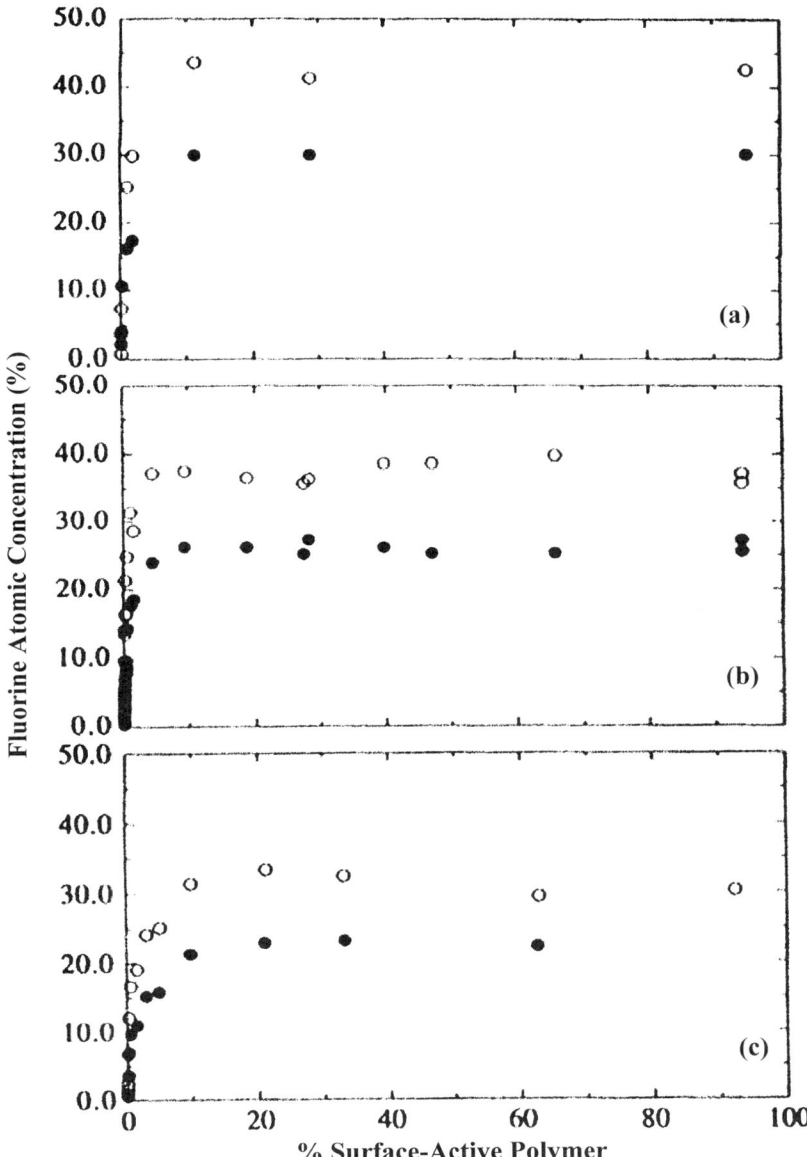

FIG. 15 Graphs relating the XPS determined fluorine atomic concentration against the concentration of (a) 3K mw PEO^{2F}, (b) 8K mw PEO^{2F}, and (c) 15K mw PEO^{2F} in blends of PEO with the same mw. Open and closed circles are from data at 15° and 75° takeoff angles respectively. (From Ref. 43.)

to extend to transport the fluorine-containing end-groups to the surface. They synthesized a range of poly(ethylene oxide) (PEO) molecules that possessed either one or two perfluorodecanoyl end-groups (final polymers were abbreviated either PEO^F or PEO^{2F}). Initially, XPS analysis was used to investigate any surface enrichment of the fluorinated end-groups of one-component cast films. The XPS data in Fig. 14 compares the surface chemistry of 3K mw PEO and 3K mw PEO^{2F}. The PEO samples displayed the expected carbon and oxygen peaks. The intensity of these peaks showed no dependence on the takeoff angle of analysis. The PEO^{2F} samples contained an additional peak for fluorine. At a takeoff angle of 75° this peak accounted for 41.7% total XPS signal (i.e., 41.7% of the atoms within the surface region were fluorine). This value decreased to 30% at a takeoff angle of 15°. The authors explored the physical meaning of this data in terms of the attenuation of photoelectron intensity at varying sampling depths. They concluded that within the surface region analyzed the top 1 nm was rich in fluorine and beneath this region there was a 3.5 nm thick region that was depleted of fluorine by the surface enrichment process.

When the perfluoroalkyl end-group PEOs were blended with normal PEOs, a very strong surface enrichment of the former occurred. This enrichment is demonstrated by large increases in fluorine signals accompanying minor additions of the surface-active PEO (see Fig. 15). This data strongly supported the hypothesis that the enthalpic component (decreasing overall material surface energy) greatly outweighed the entropic penalty.

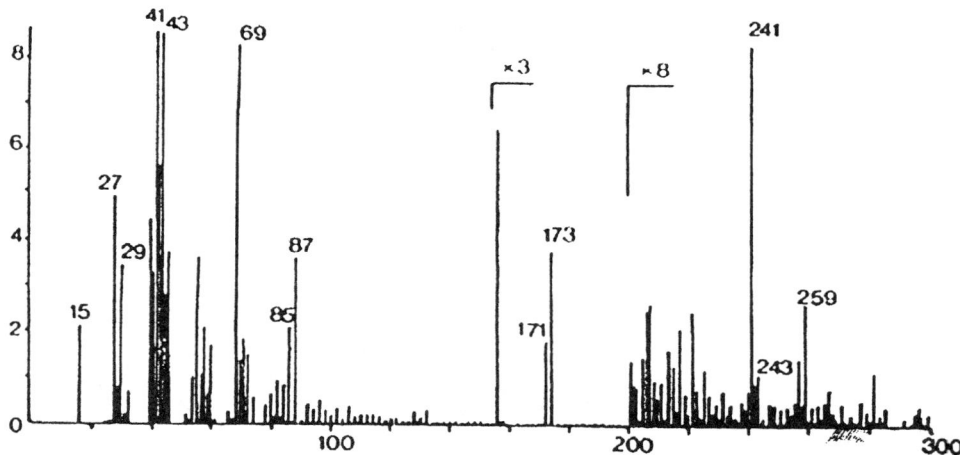

FIG. 16 Positive ion SSIMS spectrum of poly(β-hydroxybutyrate) (PHB). (From Ref. 44.)

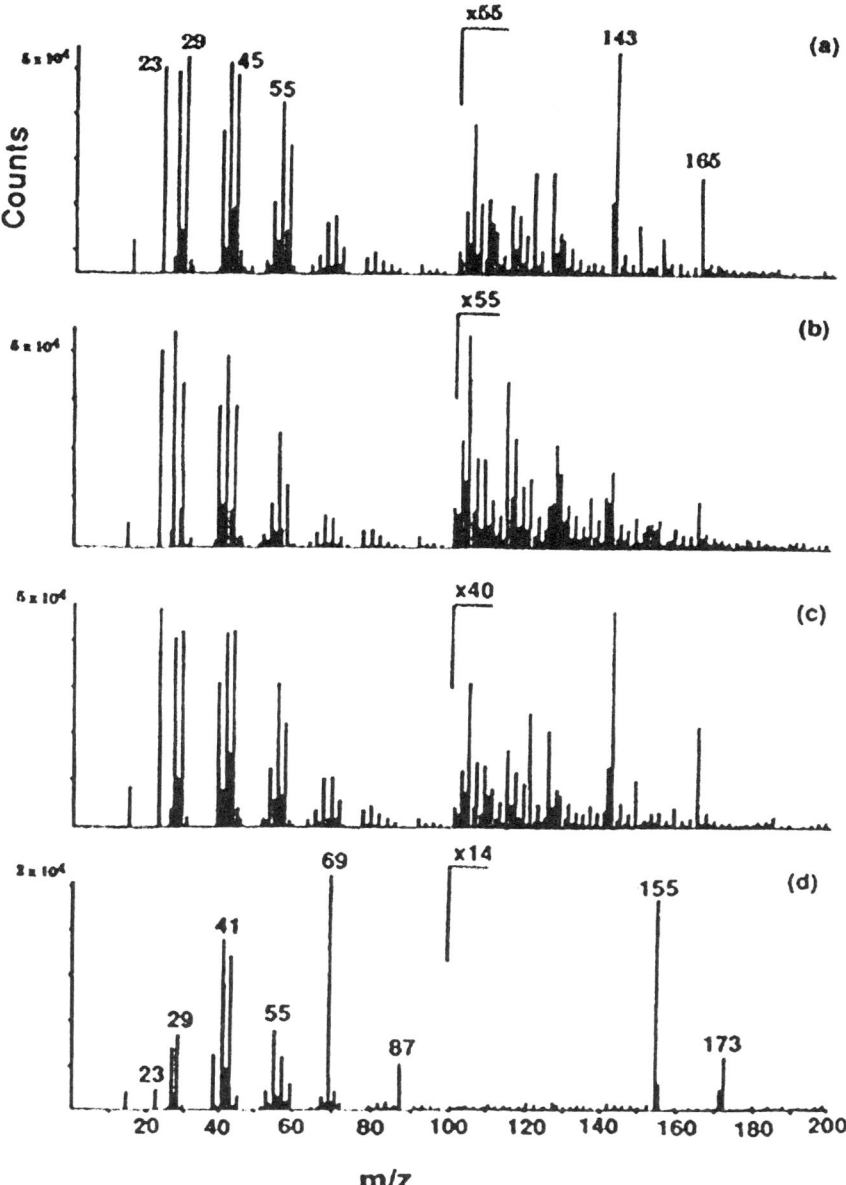

FIG. 17 Positive ion SSIMS spectra of PHB microparticles (a) uncleaned; (b) dialysis cleaned; (c) ultrafiltration cleaned; (d) GPC cleaned. (From Ref. 44.)

E. Desorption

Many technological applications involving surfactants require the removal of excess surfactant after its surface-energy reducing function has been performed. XPS and SSIMS analysis have played an important role in validating these surfactant removal and desorption events.

Koosha et al. have provided a detailed SSIMS examination of the efficiency of dialysis, ultrafiltration, and gel permeation chromatography (GPC) cleaning procedures in the removal of sodium dodecyl sulphate (SDS) from biodegradable microparticles composed of poly(β-hydroxybutyrate) (PHB) [44]. The foundation for this study was the identification of SSIMS fragments derived from the PHB component. The positive ion spectrum from PHB is shown in Fig. 16. It is dominated by series of peaks attributable to the formation of the fragments: $nM \pm H$ ($n = 1$ to 4) at m/z 85/87, 171/173, 257/259, and 343/345, and $nM - O \pm H$ ($n = 1$ to 4) at m/z 69, 155/157, 241/243, and 325/327. Series of peaks were also defined in the negative ion spectrum.

A comparison of the positive ion spectra derived from microparticles before and after cleaning is shown in Fig. 17. The most notable feature of the uncleaned sample is the absence of series of peaks observed in the spectra of PHB (e.g., m/z 69, 85, and 87). Instead, the spectrum is dominated by general hydrocarbon peaks, and evidence of the presence of SDS is indicated by the peak at m/z 23 that has been attributed to Na^+ ions originating from the ionic complex of SDS. The negative ion spectrum confirmed the absence of PHB for the surface and, in addition, confirmed the presence of SDS by the identification of peaks attributable to the sulphate ionic group of the surfactant. Cleaning by dialysis and ultrafiltration did not significantly change the distribution or intensity of peaks in either the positive or the negative SSIMS spectra. However, the size exclusion process involved in GPC cleaning was shown significantly to deplete the surface of SDS. The positive spectrum in Fig. 17 shows the appearance of peaks diagnostic of PHB (e.g., m/z 69, 87, 155, and 171). Peaks attributable to the SDS were still present, but their intensity was significantly reduced. This data proved conclusive evidence of the improved efficiency of the GPC cleaning technique.

IV. CONCLUSIONS

The self-organization of surfactant molecules and material containing components with differing surface energies offers many opportunities in polymer surface engineering. The five areas of surfactant surface characterization discussed in this review have highlighted the type of information available for XPS and SSIMS analysis. In order of increasing complexity this information ranges from elemental analysis, chemical group quantification, and fingerprint analysis to complete polymer chain mass measurement.

Advances in the design of XPS and SSIMS instruments are inherently linked to improvement in our understanding of surfactant behavior at solid surfaces. Developments including the design of instruments with imaging capabilities [45] and cryogenic platforms that allow hydrated polymer surfaces to be characterized [46] promise to continue to strengthen our knowledge of the physical chemistry underlying surfactant-based surface engineering.

REFERENCES

1. Ratner BD. Surface modification of polymers for biomedical applications: Chemical, biological, and surface analytical challenges. In: Ratner BD, Castner DG, eds. Surface Modification of Polymers, Plenum Press, New York, 1996, pp. 1–9.
2. Hoffman AS. Artificial Organs 16:43–49 (1992).
3. Hubbell JA. Bio/technology 13:565–576 (1995).
4. Langer R. Annals of Biomedical Engineering 23:101–111 (1995).
5. Park S-H, Llinás A, Goel VK, Keller JC. Hard Tissue Replacements. In: Bronzino JD, ed. The Biomedical Engineering Handbook, CRC Press, Boca Raton, FL, 1995, pp. 672–703.
6. Jantzen GM, Robinson JR. Sustained- and controlled-release drug delivery systems. In: Banker GS, Rhodes CT, eds. Modern Pharmaceutics. 3d ed., Marcel Dekker, New York, 1996.
7. Heller J. Advanced Drug Delivery Reviews 10:163–204 (1993).
8. Cima LG, Langer R. Chemical Engineering Progress June:46–54 (1993).
9. Williams DF. Biofunctionality and Biocompatibility. In: Cahn RW, Haasen P, Kramer EJ, eds. Materials Science and Technology 14. Medical and Dental Materials, VCH Germany, 1992, pp. 1–27.
10. Cook AD, Hrkach JS, Gao NN, Johnson IM, Pajvani UB, Cannizzaro SM, Langer R. Journal of Biomedical Materials Research 35:513–523 (1997).
11. Pasqualini R, Ruoslahti E. Nature 380:364–366 (1996).
12. Elbert DL, Hubbell JA. Annual Review of Materials Science 26:365–394 (1996).
13. Briggs D, Seah MP. Practical Surface Analysis 2. Ion and Neutral Spectroscopy, John Wiley, Chichester, UK, 1992.
14. Briggs D, Seah MP. Practical Surface Analysis 1. Auger and X-ray Spectroscopy, John Wiley, Chichester, UK, 1990.
15. Beamson G, Briggs D. High Resolution XPS of Organic Polymers: The Scienta 300 Database, John Wiley, Chichester, UK, 1992.
16. Davies MC, Lynn RAP. Critical Reviews in Biocompatibility 5:297–341 (1990).
17. Davies MC, Roberts CJ, Tendler SJB, Williams PM. The surface analysis of polymeric biomaterials. In: Braybrook J., ed. Biocompatibility Assessment of Medical Devices and Materials, John Wiley, 1997, pp. 65–99.
18. Vickerman JC, Briggs D. The Wiley Static SIMS library, John Wiley, Chichester, UK, 1996.
19. Bletsos IV, Hercules DM, vanLeyen D, Benninghoven A. Macromolecules 20:407–413 (1987).
20. Briggs D. British Polymer Journal 21:3–15 (1989).

21. Davies MC, Lynn RAP, Watts JF, Paul AJ, Vickerman JC, Heller J. Macromolecules 24:5508–5514 (1991).
22. Hearn MJ, Ratner BD, Briggs D. Macromolecules 21:2950–2959 (1988).
23. Shakesheff KM, Davies MC, Roberts CJ, Tendler SJB, Williams PM. Critical Reviews in Therapeutic Drug Carrier Systems 13:225–256 (1996).
24. Green RJ, Davies J, Davies MC, Roberts CJ, Tendler SJB. Biomaterials 18:405–413 (1997).
25. Malmsten M, Lassen B. Journal of Colloid and Interface Science 166:490–498 (1994).
26. Davies J, Nunnerley CS, Brisley AC, Edwards JC, Finlayson SD. Journal of Colloid and Interface Science 182:437–443 (1996).
27. Shard AG, Davies MC, Tendler SJB, Jackson DE, Lan PN, Schacht E, Purbrick MD. Polymer 36:775–779 (1995).
28. Benrashid R, Nelson GL, Linn JH, Hanley KH, Wade WR. Journal of Applied Polymer Science 49:523–537 (1993).
29. Tyler BJ, Ratner BD, Castner DG, Briggs D. Journal of Biomedical Materials Research 26:273–289 (1992).
30. Yoon SC, Ratner BD, Iván B, Kennedy JP. Macromolecules 27:1548–1554 (1994).
31. Heller J. ACS Symposium Series 567:292–305 (1994).
32. Schacht E, Vandorpe J, Dejardin S, Lenmouchi Y, Seymour L. Biotechnology and Bioengineering 102–108 (1996).
33. Briggs D, Davies MC. Surface and Interface Analysis 25:725–733 (1997).
34. Lyklema J. Fundamentals of Interface and Collid Science. Vol. 1, Fundamentals, Academic Press, London, 1991.
35. Moghimi SM, Davis SS. Critical Reviews in Therapeutic Drug Carrier Systems 11:31 (1994).
36. Illum L, Davis SS, Müller RH, Mak E, West P. Life Sciences 40:367–374 (1987).
37. Shakesheff KM, Evora C, Soriano I, Langer R. Journal of Colloid and Interface Science 185:538–547 (1997).
38. Coombes AGA, Tasker S, Lindbald M, Holmgren J, Hoste K, Toncheva V, Schacht E, Davies MC, Illum L, Davis SS. Biomaterials 18:1153–1161 (1997).
39. Brindley A, Davis SS, Davies MC, Watts JF. Journal of Colloid and Interface Science 171:150–161 (1995).
40. Chen X, Gardella JA. Macromolecules 27:3363–3369 (1994).
41. Davies MC, Shakesheff KM, Shard AG, Domb A, Roberts CJ, Tendler SJB, Williams PM. Macromolecules 29:2205–2212 (1996).
42. Hunt MO, Belu AM, Linton RW, DeSimone JM. Macromolecules 26:4854–4859 (1993).
43. Su Z, Wu D, Hsu SL, McCarthy TJ. Macromolecules 30:840–845 (1997).
44. Koosha F, Müller RH, Davis SS, Davies MC. Journal of Controlled Release 9:149–157 (1989).
45. Lhoest J-B, Detrait E, Dewez J-L, Van den Bosch de Aguilar P, Bertrand P. Journal of Biomaterial Science. Polymer Edition 7:1039–1054 (1996).
46. Ratner BD. Surface and Interface Analysis 23:521–528 (1995).

6
Evanescent Wave Scattering at Solid Surfaces

ADOLFAS K. GAIGALAS Biotechnology Division, National Institute of Standards and Technology, Gaithersburg, Maryland

I.	Introduction	173
II.	Scattering of Evanescent Waves	174
	A. Ideal interface	174
	B. Nonideal interface—scattering of evanescent waves	175
III.	Static EW Scattering	179
	A. Scattering from a single particle	181
	B. Scattering from a collection of particles	181
	C. Scattering from polymer attached to interface	185
IV.	Dynamic EW Scattering	187
	A. Restricted diffusion near interfaces	189
	B. Dynamics of polymers attached to interfaces	192
V.	Other Related Techniques	195
VI.	Conclusion	196
	References	196

I. INTRODUCTION

Evanescent electromagnetic waves can be excited at the interface of two optically transparent materials with different indexes of refraction. If a beam of light is incident on the interface from the material with the larger index of refraction and if the angle of incidence exceeds a certain critical value, then there will be no transmitted wave in the material with the smaller index of refraction, and all the incident light will be reflected back into the material with the larger index of refraction (total internal reflection). However, a wave with exponentially decaying amplitude will propagate along the interface in the material with the smaller index of refraction. This wave is called the evanescent wave (EW), and since its amplitude is confined to the interface region, the EW is an excellent probe of interface properties. The penetration length Γ, the characteristic distance over which the EW amplitude is finite, is much smaller than the wavelength of the incident light, so that EW probes over a distance much shorter than the wavelength. Fluorescence excited by EW [1], scattering of EW, and attenuated total reflection [2] have been the three major techniques for probing the interface region. In this chapter, the emphasis will be on scattering of EW from particles on or near an interface. Both static and dynamic scattering will be considered. The name "static scattering" refers to the measurement of the average scattered intensity. Static scattering is used to obtain particle distributions near or on the interface. In the case where the average scattered intensity varies sufficiently slowly and the scattering is sufficiently strong, the averaging can be performed over a short time ($\sim$ 1 ms), and it is possible to measure relatively rapid time variations of the average scattered intensity. The time dependence of the average intensity yields information about the kinetics of desorption or adsorption of particles at the interface. The term dynamic scattering refers to the measurement of the autocorrelation of the scattered intensity; this technique is used to study fast processes occurring at the interface.

In what follows, the word particle will be used to describe the source of the scattering. The word will encompass rigid structures such as latex spheres as well as macromolecules such as linear polymer chains. The particle will be viewed as a collection of smaller components each of which scatters light so that the total scattering from a particle will be written as a sum of scattered waves from the components. Such an approach permits the convenient calculation of angular dependence of scattering as well as the description of internal dynamics of linear polymer chains.

II. SCATTERING OF EVANESCENT WAVES

A. Ideal Interface

In the macroscopic theory of electromagnetic waves [3], the evanescent wave (EW) arises from the requirement that the boundary conditions be satisfied at all points on the flat (ideal) interface between two materials of different optical properties that are uniform throughout the materials. The spatial functions in the exponents describing propagation of plane waves in each material are set equal at points on the interface, and this leads to Snell's law. The requirement that the electric field be continuous across the interface leads to Fresnel's equations, which relate the magnitudes of the field on the two sides of the plane boundary. Explicit expressions [3] for the amplitudes of the EW and the internally reflected wave show that both waves depend on the dielectric properties of the material adjacent to the interface. EW exist for any polarization of the incident light.

B. Nonideal Interface—Scattering of Evanescent Waves

Fluctuations in the dielectric properties near the interface lead to scattering of the EW as well as changes in the intensity of the internally reflected wave. Changes in optical absorption can be detected in the internally reflected beam and lead to the well-known technique of attenuated total reflectance spectroscopy (ATR). Changes in the real part of the dielectric function lead to scattering, which is the main topic of this review. Polarization of the incident beam is important. For s polarization (electric field vector perpendicular to the plane defined by the incident and reflected beams or parallel to the interface), there is no electric field component normal to the interface, and the electric field is continuous across the interface. For p polarization (electric field vector parallel to the plane defined by the incident and reflected beams), there is a finite electric field component normal to the interface. In macroscopic electrodynamics this normal component is discontinuous across the interface, and the discontinuity is related to the induced surface charge at the interface. Such discontinuity is unphysical on the molecular scale [4], and the macroscopic formalism may have to be re-examined if it is applied to molecules within a few Å of the interface.

1. Exact Description of Scattering Near an Interface

There is no convenient analytical expression for the scattering of EW from a particle close to the surface. The difficulty is that the EW describes the near field of the reflection process, so that calculations of EW scattering that ignore the reflecting interface are formally inconsistent. Nevertheless, approximate expressions for scattering are obtained that satisfy boundary conditions separately for the reflection process and the scattering process. These results may be valid if the scattering particle is sufficiently far from the interface (where the intensity of the EW is small). To put the problem in perspective, the exact solution was given by Greffet [5] for the case of scattering of the incident s polarized wave from a two-dimensional rod (infinite in the y direction) near an interface (defined by the x–y plane) between two dielectrics. The solution is represented formally by an integral equation and is also applicable to the case of total internal reflection and EW:

$$E(x, z) = E_0(x, z) + \frac{k_0^2}{4\pi} \int [\varepsilon(x', z') - \varepsilon_0]$$
$$\times E(x', z') G(x - x', z, z') \, dx' \, dz' \quad (1)$$

where E_0 is the electric field for an ideal interface and consists of the incident, transmitted, and reflected components, and $G(x - x', z, z')$ is the Greens function for the ideal dielectric interface (*not* free space). The coordinate normal to the interface is z. The integral is evaluated only in the region where $\varepsilon(x, z)$ is different from ε_0. Therefore the complete solution $E(x, z)$ depends on the region where the dielectric function is not equal to ε_0 *and* the interface through the Greens function. At present only numerical solutions of Eq. (1) have been discussed [5].

2. Approximate Description of Scattering

If the particle is sufficiently far from the boundary, then the Greens function in Eq. (1) approaches the free space form, and an approximate analytical solution can be obtained. This is equivalent to the approximation that the calculation of EW scattering can be performed by neglecting the interface. Making the above approximation, Chew et al. [6] has calculated the scattering from a sphere near an interface and showed that the exponential decay of the intensity of the EW wave can introduce new effects in scattering, e.g., polarization mixing by a sphere. Liu et al. [7] has reexamined these calculations and emphasized the resonance in scattering due to particle size.

For scatterers far from the interface, the perturbation solution presented in Berne and Pecora [8, Eq. (3.2.7)] can be used after a modification of the intensity of the incident wave by an exponential function. This treatment yields an analytical result and relates the scattered intensity to spatial fluctuations of the dielectric properties. The probed fluctuations are along the direction defined by the scattering vector $\overleftarrow{q}$, which in turn is defined as the difference of the incident and scattered wave propagation vectors (the wave propagation vector is given by $2\pi e/\lambda$ where e is the unit vector in the direction of propagation of the electromagnetic wave and λ is the wavelength). For scattering in isotropic solutions, a convenient choice is made for the direction of detection, and the fluctuations along the resulting scattering vector $\overleftarrow{q}$ are representative of the sample (due to solution isotropy). The presence of the interface introduces an anisotropy, and choosing $\overleftarrow{q}$ parallel or normal to the surface may probe different fluctuations. EW always propagate parallel to the surface, but the scattered wave can be detected either in the plane of the interface ($\overleftarrow{q}$ parallel to the interface) or in a plane

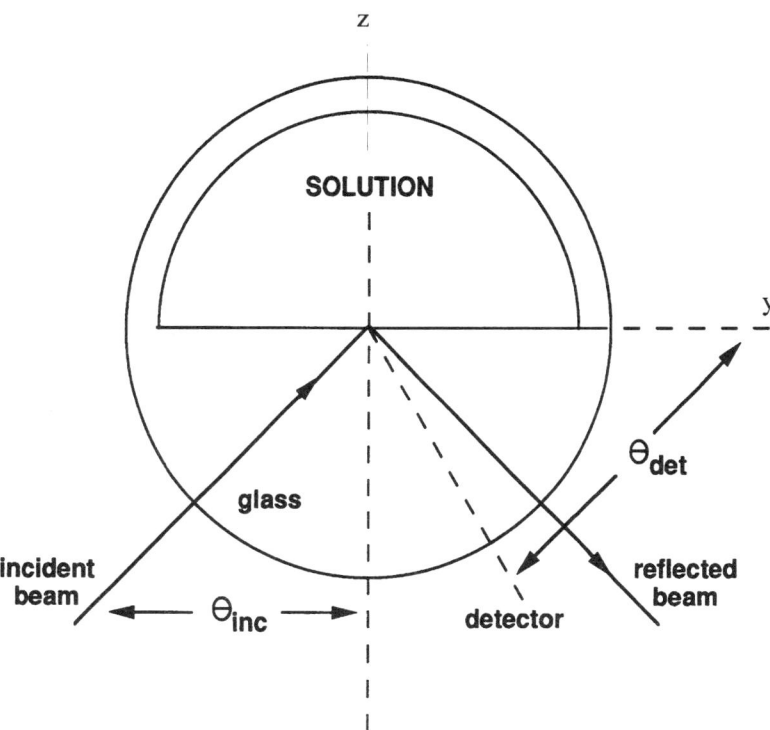

FIG. 1a A schematic of the cell used for the study of the scattering of evanescent waves. The interface between the solution and the glass can support an evanescent wave if the angle of incidence θ_{inc} exceeds a critical value. The detector is located at some angle θ_{det} not equal to the angle of reflection. The detector can view the scattering either from the side of the solution or from the side of the glass, as shown in this figure. Variations of the basic scheme are discussed in the literature, e.g., Ref. 27.

perpendicular to the interface ($\vec{q}$ has components parallel and perpendicular to the interface).

To be specific, consider a diagram of a widely used measurement geometry as shown in Fig. 1a. The scattering cell consists of half of a solid glass cylinder (or a hemisphere) to which is attached a container for the solution. The cut surface of the solid glass cylinder (or hemisphere) forms the interface at which the evanescent waves propagate. A laser beam (s or p polarization) is incident on the glass interface at an angle θ_i relative to the normal to the surface. If the angle of incidence exceeds the critical angle, then an evanescent wave propagates

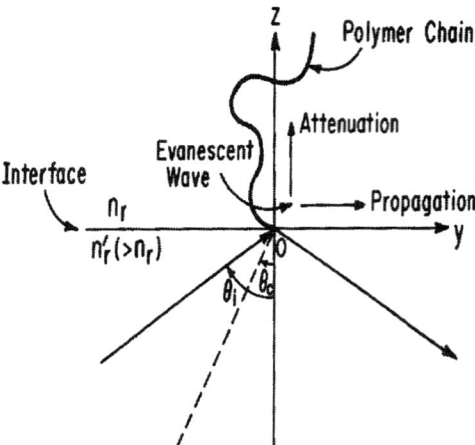

FIG. 1b A magnified view of the spot where the laser beam intercepts the interface in Fig. 1a. The evanescent wave propagates in the y direction with an amplitude that is attenuated in the z direction. A tethered polymer chain scatters the evanescent wave. From the properties of the scattered light it is possible to obtain a measure of the spatial distribution of the polymer material at the interface as well as a measure of the dynamical properties of the polymer chains. (From Ref. 9.)

along the surface (y direction) with a propagation vector k_i and an exponentially decaying amplitude in the direction normal to the surface (z direction). These concepts are summarized in Fig. 1b, which shows a magnified picture of the region where the laser beam intersects the interface. The explicit relations for the properties of the EW wave are

$$E_{ev} = E_0 e^{ik_i y} e^{-\Gamma z}$$

$$k_i = \frac{2\pi n_g}{\lambda_0} \sin(\theta_i) \tag{2}$$

$$\Gamma = \frac{2\pi n_s}{\lambda_0} \left[\frac{n_g^2}{n_s^2} \sin^2(\theta_i) - 1 \right]^{1/2}$$

where E_0 is the intensity of the EV wave, n_s and n_g are the index of refraction of the solution and glass respectively, λ_0 is the free space wavelength, and Γ is a constant that characterizes the spatial decay of the amplitude in the z direction. The evanescent wave serves as the incident beam that is scattered from particles at the interface and in the solution near the interface. The scattering process couples the EW to plane waves propagating in space with a propagation vector magnitude of $k_s = 2\pi n_s/\lambda_0$. In the context of the approximation discussed above, the scattered electric field from a scatterer located at $\overleftarrow{r}$ has the form (for s polarized incident light)

$$E_s = B'\alpha' \exp(i\overleftarrow{q} \cdot \overleftarrow{r} - \Gamma z) \tag{3}$$

where α' is the polarizability of the scatterer, $\overleftarrow{q}$ is defined as the difference between the incident (evanescent) and the scattered wave vectors ($\overleftarrow{q} = \overleftarrow{k}_i - \overleftarrow{k}_s$), z is the distance of the scatterer from the interface, and B' describes the overall propagation of the scattered wave from the scattering volume to the detector (its value is the same for all scatterers in the scattering volume). The factor Γz comes from the spatial dependence of the intensity of the EW wave. The explicit form of B' for various polarizations and scattering geometries can be found in Ref. 9. In the example shown in Fig. 1a, the scattered light is detected in a plane perpendicular to the interface at an angle of θ_d relative to the direction of propagation of the evanescent wave. The components of the scattering vector $\overleftarrow{q}$ are given in this case by

$$q_y = \frac{2\pi n_g}{\lambda_0} \sin(\theta_i) \left[1 - \frac{n_s}{n_g \sin(\theta_i)} \cos(\theta_d) \right] \approx \frac{2\pi n_g}{\lambda_0} \sin(\theta_i)(1 - \cos(\theta_d))$$

$$q_z = -\frac{2\pi n_s}{\lambda_0} \sin(\theta_d) \tag{4}$$

where the detection angle in the glass portion of the cell, θ'_d, is related to the scattering angle in solution close to the interface by $n_g \cos(\theta'_d) = n_s \cos(\theta_d)$. Changes in the angle of detection lead to changes in the magnitude of the components of q, making it possible to examine fluctuations of different length scale. Note that in the geometry described above, fluctuations both parallel and perpendicular to the interface plane are involved in the scattering. In the scattering geometry, where the scattered light is detected in the plane of the interface, only fluctuations parallel to the interface are important.

III. STATIC EW SCATTERING

By "static" is meant that a measurement is made of the average scattered intensity $\langle \overleftarrow{E}^*(t) \cdot \overleftarrow{E}(t) \rangle$, where $\overleftarrow{E}(t)$ is the scattered electric field vector at a point on the detector. The intensity can be recorded as a function of time if the intensity change is sufficiently slow to permit the acquisition of a low-noise signal. The analysis of EW static scattering data can yield the distribution of particles in the interface region as a function of the distance from the interface. Scattering data is somewhat simpler to analyze than fluorescence, since there are minimal complications from distance dependence of quantum yields, extinction coefficients, and spectral distributions, which have to be considered in fluorescence measurements [1].

Assuming that the perturbation theory of scattering applies, the detailed form of the scattered EW has been derived by Gao [9] for both in-plane and out-of-plane detection. Here the results will be summarized schematically to bring out the essential factors in interpretation. The scattered EW intensity from a collection of particles is given by

$$I(\vec{q}) = B\alpha^2 \left\langle \sum_p S(\vec{q}, z_p) \right\rangle \tag{5}$$

where B is a constant involving optical properties of the materials, and the incident light flux, α is the total polarizability of the particle (if the particle is made of n segments each with polarizability α_s, then $\alpha = n\alpha_s$), p is an index running over all particles in the scattering volume, and z_p is the distance of the center of mass of the particle from the interface. The symbol $\langle\ \rangle$ means averaging over all possible arrangements of the centers of mass of the particles in the scattering volume. It is assumed that averaging over all arrangements of the particles will yield approximately the same result as averaging the detected intensity over time. $S(\vec{q}, z_p)$ is the intensity weighted structure factor of a particle whose center of mass is located z_p from the interface:

$$S(\vec{q}, z_p) = \frac{1}{n^2} \left\langle \sum_{i,j} \exp(i\vec{q} \cdot (\vec{r}_i - \vec{r}_j) - \Gamma(z_i + z_j)) \right\rangle \exp(-2\Gamma z_p) \tag{6}$$

$$S(\vec{q}, z_p) \equiv S(\vec{q}) \exp(-2\Gamma z_p)$$

where the second line of Eq. (6) defines the function $S(\vec{q})$. The vectors $\vec{r}_i, \vec{r}_j$ are the center of mass position vectors of the components of the particle. Each particle has n components, and the averaging in Eq. (6) is over all possible positions of the components of the particle in the center of mass frame of the particle. The averaging in Eqs. (5) and (6) can be separated for the case of independent spherical particles. Recall that each particle is considered as a composite of many components, each of which scatters light (e.g., a polymer chain and the monomers comprising it). Equation (6) is obtained by summing scattered waves of the form given by Eq. (3) from all the components of a particle. The magnitude of the incident wave at the location of the components of the particle is noted explicitly. Since the detected intensity depends on the square of the magnitude of the scattered electric field, the calculation of the intensity leads to the double sum with the difference of the spatial vectors and the sum of the z coordinates. The relative differences in the path length of light scattered by the different components of the particle are determined by their physical location; hence the name structure or form factor for Eq. (6). The final expression for the average scattered intensity is

$$I(\vec{q}) = B\alpha^2 \left\langle \sum_p S(\vec{q}) e^{-2\Gamma z_p} \right\rangle = B\alpha^2 \langle N \rangle S(\vec{q}) \langle e^{-2\Gamma z_p} \rangle \quad (7)$$

where $\langle N \rangle$ is the average number of particles in the scattering volume, and the averaging of the second factor is over the positions of a single particle. If there is a particle concentration distribution along the normal to the interface (z direction) and a uniform distribution parallel to the interface, then the averaging over particle positions can be performed by using the particle number concentration to estimate the probability of finding a particle at distance z_p.

$$I(\vec{q}) = B\alpha^2 S(\vec{q}) \int_0^\infty n(z_p) e^{-2z_p \Gamma} \, dz_p \quad (8)$$

where $n(z_p)$ is the number of particles in the scattering volume per unit length normal to the interface.

A. Scattering from a Single Particle

If the scattering is due to a single particle moving in the vicinity of the interface, then $n(z_p) = \delta(z_p - z_p(t))$, where $z_p(t)$ is the trajectory of the particle over time. As the particle moves closer or farther from the interface, the intensity of the scattered light will change. By fixing the angle of observation and repeatedly sampling the intensity over a period of time (sampling time small compared to the characteristic time of intensity change) it is possible to obtain a distribution of intensities as shown in Fig. 2a [10]. Assuming that for sufficiently large measurement times (many intensity samples) the motion of the particle will take it to all possible positions, Eq. (8) can be used to relate the distribution of measured intensities to the distribution of particle positions normal to the interface. The distribution of positions in turn can be related to the characteristics of the interaction of the particle with the interface [10]. Figure 2b shows the interaction potential between the particle and the interface that yields a distribution of particle positions consistent with the intensity histogram shown in Fig. 2a. In a recent application of this technique [11], some of the variation of the sampled intensity was interpreted as originating from vibrational motion of particles (diameter $\sim 1\ \mu$m) in the potential well at the interface.

B. Scattering from a Collection of Particles

If there are many particles in the scattering volume, then Eq. (8) can be used to analyze the dependence of the measured scattering intensity on the penetration depth Γ and obtain information on the particle distribution near the interface.

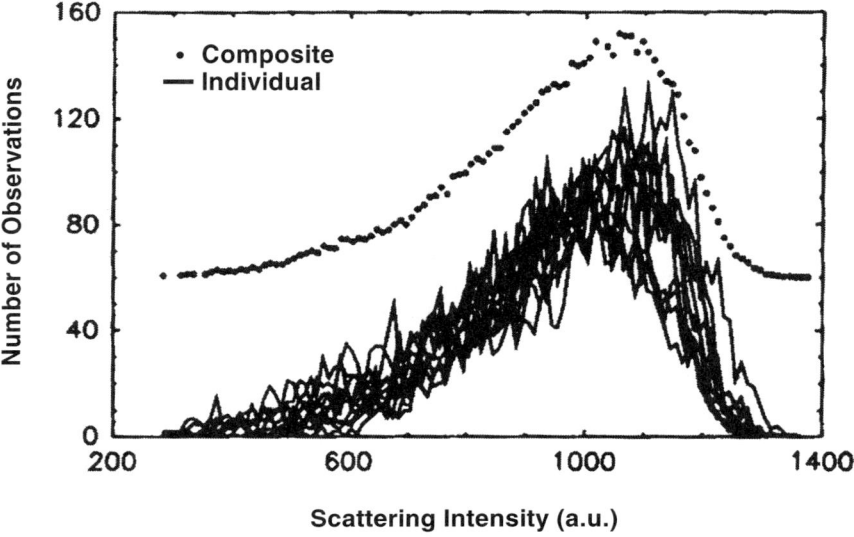

FIG. 2a Eleven 2 min histograms of scattering intensity sampled at 30 ms intervals for a 10 μm sphere (E13 in Ref. 10) in 1 mM NaCl. The points represent a composite histogram. This composite histogram is used in determining the potential energy profile for the 10 μm sphere as shown in Fig. 2b. (From Ref. 10.)

One procedure is to assume a parametrized form of the particle distribution function $n(z)$ and compare the predictions of Eq. (8) to the measured scattered intensity to estimate the values of the parameters. This procedure was used to characterize the interaction of the interface with particles in a flowing stream above an interface [12]. There was no adsorption of particles on the surface, and the particle distribution function was obtained from a solution of a mass transport equation with a term describing the interaction with the interface. The analysis yielded estimates of the parameters in the interaction potential [12].

An alternate analysis of the average scattered intensity, which does not assume any form for the particle distribution function, is based on the fact that Eq. (8) is formally a Laplace transform between the linear number density $n(z)$ and the average intensities measured as a function of the penetration depth Γ. By measuring the scattered intensity as a function of penetration depth it is possible to extract particle distribution by inverting Eq. (8). In many cases the inversion of Eq. (8) may be difficult. The analysis of Caucheteux et al. [13] (which was originally developed for small fluorescent molecules and ignores changes in fluorescent properties with distance to interface) shows that the exponential in Eq. (8) can be expanded and the integration carried out on each term in the

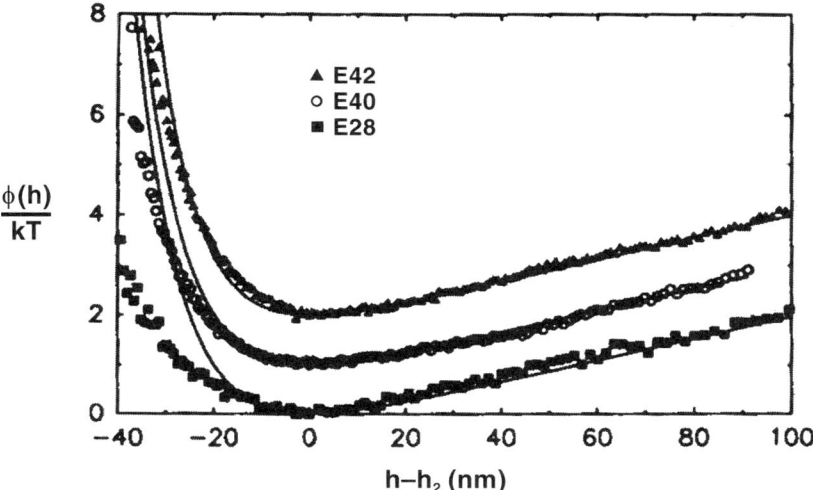

FIG. 2b Potential energy profiles for four 10 µm spheres in 1 mM NaCl. (From Ref. 10.)

expansion. The result is an expansion of $I(q)$ in terms of the moments of the particle distribution. Caucheteux's analysis was adapted by Polverari and van de Ven [14] to obtain polymer layer thickness and surface excess of adsorbed PEO on glass interface. Figure 3 shows the measured intensity as a function of penetration depth for three different polymers. The polymers were adsorbed on the interface from a jet impinging on the interface. The measurements shown in Fig. 3 were performed after the adsorbed layer stabilized and the interface was exposed to a slow flow of clean saline solution. The analysis of the measurements in Fig. 3 yields the surface excess and the "average" layer thickness. By fixing the penetration depth and measuring the change in scattering intensity with time, it is possible to obtain an estimate of the kinetics of particle adsorption (or desorption) at the interface [14].

Concentration profiles of chromophores has been obtained in an attenuated total reflection (ATR) apparatus operating between 200 nm and 300 nm with a variable incidence angle [15]. The scattering manifests as an attenuation of the internally reflected light. The scattering can be identified by the characteristic dependence on the fourth power of the wavelength. Accumulating spectra at different angles of incidence (different penetration depths) allows the measurement of the first two moments of the chromophore concentration profiles. In principle, the technique also yields the first two moments of the concentration profile of scatterers.

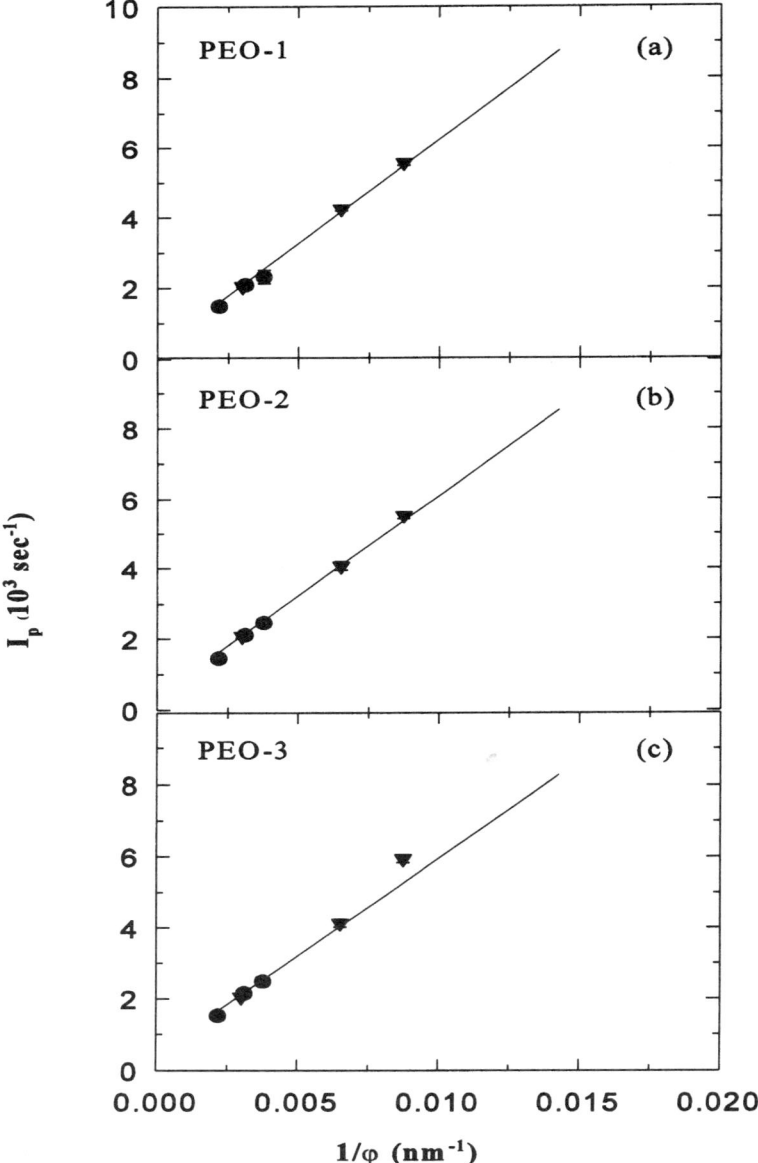

FIG. 3 Intensity of scattered light as a function of the inverse penetration depth (φ in this figure is the same as Γ in the text) for poly(ethylene oxide) molecules adsorbed at fused quartz (●) and ZnS (▼) interfaces. The polymers were adsorbed from an aqueous solution in a stagnation point flow. (From Ref. 14.)

Measurements were carried out of internally reflected light at an interface between a semitransparent electrode and a solution containing charged particles (diameter ~ 0.47 μm) [16]. It was found that the amount of internally reflected light depended on the potential applied to the semitransparent electrode. The dependence was interpreted as a variation of the particle concentration near the interface due to net electrophoretic motion of particles to or from the electrified interface. As the particles move, the particle concentration at the interface changes, leading to changes in the net EV scattering, which is manifested as a change in the amount of internally reflected light.

C. Scattering from Polymer Attached to Interface

For a system of polymer chains attached to an interface, the polymer concentration may be finite over a small portion of the penetration of EW. In this case varying penetration depth will not yield an accurate estimate of the polymer concentration profile. In addition, the distribution of the polymers on the interface may be inhomogeneous. Measurement of the angular distribution (at a fixed penetration depth) of the scattered light in a plane perpendicular to the interface yields information on the structure factor and hence on the vertical extent of the layer. Measurement of the angular distribution in the plane of the interface yields information on possible aggregation of the polymer chains.

Measurements have been performed on copolymer chains at water–air interface [17]. The scattering was observed in the plane of the interface. The copolymers consisted of polystyrene (PS) segments with a molecular mass of 880 KD and a polymethylmethacrylate (PMMA) segment with a molecular mass of 290 KD. They were deposited on the water–air interface using a Langmuir trough. The surface coverage of the copolymer layer could be controlled and the corresponding surface pressure measured. Arrays of fiber optics collectors positioned at different angles measured the angular distribution of the scattered light. Figure 4 shows the dependence of scattered light on the in-plane q for several surface pressures. The surface pressure increases from frame a in Fig. 4 to frame h. The solid lines in the frames in Fig. 4 are fits to the expected response of a collection of interacting disks. The response was calculated using a generalization of Eq. (7) to include interactions between particles (disks). In addition to the structure factor of a single disk, $S(q)$ in Eq. (7), there is now an additional term representing the structure factor of the collection of interacting disks. The good fit between the measured and the expected response is strong support for the model.

Another measurement on a system of copolymers consisting of poly(ethylene-oxide-b-styrene) (PEO-PS) chains anchored to a glass surface in toluene solution was reported by Fytas et al. [18]. The picture of the layer is that the PEO part attaches to the glass while the PS part dangles in the solution. The thickness of

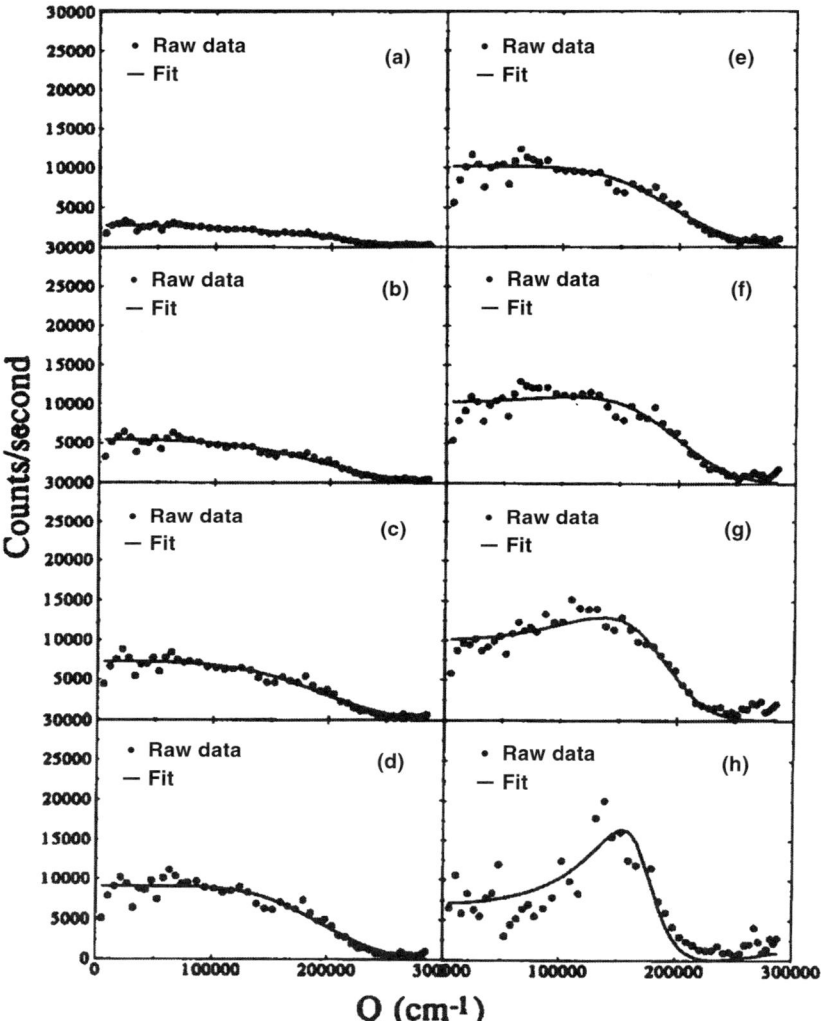

FIG. 4 The average scattered intensity as a function of the magnitude of the scattering vector q (Q in Fig. 4) in the plane of the air–water interface containing polymers. The different parts of the figure show the scattered intensity from polymers under progressively increasing compression from a to h (the interface was in a Langmuir trough). The authors assumed that the polymers on the surface aggregate into "furry disks" and wrote the structure factor [Eq. (5) in this text] as a product of the structure factor of the particles in a single "furry disk" times the structure factor of a "fluid" of hard "furry disks." The dependence on q with compression was in excellent agreement with this model. (From Ref. 17.)

the layer is mostly due to the PS portion. The measurements of angular dependence were performed in a plane normal to the interface plane. The copolymer layer thickness studied ranged from 45 nm to 130 nm. Here the angular dependence was explained by the form factor associated with variation of the copolymer density normal to the interface. The polymer chains were not free to move along the interface. The angular distribution data indicates that shorter scale fluctuations are more probable.

IV. DYNAMIC EW SCATTERING

Evanescent wave dynamic light scattering (EWDLS) can be used to study the rapid motion of submicron particles near and on interfaces. Two situations have been studied: restricted diffusion of spherical particles moving near an interface, and composition fluctuations of polymers attached to the interface.

In EWDLS, the measured quantity is the autocorrelation function of the scattered intensity, $I_2(\tau) = \langle I(\tau), I(0) \rangle = \langle |E_s(\tau)|^2 |E_s(0)|^2 \rangle$, which is related to the conditional probability that the scattered intensity is $I(\tau)$ at time τ given that the scattered intensity was $I(0)$ at an earlier time 0. In the case of scattering in solutions, the physical origin of the scattered intensity variation is the motion of scatterers, which changes the relative phase of the scattered light from the different particles at the detector. The phase differences lead to more or less constructive interference of the scattered light from the different particles, which results in changes of total detected intensity. In EWDLS there is an additional source of intensity fluctuation due to the sharp drop in EW intensity. The scatterers can move normal to the interface and enter regions of different intensity. In practice, autocorrelators are used to perform the required sampling of the intensity, forming products of the samples separated by some time interval, sorting the products into time difference bins, and averaging. Uncorrelated signal is assumed to be equal to the output of the autocorrelator at large time differences. For the purpose of comparing to model calculations it is best to convert the measured intensity autocorrelation function $I_2(\tau)$ into an electric field autocorrelation function $I_1(\tau) = \langle E_s(\tau) E_s(0) \rangle$ using the result $I_2(\tau) = |I_1(0)|^2 + |I_1(\tau)|^2$ [8, p. 40], which is a consequence of an assumption that the fluctuations in the electric field magnitude possess a gaussian distribution. The above relation is true for an ideal detection system where all of the coherence in the scattered light is preserved on detection. In actual measurements, even with pinhole apertures, there is still a large distribution of possible paths from the scattering volume to the photomultiplier (PM) surface. Hence two photons originating from the same point in the scattering volume may have an arbitrary phase difference introduced due to different paths taken to the detector. This path length distribution has been summarized by the concept of coherence area on the PM surface. Roughly, the coherence area is the area on the PM (detector) such that two photons incident on this area will maintain their initial coherence; the random phase shift due to path differences is small. The number of coherence areas on the PM surface depends on the experimental design. The net effect of the path distribution is to modify the relation between $I_2(\tau)$ and $I_1(\tau)$ to [8, p. 47]

$$I_2(\tau) = \langle I \rangle^2 \left[1 + f(A) \frac{|I_1(\tau)|^2}{|I_1(0)|^2} \right] \qquad (9)$$

where $\langle I \rangle$ is the average intensity and $f(A)$ depends inversely on the number of coherence areas on the PM surface. The larger the value of $f(A)$, the greater the sensitivity to fluctuations. The last equation is the operational equation for connecting the measured intensity autocorrelation functions $I_2(\tau)$ to the electric field autocorrelation function $I_1(\tau)$, which is used to compare to model predictions.

The scattered electric field (*not intensity*) time autocorrelation function $I_1(\tau) = \langle E_s(0)^* E_s(\tau) \rangle$ can be written for the case of noninteracting particles as (modified Eq. [8.6.1] and Eq. [8.6.2] in Ref. 8):

$$I_1(\tau) = B\alpha^2 \left\langle \sum_p S(\vec{q}, \vec{R}_p, \tau) \right\rangle$$

$$S(\vec{q}, \vec{R}_p, \tau) = \frac{1}{n^2} \left\langle \sum_{i,j}^n \exp(i\vec{q} \cdot (\vec{r}_i(\tau) - \vec{r}_j(0))) \exp(-\Gamma(z_i(\tau) - z_j(0))) \right\rangle \qquad (10)$$

$$\cdot \exp(i\vec{q} \cdot (\vec{R}_p(\tau) - \vec{R}_p(0))) \exp(-\Gamma(Z_p(\tau) - Z_p(0)))$$

Here B is an optical constant, α is the total polarizability of the particle, and n is the number of components in each particle. The indexes i and j refer to components of the same particle. If the assumption of independent particles was not made, then the indexes could refer to components of any two particles, and the autocorrelation expression could not be written as a simple sum of contributions from individual particles. The spatial vector $\vec{r}(\tau)$ refers to the center of mass of the particle, $\vec{R}(\tau)$. In the case of a *nonspherical* particle (arbitrary shape), Eq. (10) would describe the coupled motion of the center of mass and the relative arrangement of the components of the particle. For *spherical particles*, translational and rotational motion are uncoupled and we have a simplified expression for the electric field time correlation function:

$$I_1(\tau) = B\alpha^2 \langle N \rangle S(\vec{q}) \langle e^{i\vec{q} \cdot (\vec{R}_p(\tau) - \vec{R}_p(0))} e^{-\Gamma(Z_p(\tau) + Z_p(0))} \rangle \qquad (11)$$

where $S(\vec{q})$ is the intensity weighted static form factor of the particle and where it has been assumed that the particles are independent with an average number $\langle N \rangle$ in the scattering volume. The expression for $S(\vec{q})$ can be converted into an integral and evaluated for the case of a uniform spherical particle. The net effect of $S(\vec{q})$ is to introduce a variation of the scattering intensity with the angle of detection; it contributes no time variation to the autocorrelation function. Assuming the particles are rigid bodies, the entire temporal dependence comes from the center of mass motion of the particles. Combining the form factor $S(\vec{q})$, $B\langle N\rangle$, and α^2 into a constant I_{so}, the final form of the electric field time autocorrelation function becomes

$$I_1(\tau) = I_{so} \iint e^{i\vec{q}\cdot(\vec{R}-\vec{R}_0)} e^{-\Gamma(Z+Z_0)} G(\vec{R},\tau;\vec{R}_0) \, d\vec{R} \, d\vec{R}_0 \tag{12}$$

where $G(\vec{R},\tau;\vec{R}_0)$ is the joint probability density that a particle whose center of mass is at $\vec{R}_0$ at time 0 will be located at $\vec{R}$ at time τ. We neglect the subscript p in Eq. (12). The function $G(\vec{R},\tau;\vec{R}_0)$ was introduced in order to evaluate the ensemble average in Eq. (11). All possible locations $\vec{R}_0$ and $\vec{R}$ are sampled. Eq. (12) can be used to describe the motion of small spheres near an interface. Several calculations [19] have been performed of the structure factor for attached and unattached polymer chains; needless to say the procedure is much more involved than that described above for spherical particles.

A. Restricted Diffusion Near Interfaces

The presence of the interface restricts the diffusive motion to a half space. More important is the presence of particle–surface interactions, which can modify the transport properties.

The simplest case to analyze is diffusion in half space without interactions. Using $G(\vec{R},\tau;\vec{R}_0)$ appropriate for diffusion in half space with reflecting boundary conditions [20], the expected autocorrelation response is summarized as

$$I_1(\tau) = I_{so} e^{-q_y^2 D\tau} I_{zz_0} \equiv I_{so} F_s(\vec{q},\tau)$$

$$I_{zz_0} = \frac{1}{\sqrt{4\pi Dt}} \iint_0^{L_{z_0}} \cos(q_z(z-z_0)) e^{-\Gamma(z+z_0)} \tag{13}$$

$$\times \left[e^{-\frac{(z-z_0)^2}{4D\tau}} + e^{-\frac{(z+z_0)^2}{4D\tau}} \right] dz \, dz_0$$

where the upper equation defines the function $F_s(\vec{q},\tau)$, and D is the diffusion coefficient of the particle. The length L_{z_0} can be set to several decay lengths of the evanescent wave. Equation (13) shows that the electric field autocorrelation function $I_1(\tau)$ is a product of a term depending on q_y and a term depending on q_z. The first term describes the decay of concentration fluctuations parallel to the interface, and its form is the same as that obtained for unbounded solutions. The second term involves the penetration depth Γ and describes the decay of fluctuations normal to the interface; its form is more involved, since diffusive motion leads to changes in optical path length as well as changes in illumination. The integrated form of the above expression has been presented by Lan [21].

The expressions that can be used in the analysis of the measured autocorrelation response $C(\tau)$ are given below for the case of heterodyne and homodyne detection. Homodyne detection applies to the case where the interface does not contribute to the scattering. Heterodyne detection applies to cases where the interface itself contributes significantly to the scattering (perhaps due to roughness).

$$\frac{C(\tau) - B}{B} = \frac{2I_{so}}{I_{lo}}(\text{Re } F_s(\vec{q}, \tau) + b_s(\tau)) \quad \text{heterodyne}$$
$$\frac{C(\tau) - B}{B} = f\frac{(F_s(\vec{q}, \tau) + b_s(\tau))^2}{(F_s(\vec{q}, 0) + b_s(0))^2} \quad \text{homodyne}$$
(14)

Re F stands for the real part of the function F. The function $b_s(\tau)$ models a slowly varying background, which is usually present in all of the measurements. The constant background term B is measured by the autocorrelator using special time bins with extra delay. I_{lo} is the intensity of the local oscillator (may represent scattering due to the interface itself); the term $2I_{so}/I_{lo}$ indicates the relative amount of particle-scattered light and reference scattered photons and should not exceed 0.1 for heterodyne detection. The quantity f is an instrumental constant, a value around 0.5 indicating a reasonably optimized system for homodyne detection.

Lan et al. [21] published the first measurements of EWDLS. They used latex particles of diameter 90 nm and detected the scattering in a plane normal to the interface using heterodyne detection. Largest sensitivity to the normal fluctuations was obtained at small scattering angles where the decay of the fluctuations parallel to the surface is slow due to the small values of q_y. The measured autocorrelation functions were consistent with freely translating particles in the presence of a reflecting boundary. At the smallest penetration depth measured (400 nm), the measured diffusion coefficient was 8% smaller than the value in unbounded solution.

Gaigalas et al. [22] reported EWDLS for liposomes (diameter ~ 60 nm) at a water–glass interface. Large differences were observed for negative and positive liposomes. Figure 5 shows the autocorrelation function of evanescent wave scattering from negative liposomes. The incident angle θ_i was set to 66°, which gave a penetration length $1/\Gamma$ of approximately 200 nm. The analysis of the data in Fig. 5 was carried out assuming heterodyne detection. A general multiple component analysis of diffusive motion near an interface was not attempted; the author is not aware of any theoretical framework for such an analysis. There was minimal measured dependence of the effective diffusion coefficient on the

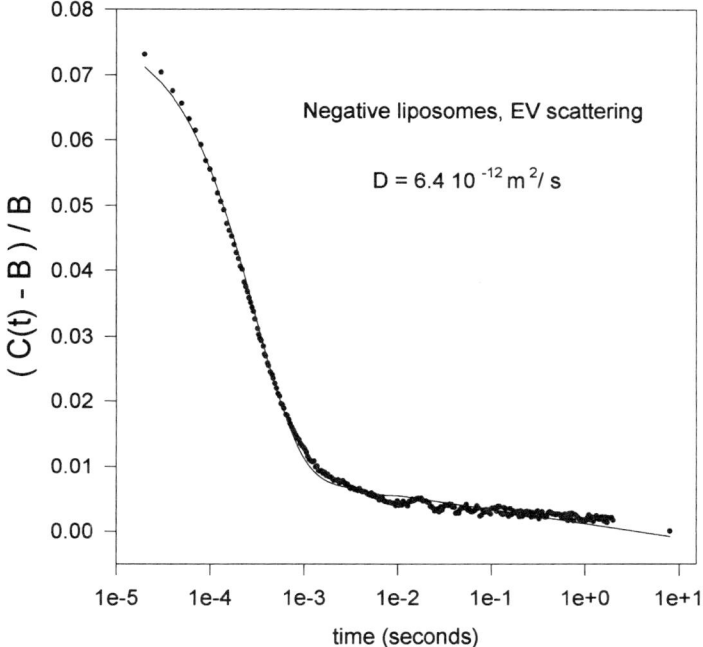

FIG. 5 Measured autocorrelation function for scattered EW from negatively charged liposomes moving in aqueous solution near a negatively charged glass interface. Comparison of the measured diffusion coefficient with the value in bulk solutions suggests that the interface leads to significant hindrance of the diffusive motion. (From Ref. 22.)

magnitude of the scattering vector $\vec{q}$. The value of D was smaller than the value measured for unbounded solution. The difference was rationalized in terms of hydrodynamic interaction with the interface. Introduction of positive liposomes into a clean cell resulted in an autocorrelation spectrum that changed in a matter of minutes. In this case, the correspondence between the model and the data was very poor. Most likely the diffusive part of the response is a minor component of the total response, and other sources of dynamic photon scattering dominate; for example, some form of association between the positive liposomes and the surface.

Marcus et al. [23] used EWDLS to study Brownian motion of dilute solutions of latex particles (diameter $\sim 1~\mu\mathrm{m}$) between two glass surfaces separated by $\sim 3~\mu\mathrm{m}$. The light was detected in a plane perpendicular to the interface, but due to the two-dimensional nature of the cell only in-plane fluctuations were expected to be important. Typical measurements are shown in Fig. 6, where the different

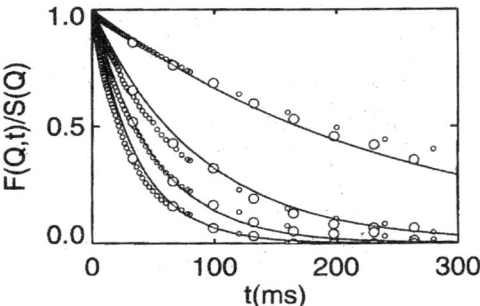

FIG. 6 The small circles show the normalized intermediate scattering function for 1 μm spheres moving between two closely spaced glass surfaces. The term intermediate scattering function is the same as $I_1(\tau)/I_1(0)$ in Eq. (9) in this text. The large spheres give the same function calculated from trajectories recorded by a video camera. The different curves correspond to different values of the magnitude of the scattering vector q. The excellent agreement between measurements taken with two different techniques suggests that the approximations made in the description of evanescent wave scattering are reasonable. (From Ref. 3.)

curves correspond to different values of the in-plane component of the scattering vector $\vec{q}$. The circles are calculated from trajectories observed with video imaging of the motion of single particles. The data was fitted to $\exp(-q^2 \tilde{D}_S(\tau))$, where $\tilde{D}_S(\tau)$ is the time-dependent self diffusion coefficient. Determination of the form of this time dependence, which is a consequence of hydrodynamic interaction between the particles, was the objective of the measurement. Care was taken to minimize persistent hydrodynamic interaction with the chamber wall (vortex propagation) by coating the wall with an alkane brush. Faucheux and Libchaber [24] used digital imaging technique to study confined Brownian motion of single silica and latex spheres (diameter ~ 2 μm) confined between two plates. They concluded that the effective diffusion coefficient is a monotonic function of the parameter $\gamma = (h - r)/r$, where r is the particle diameter and h is the average vertical position of the particle between the horizontal plates. The source of this dependence was the hydrodynamic interaction between the particle and the nearby walls.

B. Dynamics of Polymers Attached to Interfaces

Interplay of attractive surface contacts and loss of configurational entropy [25] suggest that the properties of polymers at surfaces can be different from those in bulk. The analysis of the intensity autocorrelation requires evaluation of Eq. (10) without the assumptions that particles are independent and that averaging over

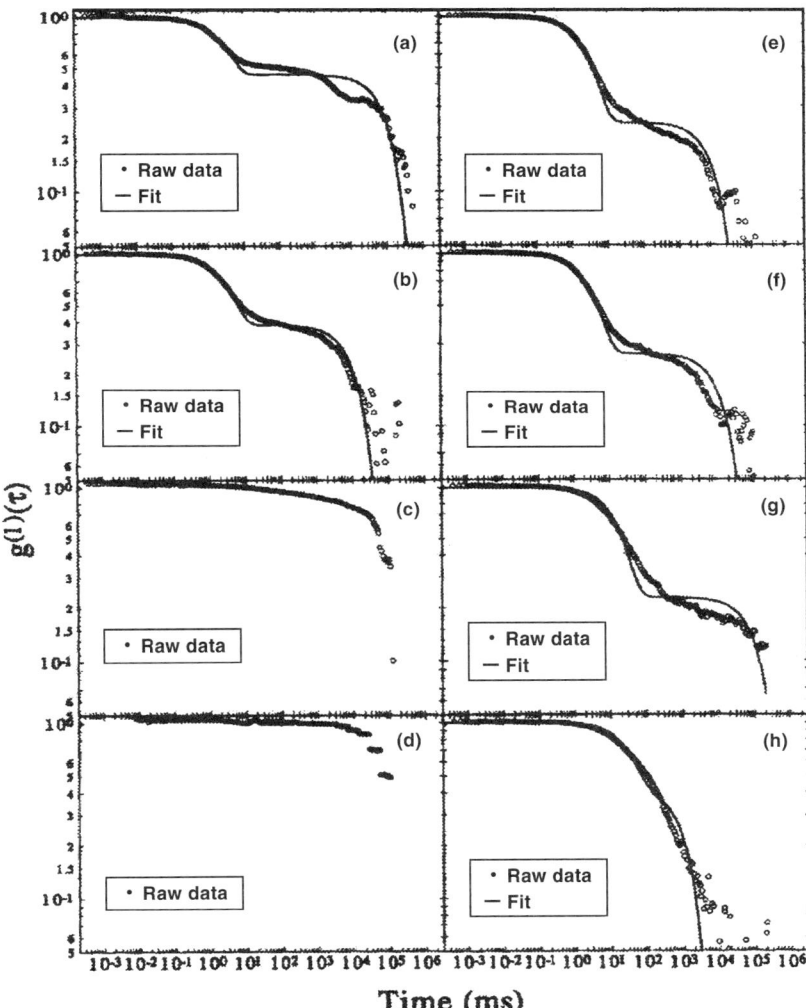

FIG. 7 The electric field autocorrelation function for scattering from polymers in the plane of the air–water interface located in a Langmuir trough. Here $g^{(1)}(\tau)$ has the same meaning as $I_1(\tau)$ used in this text. The different parts of the figure show the autocorrelation function for polymers under progressively increasing compression during the first compression cycle (a to d), and the second compression cycle (e to g). The presence of two decay times at lower compressions supports the interpretation (see Fig. 4) that the polymer on the surface consists of a fluid of interacting "furry disks." However, deviations from the model are observed at larger compressions (parts c, d, and g). (From Ref. 17.)

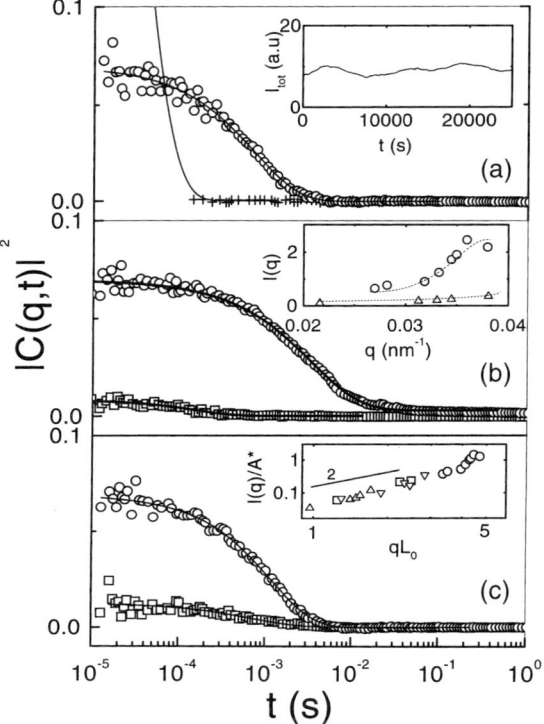

FIG. 8 (a) The square of the electric field autocorrelation function for tethered polystyrene (PS) layers on a glass surface. $C(q, t)$ in the figure corresponds to $I_1(t)$ in this text. The measurements were taken over a 7 h period. The insert shows that the average counting rate was stable over this time period. The solid line and the (+) curve are control measurements of the autocorrelation function from PS in bulk solution and the glass interface without PS, respectively. (b) Two measurements at different values of the magnitude of the scattering vector q. (c) The dependence of the measured electric field autocorrelation function on the molecular weight of PS. (From Ref. 18.)

center of mass of the particle can be uncoupled from averaging over internal coordinates. The whole undertaking is very complex. Usually the analysis of the intensity autocorrelation function is performed using one or two exponential functions. The resulting decay constants can then be rationalized in terms of the underlying dynamics of the attached polymer chains. The dynamic measurements are complementary to the static measurements discussed above.

Lin et al. [17] studied the dynamics of copolymers adsorbed on an air–water interface. These measurements complemented the static measurements described above and in Fig. 4. The extent of the polymer films perpendicular to the surface is small compared to penetration distance and wavelength so that EWDLS is most sensitive to variation of composition in the plane of the interface. Figure 7 shows the measured normalized autocorrelation $I_i(t)$ for different surface pressures. Frames a–d were taken during the first compression of the monolayer, and frames e–h were taken during the second compression. The difference between the two sets of measurements is an indication of structural changes induced by compression cycling. The frames e–g can be compared to the data in Fig. 4. The solid lines in the three frames are fits to a sum of two exponential functions, each with a characteristic decay time. The fast decay constant has a characteristic q^2 dependence and was associated with diffusive motion of the disks. The slow decay constant ($\sim$ several seconds) was ascribed to the dynamics of the associations of disks.

Fytas et al. [18] observed polymer dynamics associated with height fluctuations of PEO-PS copolymers attached to glass in toluene. Some of the measured normalized autocorrelation functions are shown in Fig. 8. The data was well represented by a single exponential function with a decay constant that exhibited minimal dependence on q. There is no calculation of the structure factor for such a system, but it is possible to rationalize the dependence of the decay constant on the 5/3 power of the chain density and the cube of the number of monomers in the PS segment.

Park et al. [26] used EWDS to measure the dynamics of nematic crystals deposited on a rubbed nylon 6/6 coated glass surface. The authors looked at scattering in a plane perpendicular to the interface and detected a polarization that was orthogonal to that of the incident wave (depolarized scattering). Both polarized (detecting the same polarization as the incident light) and depolarized dynamic light scattering were used to study the dynamics of poly(p-phenylene) adsorbed from a toluene solution on a glass surface [27].

V. OTHER RELATED TECHNIQUES

Sainov [28] discussed the possibility of EW forced Rayleigh scattering. The technique envisages the creation of an interference pattern on a surface using two coherent evanescent waves. If the material in which the EW propagates is absorbing, then a thermal grating can be produced. This thermal grating can be examined by another light beam. The decay of such a grating provides information on thermal properties of the material. In another application, scattering from the two waves can provide a measure of diffusion of the scatterers in the plane of the interface. Measurements using this technique have not appeared as of the preparation of this article.

Ramsden et al. [29] have developed a technique called optical waveguide light mode spectroscopy for the study of adsorbed biomolecules. The thin sliver of optical material acts as a waveguide passing several modes. The propagation constants of the modes depend on the index of refraction of the material in contact with the waveguide wall. This material is sampled via the EW of the propagating modes. The presence of scatterers near the waveguide boundary would be sensed as a change in the propagation properties.

VI. CONCLUSION

The interpretation of EW scattering measurements is based on an approximate theory of scattering that has an internal inconsistency. Nevertheless, the theory appears to be adequate, since measurements on model systems (e.g., latex particles) give results that are consistent with expectations. The evanescent wave scattering technique can be used fruitfully to obtain the distribution of spherical particles near optically clear interfaces. It is also relatively easy to measure transport properties of these particles near and at surfaces. Ultimately, the measurements yield a characterization of the particle interactions with the interface. Measurements of polymers at interfaces are much more difficult. The scattering signals are small, and great care must be taken to insure sample cleanliness. Data acquisition for a single scattering geometry may take hours, although multiple angle detectors speed the data acquisition for static measurements. In addition, the polymer chain attached to the interface should also be characterized by independent techniques in order to assist the interpretation of the EW data. A self-consistent description of static *and* dynamic scattering from polymer chains at interfaces is a major undertaking. At present only the first steps have been taken.

EW scattering can also be detected using internally reflected light, since any scattering of the EW results in the reduction of the intensity of the internally reflected light. In this case, the measured EW scattered intensity is the scattered intensity integrated over a large range of scattering angles.

REFERENCES

1. V. Hlady, D. R. Reinecke, and J. D. Andrade. Journal of Colloid and Interface Science *111*:555 (1986).
2. Marek W. Urban, *Attenuated Total Reflectance Spectroscopy of Polymers: Theory and Practice*, American Chemical Society, Washington, D.C., 1996.
3. John David Jackson, *Classical Electrodynamics*, 2d ed., John Wiley, New York, 1975.
4. Peter J. Feibelman. Progress in Surface Science *12*:287 (1982).
5. Jean-Jacques Greffet. Optics Communications *72*:274 (1989).

6. Herman Chew, Dau-Sing Wang, and Milton Kerker. Applied Optics *18*:2679 (1979).
7. C. Liu, T. Kaiser, S. Lange, and G. Schweiger. Optics Communications *117*:521 (1996).
8. Bruce J. Berne and Robert Pecora, *Dynamic Light Scattering*, John Wiley, New York, 1976.
9. Jun Gao and Stuart A. Rice. J. Chem. Phys. *90*:3469 (1989).
10. Dennis C. Prieve and Nasser A. Frej. Langmuir *6*:396 (1990). R. Burchett and Dennis C. Prieve. Biophysical Journal *69*:66 (1995).
11. S. Tanimoto, H. Matsuoka, and H. Yamaoka. Colloid Polym. Sci. *273*:1201 (1995).
12. Marc Polverari and Theo G. M. van de Ven. Langmuir *11*:1870 (1995).
13. I. Caucheteux, H. Hervet, R. Jerome, and F. Rondelez. J. Chem. Soc. Faraday Trans. *86*:1369 (1990).
14. Marco Polverari and Theo G. M. van de Ven. Journal of Colloid and Interface Science *73*:343 (1995).
15. Mathias Trau, Franz Grieser, Thomas W. Healy, and Lee R. White. J. Chem. Soc. Faraday Trans. *90*:1251 (1994).
16. J. T. Remillard, J. M. Ginder, and W. H. Weber. Applied Optics *34*:3777 (1995).
17. Binhua Lin, Stuart A. Rice, and D. A. Weitz. J. Chem. Phys. *99*:8308 (1993).
18. George Fytas, Spiros H. Anastasiadis, Rachid Seghrouchni, Dimitris Vlassopoulos, Junbai Li, Bradford J. Factor, Wolfgang Theobald, and Chris Toprakcioglu. Science *274*:2041 (1996).
19. Jun Gao, Karl F. Freed, and Stuart A. Rice. J. Chem. Phys. *93*:2785 (1990). Epaminondas Rosa, Jr., and John S. Dahler. J. Chem. Phys. *97*:1 (1992).
20. S. Chandrasekhar. Rev. Mod Phys. *15*:1 (1943).
21. K. H. Lan, N. Ostrowsky, and D. Sornette. Phys. Rev. Lett. *57*:17 (1986).
22. A. K. Gaigalas, J. B. Hubbard, A. L. Plant, and V. Reipa. Journal of Colloid and Interface Science *175*:181 (1995).
23. Andrew H. Marcus, Binhua Lin, and Stuart A. Rice. Physical Review E *53*:1765 (1996).
24. Luc P. Faucheux and Albert J. Libchaber. Phys. Rev. E *49*:5158 (1994).
25. Jack F. Douglas, Adolfo M. Nemirovsky, and Karl F. Freed. Macromolecules *19*:2041 (1986).
26. C. S. Park, M. Copic, R. Mahmood, and N. A. Clark. Liquid Crystals *16*:135 (1994).
27. B. Loppinet, G. Petekidis, and G. Fytas. Langmuir *14*:4958 (1998).
28. S. Sainov. J. Chem. Phys. *104*:6901 (1996).
29. J. J. Ramsden. Phys. Rev. Lett. *71*:295 (1993). R. Kurrat, M. Textor, J. J. Ramsden, P. Böni, and N. D. Spencer. Rev. Sci. Instrum. *68*:2172 (1997).

7
Characterizing Colloidal Materials Using Dynamic Light Scattering

LEO H. HANUS and HARRY J. PLOEHN Department of Chemical Engineering, University of South Carolina, Columbia, South Carolina

I.	Introduction	199
	A. Colloidal characterization via dynamic light scattering	200
	B. Light scattering theory	201
	C. Light scattering instrumentation	208
II.	Dynamic Light Scattering	208
	A. Theory	209
	B. Data analysis methods	216
	C. Applications	226
III.	Electrophoretic Light Scattering	228
	A. Theory	229
	B. Experimental methods	237
	C. Applications	241
	References	242

I. INTRODUCTION

This chapter reviews the theoretical and experimental foundations of light scattering as applied in the characterization of colloidal materials. First we introduce the basic principles of light scattering. Then we provide a detailed review of dynamic light scattering, emphasizing the theoretical considerations necessary for reliable interpretation of experimental data and accurate determination of average particle size and size distribution. Finally we review electrophoretic light scattering—the application of dynamic light scattering to measure the collective motion of colloidal particles under the influence of an applied electric field. From the measured electrophoretic velocity, one can calculate the particle electrophoretic mobility, providing the starting point for investigation of particle electrostatic and electrokinetic properties.

A. Colloidal Characterization via Dynamic Light Scattering

1. Particle Size

Colloids are multiphase mixtures of matter wherein one of the phases is continuous and one or more phases are dispersed with a length scale in the nanometer to micron size range. Each phase may be solid, liquid, or gas, resulting in different colloidal classifications [1,2] (aerosols, emulsions, foams, and suspensions). Knowledge of the length scales associated with the different phases (or regions) is crucial for understanding and controlling the structure and properties of colloidal materials.

For the different classes of colloidal materials, the characteristic length scale is generally termed particle size. Many techniques [3,4,5,6] have been developed for characterizing particle size and shape, including optical and electron microscopy, sedimentation, and various kinds of chromatography. Light scattering techniques have been utilized extensively because of their speed, versatility, and ease of use relative to the other methods.

2. Particle Electrostatics

When the relevant length scales of a material fall within the colloidal domain, colloidal forces can strongly influence the material's bulk properties. The ability to control and tailor material properties to meet specific needs requires knowledge of the underlying interfacial chemistry and structure. For example, attractive colloidal forces (van der Waals, electrostatic, polymer mediated) can promote particle aggregation in suspensions and drop coalescence in emulsions. Repulsive forces (electrostatic, polymer mediated) can be interposed to stabilize particles against aggregation. Because these colloidal forces perturb the size, shape, and spatial distribution of the dispersed phase, light scattering techniques can be employed to probe the interfacial features of colloidal phases.

The characterization and control of electrostatic forces are of particular interest. Electrostatic forces depend on the electric charge and potential at the particle surfaces. When subjected to a uniform, unidirectional electric field **E**, charged colloidal particles accelerate until the electric body force balances the hydrodynamic drag force, so that the particles move at a constant average velocity **v**. This motion is known as electrophoresis, and **v** is the electrophoretic velocity.

If the magnitude of the field **E** is not too large, then **v** is directly proportional to **E** with

$$\mathbf{v} = \mu \mathbf{E} \tag{1}$$

The proportionality factor μ is known as the electrophoretic mobility. In general, μ depends on the particle shape, size, surface charge density, and surface electric potential, and on the suspension medium composition. For spherical particles, μ is a scalar. Additionally, for nonconducting particles, μ does not depend on the particle's bulk properties (such as its density or dielectric permittivity).

Electrophoretic light scattering (ELS) is commonly used to measure **v**. The electrophoretic mobility μ can be calculated from **v** and the known value of **E** according to Eq. (1). Theoretical models [1,7–10] that describe colloidal electrostatics and hydrodynamics can then be used to relate the measured values of μ to particle electrical characteristics including surface charge density and surface electric potential. Because μ depends on the surface electrostatic properties but not particle bulk properties, ELS can characterize surface electrostatic properties exclusively for a wide range of colloidal materials.

B. Light Scattering Theory

Figure 1 provides an overview of the relationships among different light scattering methodologies. This chapter focuses on quasielastic light scattering (QELS) in which the incident and scattered photons have essentially the same energy ($\hbar\Delta\omega \approx 0$). QELS can be divided into static and dynamic techniques. Static light scattering (SLS) techniques [3,6,11–15] measure the angular dependence of the average scattered light intensity. SLS provides information on the size, shape, and spatial orientation of the dispersed colloidal particles. SLS will not be discussed further, but the cited references contain more information.

Dynamic light scattering (DLS) techniques measure the fluctuations in the scattered light intensity caused by the random Brownian motion of the dispersed particles. The use of a theoretical model of particle Brownian motion enables us to extract particle size from DLS data. Other dynamic light scattering techniques such as electrophoretic light scattering (ELS) study collective particle motions. Theoretical interpretation of ELS data leads to other particle properties such as electrophoretic mobility μ and zeta potential ζ. These techniques will be discussed in more detail in subsequent sections.

Most light scattering techniques use a similar experimental setup. Typically, one or more beams of light are projected through a sample of a colloidal material. Laser light is preferred because it is coherent (in-phase) and nearly monochromatic. The particles of the dispersed phase scatter light in all directions. A photodetector (generally a photomultiplier tube or PMT), positioned in the same plane and at known angles relative to the incident laser beams, collects

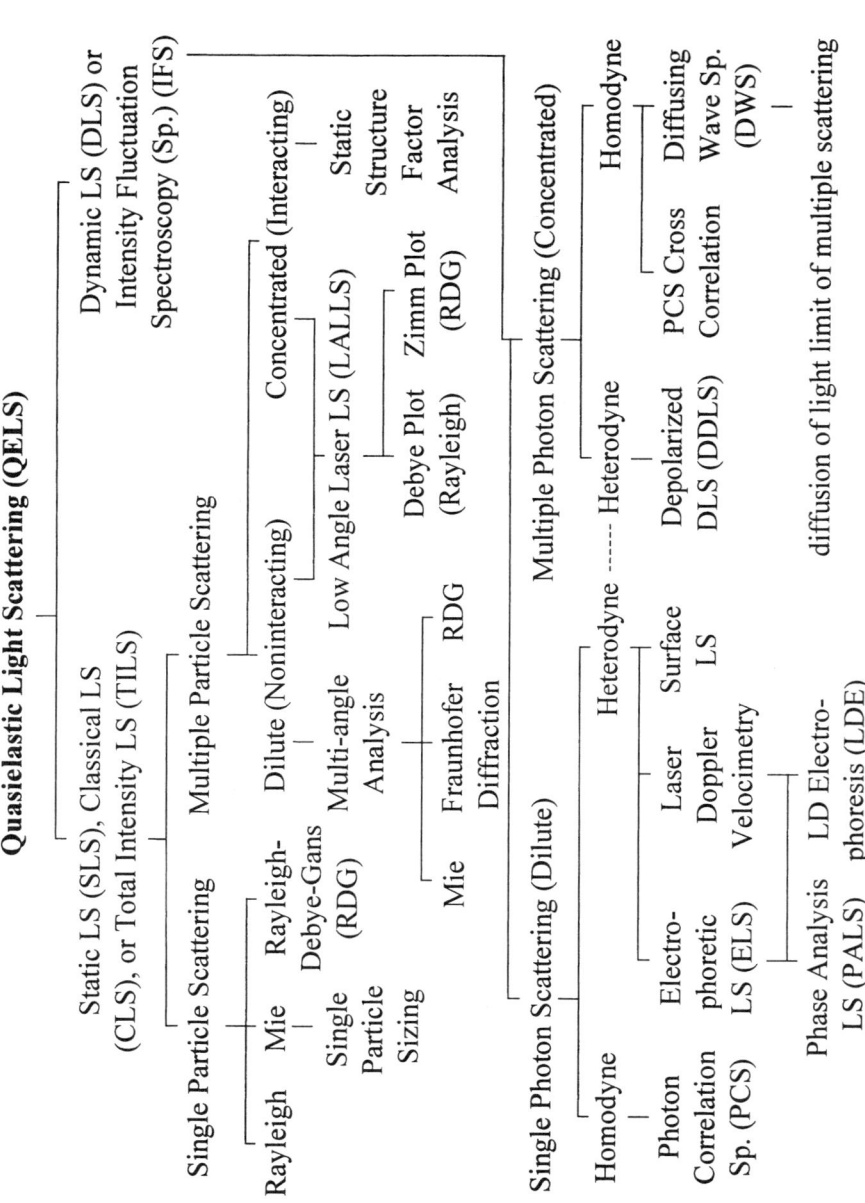

FIG. 1 Overview of quasielastic light scattering techniques.

some of the scattered light and provides a signal proportional to its intensity. A dedicated microprocessor averages the intensity signal or computes its autocorrelation function (ACF). An appropriate theoretical model can then be used to extract the desired properties of the colloidal particles from the averaged or autocorrelated intensity function.

The critical event is the scattering of the light. When light passes through a material, its molecules absorb photons, causing promotion of electrons to higher energy states. Upon relaxation of the electrons to lower energy states, photons are emitted in all directions as scattered light. In terms of the electric field (magnitude E) of the incident light and the isotropic polarizability α of the material, the incident light induces dipole moments of magnitude $\mu = \alpha E$ in the molecules of the material. These induced dipoles store additional energy [8] equal to $\alpha E^2/2$. Some of this energy is converted to thermal energy, but the rest is subsequently reradiated as scattered light. If the material has a uniform polarizability over a length scale greater than the wavelength of the incident light, then the light scattered from different regions interferes destructively, and only transmitted (unscattered) light leaves the material. For a material with a nonuniform polarizability on the scale of the light's wavelength, the interference of scattered light from regions with different polarizabilities depends on the relative positions, shapes, and sizes of the regions. Therefore analysis of scattered light can provide information about the relative positions, shapes, and sizes of the dispersed particles in a colloidal material [9,16].

Rigorous analysis of light scattering requires solution of Maxwell's equations [17] governing the propagation of electromagnetic fields in dielectric materials. Various texts [18,19] provide a thorough discussion of this topic. The use of light scattering to characterize colloidal particles of arbitrary size, shape, and concentration requires a thorough theoretical analysis of the changes in the polarization, phase, and intensity of light as it undergoes scattering. In a light scattering experiment, laser light impinges upon a sample and scatters in all directions. The scattering event changes the polarization, phase, and intensity of the light. The purpose of the photodetection system is to quantify these changes. Light scattering theory provides the means for interpreting the data in terms of the physical characteristics of the scatterers.

The orientation of the laser and photodetector relative to the sample establishes the scattering geometry. The orientation of the photodetector determines the wave vector k_S of the scattered light that is sampled (Fig. 2). The wave vectors k_I and k_S define the scattering plane. In practice, the scattering plane is usually horizontal in the laboratory frame of reference. The polarization of the incident and scattered light can be resolved into vector components perpendicular and parallel to the scattering plane. Commercial lasers used in light scattering experiments produce vertically polarized light (i.e., linearly polarized light with E_0 oriented perpendicular to the scattering plane).

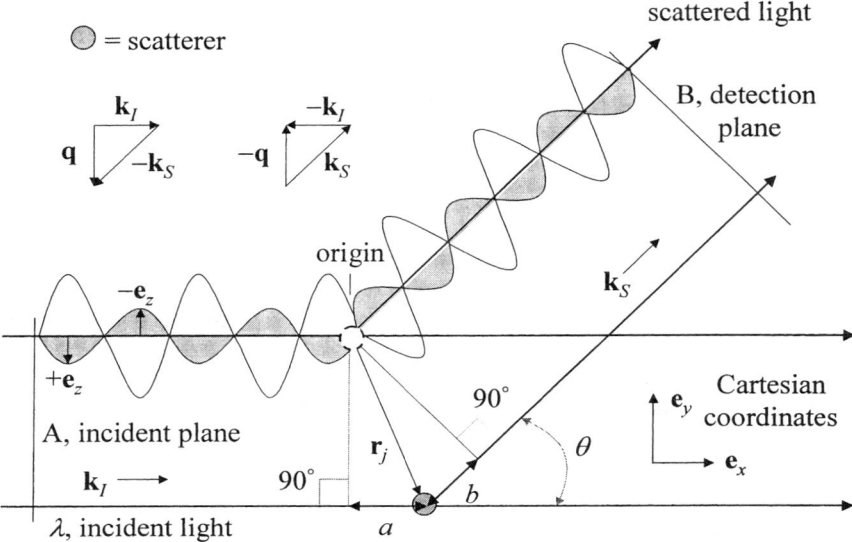

FIG. 2 Vector representation of light scattered by a moving scatterer.

For the quasielastic light scattering techniques reviewed here, the change in the frequency of the radiation upon scattering is approximately equal to zero ($\Delta\omega \approx 0$ and $\omega_I \approx \omega_S \equiv \omega$). Thus magnitudes of the wave vectors are given as $k_I \approx k_S \equiv k = 2\pi n/\lambda_0$, where λ_0 is the laser wavelength in vacuum and n is the refractive index of the medium. Additionally, conservation of linear momentum requires

$$\hbar(\mathbf{k}_I - \mathbf{k}_S) = \hbar\mathbf{q} \tag{2}$$

indicating that the difference between the linear momenta of an incident and a scattered photon equals the force exerted on the scatterer, $\hbar\mathbf{q}$. This phenomenon provides the basis for radiation pressure techniques for manipulating the positions of colloidal particles [20,21]. The difference between the incident and scattered wave vectors, $\mathbf{q}$, is known as the scattering vector. Simple vector analysis leads to

$$q \equiv |\mathbf{q}| = \frac{4\pi n}{\lambda_0} \sin\left(\frac{\theta}{2}\right) \tag{3}$$

where θ is the scattering angle.

Now that we have defined the scattering geometry and considered the implications of conservation principles, we turn to the scattered light. Consider the light scattered from a volume element located at $\mathbf{L} = \mathbf{0}$ (i.e., the origin of a coordinate system). If the volume element is small compared to the wavelength of light, then the scattered electric field can be treated as though the volume element were a dipole [17]. We shall also assume that the material in the volume element is isotropic, so that the permittivity and excess polarizability and scalars. At a position $L = |\mathbf{L}|$ far from the scattering volume element, the scattered field has the form (Ref. 17, pp. 136–140; Refs. 22, 23)

$$\mathbf{E}_S(\mathbf{L}, t) = \mathbf{k}_S \times (\mathbf{k}_S \times \mathbf{E}_0) \exp[i(\mathbf{k}_S \cdot \mathbf{L} - \omega t)] \frac{\alpha}{4\pi \varepsilon L} \tag{4}$$

where $\mathbf{E}_0$ is the magnitude of the incident laser light electric field, α is the excess polarizability of the material in the scattering volume, and ε is the dielectric permittivity of the medium. The scattered field magnitude (also called its amplitude),

$$E_S(\mathbf{L}, t) = \frac{E_0 k^2 \sin(\gamma) \alpha}{4\pi \varepsilon L} \exp[i(\mathbf{k}_S \cdot \mathbf{L} - \omega t)] \tag{5}$$

follows since $|\mathbf{k}_S \times (\mathbf{k}_S \times \mathbf{E}_0)| = k^2 E_0 \sin(\gamma)$, where γ is the angle between $\mathbf{E}_0$ and $\mathbf{k}_S$. Equation (4) clearly shows the changes in the polarization and phase of the scattered light relative to the incident light. The scattered intensity can be calculated by

$$I_S \equiv |\langle \mathbf{S}_S \rangle| = \frac{1}{2} \text{Re} \left[\left(\frac{\varepsilon}{\mu} \right)^{1/2} E_S E_S^* \right] = \frac{1}{2} c \, \text{Re} \left(\frac{\varepsilon}{n} \right) E_S^2 = \frac{1}{2} c n E_S^2 \tag{6}$$

where $\mathbf{S}_S$ is the Poynting vector [17], μ is the magnetic permeability of the medium, c is the speed of light in the medium, n is the complex refractive index of the medium equal to $c(\varepsilon\mu)^{1/2}$, Re denotes the real part, and the asterisk signifies the complex conjugate of the function. The last equality in Eq. (6) assumes $\varepsilon = n^2$ for the optical frequencies of interest. Insertion of Eq. (5) into Eq. (6) leads to

$$I_S = \frac{cn E_0^2}{2} \left[\frac{k^2 \sin(\gamma) \alpha}{4\pi \varepsilon L} \right]^2 = I_I \left[\frac{k^2 \sin(\gamma) \alpha}{4\pi \varepsilon L} \right]^2 \tag{7}$$

for single particles, where $I_I = cn E_0^2/2$ is the incident laser intensity. For vertically polarized incident light, $\gamma = \pi/2$; for horizontal polarization, $\gamma = \pi/2 - \theta$. The use of vertically polarized light in experiments conveniently eliminates the $\sin(\gamma)$ term in Eqs. (5) and (7).

Suppose that the scattering volume element is located at a position $\mathbf{r}$ rather than at the origin. The path length traveled by the scattered photons varies with the position of the volume element (Fig. 2). This shifts the phase of the scattered light. Vector analysis identifies the path length difference as

$$a + b = \mathbf{r} \cdot \frac{\mathbf{k}_I}{k_I} - \mathbf{r} \cdot \frac{\mathbf{k}_S}{k_S} = \mathbf{r} \cdot \frac{(\mathbf{k}_I - \mathbf{k}_S)}{k} = \frac{\mathbf{q} \cdot \mathbf{r}}{k} \tag{8}$$

where a and b are lengths shown in Fig. 2 and $k_I \approx k_S \equiv k$. The phase shift

$$\phi = k(a+b) = \mathbf{q} \cdot \mathbf{r} \tag{9}$$

clearly varies with the position of the scattering volume element. The scattered electric field amplitude becomes

$$E_S(q,t) = \frac{E_0 k^2 \sin(\gamma) \alpha}{4\pi \varepsilon L} \exp[i(\mathbf{k}_S \cdot \mathbf{L} - \omega t)] \exp(i\mathbf{q} \cdot \mathbf{r}) \tag{10}$$

Substitution into Eq. (6) shows that relocation of the particle produces no change in the intensity of the scattered light. Therefore the phase provides a direct indication of the position of the scattering volume element, but the phase of a single scatterer cannot be determined directly through intensity measurements employing only one incident beam.

Regardless of the information provided by intensity measurements, practical light scattering experiments never measure the scattering due to a single, small scatterer. Even in dilute suspensions, many colloidal particles will be present in the scattering volume defined by the intersection of the incident beam and the solid angle observed by the detector. The size of the particles or their average separation distance may be comparable to the wavelength of the light. Generalization of Eq. (10) provides the starting point for analyzing the scattering from a collection of particles of arbitrary size, shape, and concentration. The total scattered electric field can be represented as the sum of the fields scattered from all volume elements. We assume that the position $\mathbf{L}$ is far from the scattering volume, so that $\mathbf{k}_S$ and $\mathbf{L}$ have the same values for all scattering volume elements. Integration over all differential scattering volume elements yields the field amplitude

$$E_S(q,t) = \frac{E_0 k^2 \sin(\gamma)}{4\pi \varepsilon L} \exp[i(\mathbf{k}_S \cdot \mathbf{L} - \omega t)] \int_V \alpha'(\mathbf{r},t) \exp(i\mathbf{q} \cdot \mathbf{r}) \, dV \tag{11}$$

in which the excess polarizability per unit volume is defined by $\alpha' \equiv 3\varepsilon((\varepsilon_r - 1)/(\varepsilon_r + 2))$. The excess polarizability may vary in time owing to fluctuations in the density of the medium or the motion of particles. Substitution into Eq. (6) leads to the scattered intensity

$$I_S(q,t) = I_1 \left[\frac{k^2 \sin(\gamma)}{4\pi \varepsilon L} \right]^2 \int_V \int_V \alpha'(\mathbf{r}_1,t) \alpha'(\mathbf{r}_2,t) \tag{12}$$
$$\times \exp[i\mathbf{q} \cdot (\mathbf{r}_1 - \mathbf{r}_2)] \, dV_1 \, dV_2$$

involving a double integral over the scattering volume. The scattered intensity is incoherent, i.e., composed of many contributions that are not in phase and thus interfere. A photodetector would observe a speckle pattern that fluctuates in space and time.

Temperature and pressure fluctuations in the solvent medium create density fluctuations with nonzero excess polarizability. However, the contributions of these fluctuations to the scattered intensity can generally be neglected compared to those of molecular or colloidal solutes. As a result, the volume integral in Eq. (11) is dominated by contributions from solute particles. Suppose that the jth solute particle has volume V_j and center-of-mass located at position $\mathbf{r}_j(t)$. Then the contribution of this particle to the integral in Eq. (11) can be expressed as an integral

$$\int_{V_j} \alpha'_j(\mathbf{r},t) \exp(i\mathbf{q}\cdot\mathbf{r})\, dV = \int_{V_j} \alpha'_j(\delta\mathbf{r},t) \exp(i\mathbf{q}\cdot\delta\mathbf{r}) \exp[i\mathbf{q}\cdot\mathbf{r}_j(t)]\, dV \tag{13}$$

over the internal coordinate $\delta\mathbf{r}$ defined via $\mathbf{r} = \mathbf{r}_j(t) + \delta\mathbf{r}$. The subscript j on α'_j recognizes that the polarizability may differ among the particles. The second phase factor comes out of the integral, giving

$$\int_{V_j} \alpha'_j(\mathbf{r},t) \exp(i\mathbf{q}\cdot\mathbf{r})\, dV = a_j(q,t) \exp[i\mathbf{q}\cdot\mathbf{r}_j(t)] \tag{14}$$

with a new amplitude function

$$a_j(q,t) = \int_{V_j} \alpha'_j(\delta\mathbf{r},t) \exp(i\mathbf{q}\cdot\delta\mathbf{r})\, dV \tag{15}$$

describing interference of light scattered from different portions of the jth particle. Although the particle has constant shape, $a_j(q,t)$ depends on time, because the orientation of the particle may vary.

The volume integral in Eq. (11) can be replaced by a sum over all N solute particles

$$E_S(q,t) = E_0 \exp[i(\mathbf{k}_S\cdot\mathbf{L} - \omega t)] \sum_{j=1}^{N} b_j(q,t) \exp[i\mathbf{q}\cdot\mathbf{r}_j(t)] \tag{16}$$

with

$$b_j(q,t) = \frac{k^2 \sin(\gamma)}{4\pi\varepsilon L} \int_{V_j} \alpha'_j(\delta\mathbf{r},t) \exp(i\mathbf{q}\cdot\delta\mathbf{r})\, dV = \frac{k^2 \sin(\gamma)}{4\pi\varepsilon L} a_j(q,t) \tag{17}$$

defining another amplitude function. The scattered intensity can also be written as

$$I_S(q,t) = I_I \sum_{j=1}^{N} \sum_{l=1}^{N} b_j(q,t) b_l(q,t) \exp\{i\mathbf{q} \cdot [\mathbf{r}_j(t) - \mathbf{r}_l(t)]\} \tag{18}$$

Equations (16) and (18) discriminate between intraparticle and interparticle interference effects embodied in $b_j(q,t)$ and $\exp\{i\mathbf{q}\cdot[\mathbf{r}_j(t)-\mathbf{r}_l(t)]\}$, respectively. The amplitude function $b_j(q,t)$ contains information on the internal structure, shape, orientation, and composition of individual particles. Variations of $b_j(q,t)$ across the particle population reflect the polydispersity of particle size, shape, orientation, and composition. The phase function $\exp\{i\mathbf{q}\cdot[\mathbf{r}_j(t)-\mathbf{r}_l(t)]\}$ carries information on the random motion of individual particles, the collective motion of many particles, and the equilibrium arrangement of particles in the suspension medium.

Clearly, Eqs. (16) and (18) are rich in information. However, their generality precludes any possibility of extracting information from them without invoking simplifying assumptions. By combining different types of intensity measurements with judicious simplifying approximations, we can begin to dissect Eq. (18) in order to characterize the individual and collective properties of colloidal particles. The sections that follow describe dynamic light scattering techniques with emphasis on their relationship to the fundamental theory reviewed above.

C. Light Scattering Instrumentation

Technological advances in semiconductors, micro-optics, and microelectronics have promoted the development of more compact light scattering instrumentation [24]. The traditional setup consists of a gas ion laser, a goniometer, and a photomultiplier tube (PMT) detector mounted on a large vibration isolation table and connected to a PC-based correlator. Compact solid-state and semiconductor lasers [13], single and multimode optical fibers [25–31], and avalanche photodiode detectors [32] have led to improved designs consisting of smaller components enclosed in a single, transportable unit. Advances in correlator design [33–35], incorporating multitau and symmetric normalization schemes, provide enhanced data acquisition capabilities. Recent reviews [13,36,37] contain more detailed discussions of these developments with additional links to relevant literature. Additionally, Ref. 38 provides information about many of the commercial light scattering instruments currently available.

II. DYNAMIC LIGHT SCATTERING

Dynamic light scattering (DLS) is most commonly used to determine the average size of colloidal particles. Normally, this application requires sample dilution to obtain a particle concentration that eliminates particle interactions while maintaining a sufficient amount of light scattering. Under these conditions, the average particle size measured by DLS is independent of particle concentration. DLS yields an intensity-weighted average, which complicates comparisons with the number-weighted averages obtained via electron microscopy or other measurement techniques [39].

The lower limit of the particle size range accessible to DLS depends on the scattered intensity which, according to Eqs. (17) and (18), is proportional to the square of the particle volume. Thus scattering decreases rapidly as particle size decreases. Sedimentation sets the upper particle sizing limit. Gravity can cause settling or flotation of large particles with densities that differ from the suspending medium. The underlying theoretical assumptions of DLS are violated when the sedimentation velocity [10] becomes comparable to the root-mean-square velocity associated with Brownian motion.

A. Theory

Scattered light carries information about colloidal particles in both amplitude and phase functions, represented by $b_j(q,t)$ and $\exp\{i\mathbf{q}\cdot[\mathbf{r}_j(t)-\mathbf{r}_l(t)]\}$, respectively. DLS techniques are based on exploiting the phase function, or more precisely, its variation due to particle motion. The utility of DLS for the characterization of colloidal particles relies upon our ability to describe the mechanisms of colloidal particle motions (Brownian, electrophoretic, etc.) and their dependence on size or other particle properties.

Most DLS systems employ a single beam (known as the homodyne arrangement) with a photodetector that quantifies scattered intensity represented by Eq. (18). These systems cannot measure phase changes directly. Instead, they detect changes in the amplitude functions $b_j(q,t)$ that are, in effect, weighted by phase factors $\exp\{i\mathbf{q}\cdot[\mathbf{r}_j(t)-\mathbf{r}_l(t)]\}$. This presents several limitations. For dilute suspensions for which $\mathbf{q}\cdot\mathbf{r}_{lj}(t) \gg 1$, Eq. (18) reduces to

$$I_S(q,t) = I_I \sum_{j=1}^{N} b_j^2(q,t) \tag{19}$$

which carries no phase information and has little utility unless we make further assumptions. For identical spherical particles, Eq. (19) further reduces to

$$I_S(q,t) = I_I N(t) b^2(q) \tag{20}$$

which provides the basis for number fluctuation spectroscopy [40]. This technique detects particle motions over length scales comparable to the width of the laser beam. Due to the time required for this motion as well as the limiting assumptions, this technique has not been used widely.

Collective particle motions cannot be detected using a homodyne arrangement. Displacement of every particle in a scattering volume by the same vector **R** multiplies the scattered electric field, Eq. (16), by $\exp(i\mathbf{q} \cdot \mathbf{R})$, but the scattered intensity, Eq. (18), remains unchanged. However, random motions that change the distribution of interparticle separations ($\mathbf{r}_{lj}$) can be detected, because they change the distribution of phase-weighted amplitudes and thus the total scattered intensity. For this reason, characterization of colloidal particles by DLS relies heavily on the theory of Brownian motion for data interpretation.

The random Brownian motion of colloidal particles creates temporal fluctuations in the intensity of the scattered light. The fluctuating intensity signal cannot be readily interpreted because it contains too much detail. Instead, the fluctuations are commonly quantified by constructing an intensity autocorrelation function (ACF) [41]. For this reason, DLS often goes by the name photon correlation spectroscopy (PCS).

A suitable photodetector reports a signal $n(t)$ that is proportional to the intensity $I(t)$ and represents the number of photons detected over a short sampling interval centered on time t. The intensity ACF is defined as

$$G^{(2)}(q, \tau) = a^2 \langle n(t)n(t+\tau) \rangle = \langle I(0)I(\tau) \rangle \qquad (21)$$
$$= \langle E_S(0) E_S^*(0) E_S(\tau) E_S^*(\tau) \rangle$$

where a is a proportionality constant and we have arbitrarily set the starting time t to zero. Assuming that the random processes $n(t)$ and $I(t)$ display Gaussian statistics [40,41], we can show that

$$G^{(2)}(q, \tau) = \langle E_S(0) E_S^*(0) \rangle^2 + \langle E_S^*(0) E_S(\tau) \rangle^2 \qquad (22)$$
$$= [G^{(1)}(q, 0)]^2 + [G^{(1)}(q, \tau)]^2$$

indicating that the intensity ACF can be expressed in terms of electric field ACF, defined by

$$G^{(1)}(q, \tau) \equiv \langle E_S(q, t) E_S^*(q, t+\tau) \rangle = \langle E_S(q, 0) E_S^*(q, \tau) \rangle \qquad (23)$$

Normalized forms of the field and intensity ACFs are defined by

$$g^{(1)}(q, \tau) \equiv \frac{G^{(1)}(q, \tau)}{G^{(1)}(q, 0)} = \frac{\langle E_S(q, 0) E_S^*(q, \tau) \rangle}{\langle E_S(q, 0) E_S^*(q, 0) \rangle} \qquad (24)$$

and

$$g^{(2)}(q, \tau) \equiv \frac{\langle I(0)I(\tau) \rangle}{\langle I(0) \rangle^2} = \frac{\langle E_S(0) E_s^*(0) E_S(\tau) E_S^*(\tau) \rangle}{\langle E_S(q, 0) E_S^*(q, 0) \rangle^2} \qquad (25)$$
$$= 1 + |g^{(1)}(q, \tau)|^2$$

The latter equation, known as the Siegert relation, is valid for a point detector. In practice, the finite area of a photocathode detector [29,42] necessitates a correction factor [13] so that

$$g^{(2)}(q,\tau) = 1 + \beta |g^{(1)}(q,\tau)|^2 \qquad (26)$$

or equivalently

$$G^{(2)}(\tau,\theta) = A + B|G^{(1)}(\tau,\theta)|^2 \qquad (27)$$

where $A = [G^{(1)}(q,0)]^2$ and $B \equiv A\beta$ [23].

The normalized electric field ACF can also be expressed in terms of structure factors. Employing Eq. (16), we can define the dynamic structure factor $S(q,\tau)$ through

$$\langle E_S(q,0) E_S^*(q,\tau) \rangle = E_0^2 \sum_{j=1}^{N} \sum_{l=1}^{N} \langle b_j(q) b_l(q) \exp\{i\mathbf{q} \cdot [\mathbf{r}_j(0) - \mathbf{r}_l(\tau)]\} \rangle \qquad (28)$$

$$= E_0^2 N \overline{b^2(q)} S(q,\tau)$$

so that

$$S(q,\tau) \equiv \frac{1}{N \overline{b^2(q)}} \sum_{j=1}^{N} \sum_{l=1}^{N} \langle b_j(q) b_l(q) \exp\{i\mathbf{q} \cdot [\mathbf{r}_j(0) - \mathbf{r}_l(\tau)]\} \rangle \qquad (29)$$

When $\tau = 0$, $S(q,\tau)$ becomes the static structure factor [13]. Thus the normalized electric field ACF is simply

$$g^{(1)}(q,\tau) = \frac{S(q,\tau)}{S(q,0)} \qquad (30)$$

$$= \frac{\sum_{j=1}^{N} \sum_{l=1}^{N} \langle b_j(q) b_l(q) \exp\{i\mathbf{q} \cdot [\mathbf{r}_j(0) - \mathbf{r}_l(\tau)]\} \rangle}{\sum_{j=1}^{N} \sum_{l=1}^{N} \langle b_j(q) b_l(q) \rangle}$$

$$= \frac{1}{N \overline{b^2(q)}} \sum_{j=1}^{N} \sum_{l=1}^{N} \langle b_j(q) b_l(q) \exp\{i\mathbf{q} \cdot [\mathbf{r}_j(0) - \mathbf{r}_l(\tau)]\} \rangle$$

The summation in Eq. (30) can be written as

$$g^{(1)}(q,\tau) = \frac{1}{N\overline{b^2(q)}} \sum_{j=1}^{N} \left\langle \begin{array}{l} b_j^2(q)\exp\{i\mathbf{q}\cdot[\mathbf{r}_j(0)-\mathbf{r}_j(\tau)]\} \\ + \sum_{\substack{l=1 \\ l\neq j}}^{N} b_j(q)b_l(q)\exp\{i\mathbf{q}\cdot[\mathbf{r}_j(0)-\mathbf{r}_l(\tau)]\} \end{array} \right\rangle \quad (31)$$

which separates single-particle (self) and two-particle (distinct) contributions.

1. Dilute Suspensions of Identical Particles

In dilute suspensions, particle positions are uncorrelated. That is, $\mathbf{q}\cdot[\mathbf{r}_l(\tau) - \mathbf{r}_j(0)] \gg 1$, so that the cross term in Eq. (31) can be neglected, leaving

$$g^{(1)}(q,\tau) = \frac{1}{N\overline{b^2(q)}} \sum_{j=1}^{N} \langle b_j^2(q)\exp\{i\mathbf{q}\cdot[\mathbf{r}_j(0)-\mathbf{r}_j(\tau)]\}\rangle \quad (32)$$

describing the autocorrelation of the scattered electric fields from an ensemble of uncorrelated single particles.

For identical particles, we have simply

$$g^{(1)}(q,\tau) = \frac{1}{N}\sum_{j=1}^{N}\langle\exp\{i\mathbf{q}\cdot[\mathbf{r}_j(0)-\mathbf{r}_j(\tau)]\}\rangle = \langle\exp\{i\mathbf{q}\cdot[\Delta\mathbf{r}(\tau)]\}\rangle \quad (33)$$

where $\Delta\mathbf{r} = \mathbf{r}(\tau) - \mathbf{r}(0)$ denotes the change of position of a typical particle over the time interval τ. To compute the ensemble average in Eq. (33), we require a theoretical description of the Brownian motion of noninteracting colloidal particles.

Brownian motion of a single noninteracting particle can be described in terms of self-diffusion characterized by D_0, the particle self-diffusion coefficient in the infinite dilution limit. The probability $p(\Delta\mathbf{r},\tau)$ of a particle displacement $\Delta\mathbf{r}$ in time τ satisfies the diffusion equation

$$\frac{\partial p(\Delta\mathbf{r},\tau)}{\partial \tau} = D_0 \nabla^2 p(\Delta\mathbf{r},\tau) \quad (34)$$

with the well-known solution [10,41]

$$p(\Delta\mathbf{r},\tau) = (4\pi D_0\tau)^{-3/2}\exp\left[-\frac{|\Delta\mathbf{r}|^2}{4D_0\tau}\right] \quad (35)$$

From the definition of the ensemble average, we have [10]

$$\langle\exp(i\mathbf{q}\cdot\Delta\mathbf{r})\rangle = \int_V p(\Delta\mathbf{r},\tau)\exp(i\mathbf{q}\cdot\Delta\mathbf{r})d(\Delta\mathbf{r}) = \exp(-q^2 D_0\tau) \quad (36)$$

Characterizing Colloidal Materials Using DLS

Thus Eq. (33) becomes

$$g^{(1)}(q, \tau) = \exp(-q^2 D_0 \tau) = \exp(-\Gamma \tau) \tag{37}$$

where $\Gamma = q^2 D_0$ represents the reciprocal of the characteristic time of the diffusion process. Substitution of Eq. (23) into the Siegert relation Eq. (26) gives

$$g^{(2)}(q, \tau) = 1 + \beta \exp(-2\Gamma \tau) \tag{38}$$

Under these conditions, the normalized intensity ACF can be measured at a single angle (q) and fitted with a single exponential function to determine the characteristic time Γ and thus D_0. For spherical particles, the Stokes–Einstein equation [10]

$$D_0 = \frac{k_B T}{6\pi \eta R} \tag{39}$$

relates D_0 to the solvent viscosity η and the particle radius R. DLS therefore provides a simple method for determining the size of monodisperse, spherical colloidal particles. For nonspherical particles, DLS provides an effective intensity-averaged spherical radius. In practice, we must account for instrument artifacts as well as particle size and shape polydispersity. Considerable effort has been devoted to addressing these issues.

2. Size and Shape Polydispersity

All real colloidal suspensions contain a distribution of particle sizes and shapes and thus a distribution of diffusivities D_0 and decay constants Γ. Extracting the particle size distribution (PSD) from DLS data is a considerable challenge and the subject of ongoing research.

In the absence of interactions, particles of differing sizes and shapes are statistically independent. For this reason, we can treat the statistical properties of light scattered from a dilute polydisperse suspension as the sum of contributions of many dilute monodisperse suspensions of particles with characteristic shape and size. Suppose that each characteristic shape/size combination is labeled with the index s. Let N_s represent the number of particles having a particular shape and size. Clearly, we require $N = \sum_s N_s$. All sums over the N particles in a suspension can be expressed in terms of sums over the shape/size distribution. Thus we have

$$N\overline{b^2} \equiv \sum_{j=1}^{N} b_j^2 = \sum_s N_s \overline{b_s^2} \tag{40}$$

where b_s is the amplitude function for all particles of shape/size s. The overbar denotes an orientational average for nonspherical shapes. Likewise, Eq. (32) can be written as

$$g^{(1)}(q,\tau) = \frac{\sum_s N_s \overline{b_s^2} \langle \exp\{i\mathbf{q} \cdot [\mathbf{r}_s(0) - \mathbf{r}_s(\tau)]\}\rangle}{\sum_s N_s \overline{b_s^2}} \quad (41)$$

assuming that the average orientation of nonspherical particles is independent of their position.

The translational diffusion of particles of each shape/size combination can be treated independently, as mentioned before. Equation (36) describes each ensemble average, leading to

$$g^{(1)}(q,\tau) = \frac{\sum_s N_s \overline{b_s^2} \exp(-\Gamma_s \tau)}{\sum_s N_s \overline{b_s^2}} = \sum_s f(\Gamma_s) \exp(-\Gamma_s \tau) \quad (42)$$

with $f(\Gamma_s)$ equal to the intensity-average fraction of particles having the shape/size combination s and with characteristic decay constant Γ_s. In other words, $f(\Gamma_s)$ represents a generalized particle shape/size distribution.

The field ACF can also be written as an integral over a number distribution. From Eq. (42) we have

$$f(\Gamma_s) = \frac{N_s \overline{b_s^2}}{\sum_s N_s \overline{b_s^2}} = \frac{N_s \overline{b_s^2}}{N \overline{b^2}} = f_\#(\Gamma_s) \frac{\overline{b_s^2}}{\overline{b^2}} \quad (43)$$

where the last equality defines the number fraction $f_\#(\Gamma_s)$ of particles in shape/size configuration s. From Eq. (17), we have

$$\overline{b_s^2} = \left[\frac{k^2 \sin(\gamma)}{L}\right]^2 \overline{\alpha_s^2 P_s(q)} = \left[\frac{k^2 \sin(\gamma)}{\rho L}\right]^2 \overline{(\alpha_s')^2 M_s^2 P_s(q)} \quad (44)$$

where α_s, α_s', $P_s(q) \equiv V_p^{-2} \int_{V_p} \int_{V_p} \langle \exp[i\mathbf{q} \cdot (\delta\mathbf{r}_1 - \delta\mathbf{r}_2)]\rangle \, dV_1 \, dV_2$, and M_s are the excess polarizability, excess polarizability per unit volume, form factor [43–45], and particle mass for particles of shape/size configuration s. The overbar indicates an orientation average for nonspherical particles. A similar equation exists for $\overline{b^2}$ with $M_s^2 P_s(q)$ replaced by $\langle M_s^2 P_s(q)\rangle$, the average over shape/size configurations. Employing Eqs. (43) and (44), Eq. (42) becomes

$$g^{(1)}(q,\tau) = \sum_s f_\#(\Gamma_s) \frac{M_s^2 P_s(q,\Gamma_s)}{\langle M_s^2 P_s(q,\Gamma_s)\rangle} \exp(-\Gamma_s \tau) \quad (45)$$

The denominator of the fraction is a constant that can be absorbed in a renormalization of $g^{(1)}$.

If we restrict our attention to particles having identical shapes, $f(\Gamma_s)$ is the intensity-weighted PSD. In the limit of a continuous size distribution, Eq. (42) has the form of a Laplace transform

$$g^{(1)}(q,\tau) = \int_0^\infty f(\Gamma) \exp(-\Gamma\tau) \, d\Gamma \tag{46}$$

(i.e., a Fredholm integral of the first kind). In terms of the number distribution, we have

$$g^{(1)}(q,\tau) = \int_0^\infty f_\#(\Gamma) \frac{M^2(\Gamma) P(q,\Gamma)}{\langle M^2(\Gamma) P(q,\Gamma) \rangle} \exp(-\Gamma\tau) \, d\Gamma \tag{47}$$

Again, the constant in the denominator of the fraction can be absorbed in renormalization of $g^{(1)}$. Although f and $f_\#$ have been expressed as explicit distributions in Γ, these also represent distributions in D_0, since $\Gamma = q^2 D_0$.

Determination of the PSD requires knowledge of the relationship between D_0 and particle size analogous to the Stokes–Einstein Eq. (39). The number distribution of particle size also requires an expression for $P(q)$. For roughly spherical particles that do not absorb light at the laser wavelength, the Rayleigh–Debye–Gans (RDG) [43,44] approximation for spheres

$$P(q,R) = \frac{9\pi J_{3/2}^2(qR)}{2(qR)^3} = 9 \left[\frac{\sin(qR) - qR\cos(qR)}{(qR)^3} \right]^2 \tag{48}$$

can be used to approximate $P(q)$. RDG theory requires $|n_r - 1| \ll 1$, so that the incident light has uniform magnitude and phase within each particle. The phase condition also depends on the characteristic particle size R: when $(4\pi n R/\lambda_0)|n_r - 1| \ll 1$, the incident light experiences no phase shift. If these conditions are not met, then complete Mie theory [17] (which is general for spheres) can be used to calculate $P(q)$.

Accurate determination of suspension PSD via DLS represents a challenging problem because solution of Eqs. (46) or (47) for $f(\Gamma)$ or $f_\#(\Gamma)$ requires numerical Laplace inversion [46]. Noise in the ACF data makes this problem mathematically ill-conditioned with no unique solution. As a result, only a limited amount of useful information can be obtained from any measured ACF. Much research has been devoted to this problem, and several techniques have been shown to produce encouraging results [23,36,47–49]. Nevertheless, as the suspension PSD becomes increasingly complex, it becomes increasingly difficult to resolve the correct PSD using DLS.

B. Data Analysis Methods

Figure 3 classifies many of the data analysis techniques that have been reported for extracting PSDs from DLS data. Detailed discussions of the theory and application of these methods can be found in several reviews [23,36,48,49]. Some of the most commonly used methods will be summarized in the sections that follow.

1. ACF Selection

The analysis methods in Fig. 3 can be used to fit scattering data in the form of $G^{(2)}$, $g^{(2)}$, or $g^{(1)}$ ACFs. Each of these forms has advantages and disadvantages with respect to data analysis. Most DLS experiments employ homodyne detection and therefore measure $G^{(2)} = Ag^{(2)}$. We can fit homodyne $G^{(2)}$ data with a functional form derived from a combination of Eqs. (26) and (46):

$$\frac{G^{(2)}(\tau)}{A} - 1 = g^{(2)}(\tau) - 1 = \beta \left[\int_0^\infty f(\Gamma) \exp(-\Gamma \tau) \, d\Gamma \right]^2 \tag{49}$$

The baseline constant A can be determined from measured values of $G^{(2)}$ at long delay times τ or can be treated as an additional fitting parameter. Unfortunately, the nonlinearity of this regression makes the analysis rather complicated except for the simplified cases of single or double exponential fits. Alternatively, homodyne-measured $G^{(2)}$ data can be converted into $g^{(1)}$ data using either Eq. (26) or Eq. (27) prior to analysis:

$$\sqrt{\frac{G^{(2)}(\tau, \theta)}{A} - 1} = \sqrt{g^{(2)}(\tau, \theta) - 1} = \sqrt{\beta} g^{(1)}(\tau, \theta) \tag{50}$$

Heterodyne experiments measure a linear function of $g^{(1)}$ and therefore produce $g^{(1)}$ data with improved signal-to-noise ratio due to removal of the errors [13] introduced by the square root in Eq. (50). However, particle sizing applications rarely employ this approach because of experimental difficulties [13,50] and the sensitivity of heterodyne experiments to collective particle motions.

Consequently, most data analysis methods fit $g^{(1)}$ calculated from $G^{(2)}$ data measured in the homodyne mode. Difficulties in measuring the baseline complicate this conversion. Noise sometimes produces values of $G^{(2)}$ that are less than A for large values of τ, creating a negative number under the square root in Eq. (50). Removal of these data points, setting them to zero, or setting $\sqrt{\beta} g^{(1)} = -\sqrt{1 - g^{(2)}}$ when $g^{(2)} < 1$ remedy the problem but can bias the regression. Another approach truncates the data set at the first negative data point at τ_-, thereby removing all the data at $\tau \geq \tau_-$. However, it is not clear that this approach produces any less bias in the data regression.

Even if the baseline has relatively low levels of noise, run-to-run variations in the baseline photon count can have a significant impact on the predicted PSD.

Characterizing Colloidal Materials Using DLS

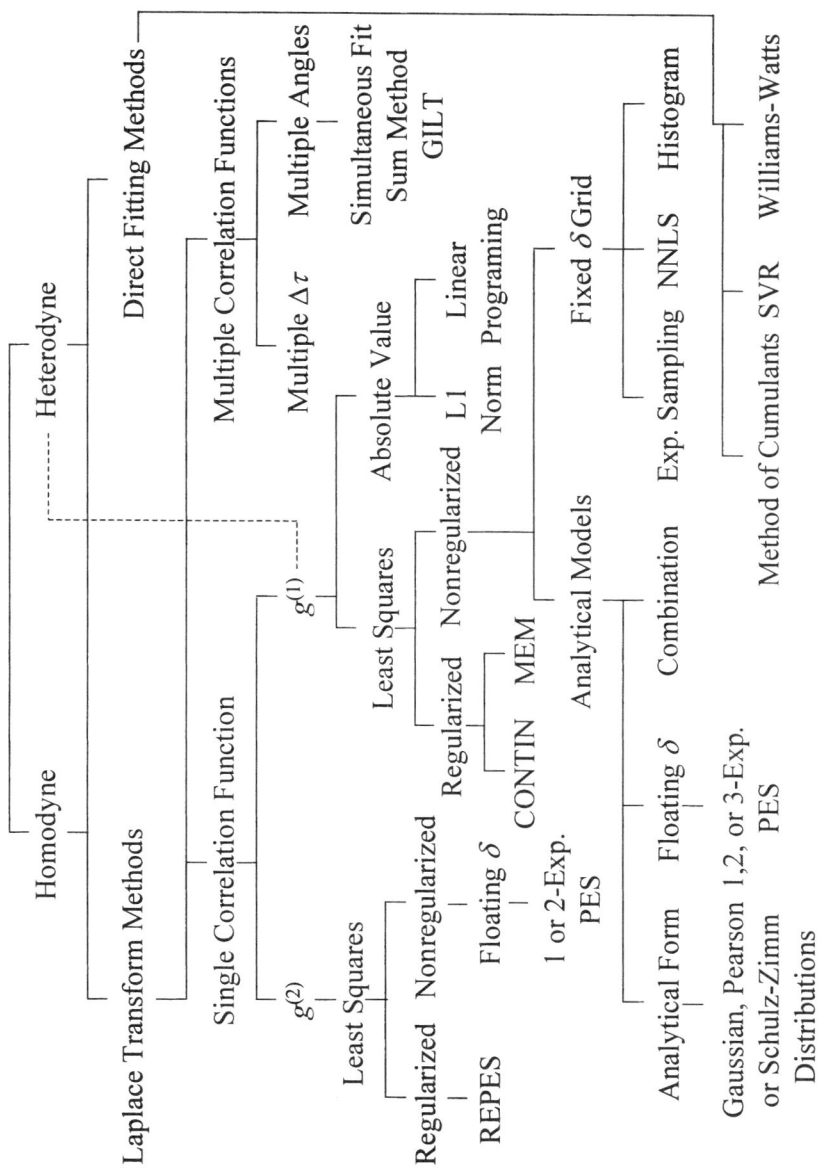

FIG. 3 DLS analysis methods. (Adapted from Ref. 49.)

The propagation of baseline error is especially significant for the regression of $g^{(2)}$ or $g^{(1)}$ data because each data point is normalized with the baseline value A. The error in $g^{(1)}$ due to a deviation ΔA between the measured and expectation values of A can be approximated by [51]

$$\Delta g^{(1)}(\tau) \approx -\frac{1}{2}\left[\frac{1}{g^{(1)}(\tau)} + g^{(1)}(\tau)\right]\frac{\Delta A}{A} \tag{51}$$

The effect of normalization errors on PSD prediction has been studied in the context of one Laplace inversion method [51], but the implications for the general problem of data weighting (Section II.B.2) have not been addressed.

2. Data Weighting

All the data analysis methods shown in Fig. 3 involve linear or nonlinear regression of ACF data, y_j^{data} (representing data point j of $G^{(2)}$, $g^{(2)}$, or $g^{(1)}$), to fit a proposed model, y_j^{model}. The model parameters or amplitudes of a proposed distribution are adjusted until a characteristic function is minimized or maximized. The characteristic function is often the chi-square (χ^2) statistic [52],

$$\chi^2 \sum_{j=1}^{N} \frac{1}{\sigma_j^2}(y_j^{\text{data}} - y_j^{\text{model}})^2 = \sum_{j=1}^{N} w_j(y_j^{\text{data}} - y_j^{\text{model}})^2 \tag{52}$$

that is, the variance-weighted sum of the squares of the residuals $y_j^{\text{data}} - y_j^{\text{model}}$ with N as the number of data points. The second equality defines $w_j = \sigma_j^{-2}$ as the data weighting used in the regression.

Alternative characteristic functions have also been used. For example, the sum of the absolute values of the residuals [53] has been evaluated for fitting $g^{(1)}$ data and produced results similar to those based on the chi-square statistic [49]. When one sets $w_j = 1$, the characteristic function in Eq. (52) is known as the L2 norm.

Most data analysis schemes employ minimization of the χ^2 statistic. However, the methods differ in their choice of the functional forms for y_j^{model} and in their estimates of the variance σ_j^2. Various schemes have been discussed in the literature [23,54–62]. Most of the data weighting expressions stem from the work of Jakeman et al. [61] for estimating noise effects on ACF data. This work is only strictly applicable for weak single exponential signals [58,62]. More recent research [40,63–65] has produced more accurate models of σ_j^2 for ACF data, leading to their use to compute weighting factors in data analysis routines [65,66]. Accurate models for estimating σ_j^2 are needed because calculated PSDs may be sensitive to the data weighting factors [54].

3. Direct Fitting Methods

The previous sections address data analysis issues relevant for all the methodologies summarized in Fig. 3, including selection of ACF form, the regression statistic, and data weighting factors. Figure 3 divides the data analysis methods into two categories: direct fitting methods and Laplace transform methods. Direct fitting methods include the method of cumulants (MC) [60,62,67], the William–Watts (WW) method [49], and the singular value analysis and regression method (SVR) [36,48].

The method of cumulants and the other direct fitting methods can be used to size particles accurately with smooth, unimodal PSDs. However, for broad or multimodal PSDs, these methods generally produce less accurate, sometimes misleading results. In a sense, one or two parameters regressed via direct fitting cannot capture most of the information inherent in a complex PSD. For more complicated PSDs, numerical Laplace inversion methods may provide improved accuracy and reliability. Nevertheless, direct fitting methods can be fruitfully employed to provide information that constrains an otherwise ill-conditioned Laplace inversion. For example, a direct fitting method can be used to estimate the limits of the particle size range.

Figure 4 shows an example of weighted and unweighted ($w_j = 1$) methods of cumulants regressions of PCS data for a Stöber [68,69] silica suspension. This example employs a quadratic cumulant (QC) expansion given by

$$\ln[\sqrt{B}g^{(1)}(\tau)] = K_0 - K_1\tau + \frac{K_2\tau^2}{2} \qquad (53)$$

For the weighted regression, the methodology of Brookhaven's light scattering software [70] with $w_j = [g^{(2)}(\tau_j) - 1]^2 = \beta^2[g^{(1)}(\tau_j)]^4$ was used which, in the context of the method of cumulants, is equivalent to other weighting methods found in the literature [61,62]. As Figure 4 shows, the QC expansion fits the measured PCS ata remarkably well despite the width of the PSD (25% of the average diameter as measured by TEM, i.e., $Q = 6.1\%$). Intensity-averaged diameter $\overline{d_{PCS}}$ and polydispersity Q values of 193.2 nm and 9.6% (± 59.9 nm) and 181.2 nm and 14.3% (± 68.5 nm) were obtained from the weighted and unweighted QC fits, respectively. From a data analysis perspective, the weighted regression results are more accurate and therefore are used in the discussion that follows. TEM analysis [71] of the particles (shown in the inset of Fig. 4) yielded a number-averaged diameter $\overline{d_{TEM}}$ of 144.8 ± 35.9 nm, which is 33% less than the intensity-averaged $\overline{d_{PCS}}$ value.

For better comparison with $\overline{d_{TEM}}$, the intensity-averaged $\overline{d_{PCS}}$ value can be converted to a number-averaged basis $\overline{d_\#}$ using the formula derived by Thomas [72], $\overline{d_\#} = \overline{d_{PCS}}/(1+Q)^5$. This formula assumes a log-normal PSD and Rayleigh scattering and corrects for noise in the ACF by subtracting 3% from the mea-

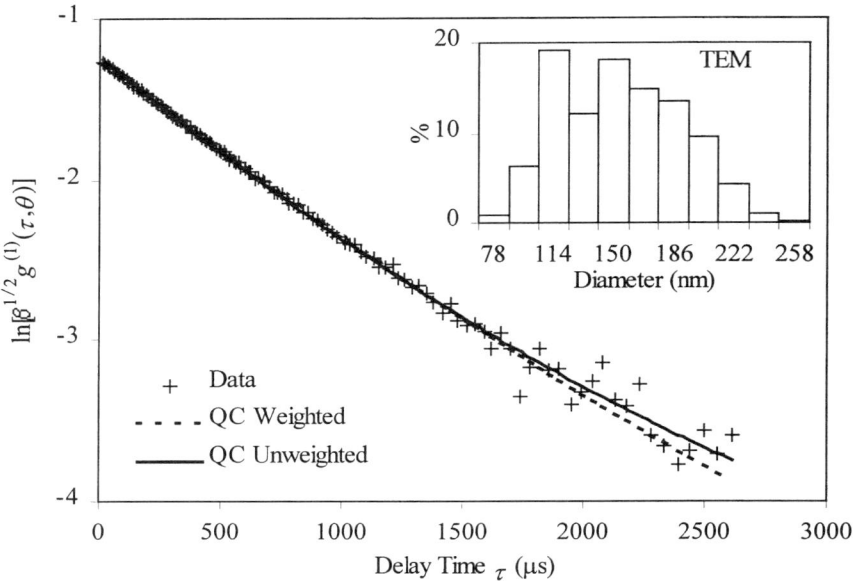

FIG. 4 Weighted and unweighted quadratic cumulants (QC) fits of ACF data collected at $\theta = 90°$ for Stöber [68] silica particles (inset shows the number-weighted TEM histogram of particle sizes [71]).

sured Q. Using this formula, the weighted regression $\overline{d_{PCS}}$ value converts to a number-averaged diameter of 140.2 nm, which is 3% less than the $\overline{d_{TEM}}$ value. The corrected Q of 6.6% is 8% greater than the TEM value. Conversion formulas for normal (Gaussian) and Schultz–Zimm PSDs have also been developed, and alternative Q correction heuristics have been found to produce closer comparisons between converted $\overline{d_{PCS}}$ and $\overline{d_{TEM}}$ values [73].

4. Laplace Inversion Methods

Laplace inversion techniques (the second category in Fig. 3) involve the inversion of Eq. (46), or more generally,

$$g^{(1)}(\tau) = \int_0^\infty f(\Gamma) K(\Gamma, \tau) \, d\Gamma \tag{54}$$

where $K = \exp(-\Gamma \tau)$ is the kernel of the transform. Although analytical inversion is possible in certain special cases, the general problem requires numerical inversion. A digital correlator collects ACF data as a set of N discrete values at delay times τ_j, where $1 \leq j \leq N$. The particle size distribution is expressed in terms of "bins" with index k, with $1 \leq k \leq M$. With these definitions, Eq. (54) for $g_j^{(1)} = g^{(1)}(\tau_j)$ can be written in discrete form as

$$g_j^{(1)} = \sum_{k=1}^{M} f(\Gamma_k) K(\tau_j, \Gamma_k) = \sum_{k=1}^{M} K_{jk} f_k \tag{55}$$

with the second expression defining a more compact notation. This expression can be inverted to determine f_k through least-squares minimization of

$$\chi^2 = \sum_{j=1}^{N} \frac{1}{\sigma_j^2} \left[g_j^{(1)} - \sum_{k=1}^{M} K_{jk} f_k \right]^2 \tag{56}$$

where σ_j^2 is the (perhaps unknown) variance in the ACF data. Most Laplace inversion methods solve the linear problem of fitting $g^{(1)}$ data (converted from measured $G^{(2)}$ data) rather than the nonlinear problem of fitting $G^{(2)}$ data directly (Section II.B.1). To our knowledge, the latter approach has been followed only by Jakes [48,49] and Honerkamp et al. [65] . In all cases, the χ^2 statistic (either alone or with a regularization constraint) serves as the fitting criterion.

The user sets the number of bins, the range of particle sizes over which they are distributed, and the functional form of this distribution (e.g., linear, logarithmic, etc.). Computational resources set the upper limit on the number of bins. The ill-conditioned nature of Laplace inversion of noisy ACF data necessitates some sort of prior knowledge for setting the distribution of bins. In fact, all Laplace inversion methodologies involve some form of constraint (based on prior knowledge) for dealing with the ill-conditioned inversion of Eqs. (54) or (55). Typical constraints include (1) specifying a limited number of bins and their spacing within the range of expected particle sizes; (2) assuming the shape of the distribution (e.g., normal, Pearson, single or double exponential) or features of the distribution (e.g., the number of peaks); or (3) requiring that the values of f_k remain nonnegative.

Much effort has been devoted to optimizing the spacing of the bins. Some approaches employ equal, quadratic [83], or exponential [74] spacing of the bins over a specified range of Γ values. Equal spacing in Γ has an advantage because it introduces no sampling bias into the final results. The eigenvalue analysis of the Laplace inversion by Ostrowsky et al. [75] indicates that exponential spacing in Γ is preferred over equal spacing. Unequal spacing procedures must be used carefully because they can bias the final results unless the grid points are weighted properly [36]. "Binning" is another issue: exactly what range of sizes should be grouped together in a single bin? It is common practice [49] to employ δ functions at the grid points in order to represent a PSD as a set of discrete sizes. However, other studies have defined bins having specified functional forms that cover a range of particle sizes. For example, bins represented by triangle functions [74] have been explored.

Prior knowledge (such as images from electron microscopy) may motivate the assumption that the PSD has an analytical form, such as normal, Pearson, or Schultz–Zimm distributions [23,76,77]. Laplace inversion under this assumption has produced accurate results in certain circumstances. In some cases, combinations of assumed distribution functions or combinations of distribution functions with δ functions have been used [49].

Another approach to dealing with the ill-conditioned nature of Laplace inversion is regularization, also known as parsimony. Regularization involves the imposition of additional constraints designed to favor some distributions over others, consistent with the measured data. For example, Tikhonov's regularization [49,65] adds a smoothing constraint to the least squares minimization, so that

$$\chi^2 = \sum_{j=1}^{N} \frac{1}{\sigma_j^2} \left[g_j^{(1)}(\tau_j) - \sum_{k=1}^{M} K_{jk} f_k \right]^2 + \alpha \sum_{k=1}^{M} [L f_k]^2 \tag{57}$$

where L is an operator (generally the second derivative $L = d^2/d\Gamma^2$) and α is a constant that controls the strength of the smoothing. This regularization favors smooth distributions over more complex distributions. A different approach, known as maximum entropy regularization [23,49], favors distributions that maximize the entropy

$$S = -\sum_{j=1}^{N} p_k \ln \left[\frac{p_k}{m_k} \right] \tag{58}$$

where p_k is proportion of the distribution $f(\Gamma)$ at decay rate Γ and m_k is an a priori guess of the distribution. In this case, the regularization parameter in Eq. (57) becomes $\alpha \sum_{k=1}^{M} [L f_k]^2 = -S$.

It is useful to distinguish between floating grid and fixed grid methods. Floating grid methods allow the locations of the bins (R_k or Γ_k) as well as the bin amplitudes (f_k) to vary. Fixed grid methods place bins at fixed values of R_k or Γ_k and allow variations only in the amplitudes f_k. The user sets the bin locations (fixing the size range and bin distribution within that range) based on a priori knowledge about the sample, trial and error, or information obtained from direct fitting methods.

Table 1 lists several Laplace inversion techniques together with the constraints that they place on the inversion. Details can be found in the cited references. The review of Stepanek [49] provides a good discussion of the difficulties encountered in comparing the sizing techniques in Table 1 and assessing their relative merits. Included are citations of some of the comparisons that have been made

TABLE 1 Laplace Inversion Methods

Method	Constraints	Refs.
Nonregularized methods		
Single exponential	$g^{(1)} = \exp(-\Gamma\tau)$, floating grid, fits $G^{(2)}$	62,67
Double exponential	$g^{(1)} = \sum_{k=1}^{2} A_k \exp(-\Gamma_k\tau)$, floating grid, fits $G^{(2)}$	49,55,67
Exponential sampling	Fixed grid, fits $g^{(1)}$	48,49
PES	$f(\Gamma)$ nonnegative (NN), floating grid, fits $g^{(1)}$ or $G^{(2)}$	49
DISCRETE	Floating grid	49,80
Histogram	NN, fixed grid, fits $g^{(1)}(\tau_k) = \sum_{k=1}^{N} f(\Gamma_k) \int_{\Gamma_k-\Delta\Gamma/2}^{\Gamma_k+\Delta\Gamma/2} \exp(-\Gamma\tau_k)\,d\Gamma$	48,49
NNLS	NN, fixed grid, fits $g^{(1)}$	49
MNNLS	multiangle NNLS analysis, fits $g^{(1)}$	47
Linear programming	NN, minimizes $\sum_j |g^{(1)}(\tau_j) - \sum_k A + B_k \exp(-\Gamma_k\tau_j)|$	49
Regularized methods		
CONTIN, RILIE	NN, fixed grid, fits $g^{(1)}$	48,49,81
CONTIN-multiq	Multiangle data CONTIN analysis, fits $g^{(1)}$	82
REPES	NN, floating grid, fits $G^{(2)}$	48,49
MAXENT	NN, fixed grid, fits $g^{(1)}$	48,49
Honerkamp et al.	NN, fixed grid, improved σ^2 formula, fits $G^{(2)}$	65
Ruf et al.	CONTIN, with correction for normalization errors	51

NN denotes nonnegativity constraint on $f(\Gamma)$.

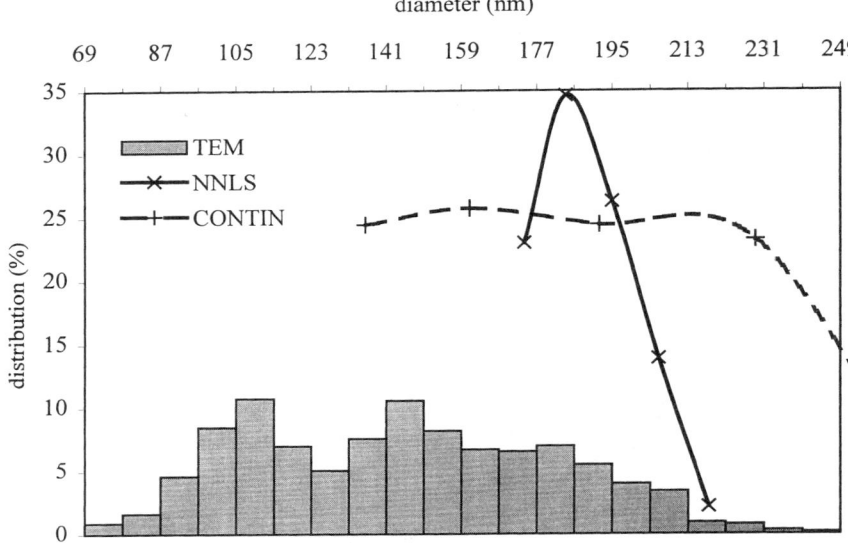

FIG. 5 PSDs generated from NNLS and CONTIN analysis of the ACF data shown in Fig. 4 for Stöber [68] silica particles. For comparison, the bar graph is the number-weighted TEM histogram of particle sizes [71].

among various methods. The review of Finsy [36] also offers a comparison of results obtained for monomodal and bimodal samples sent to different labs and analyzed using different methods (SVR, NNLS, CONTIN, and MAXENT).

It is worth noting that Laplace inversion methods can be used for particle size analysis only under certain conditions. The Siegert relation, Eq. (27), assumes that a Gaussian random process describes the scattering events. Equation (27) is invalidated by non-Gaussian behavior in suspensions with strongly interacting particles [49,78], large fluctuations in the number of particles in the scattering volume [49], or nonergodic behavior (e.g., glasses or gels) [23,49,79]. The expression of the electric field ACF as a Laplace transform, Eq. (46), assumes that particle motion occurs due to Brownian translational diffusion. Laplace inversion methods are therefore inapplicable for characterizing the sizes of particles manifesting other types of dynamic behavior, such as internal relaxation modes in large polymer coils [49], combined translation and rotational diffusion (Section II.C.2), or collective particle motion (Sections II.C.3 and III).

Figure 5 shows an example of NNLS and CONTIN analyses of PCS data for a Stöber [68] silica suspension (Fig. 4). In this example, the NNLS and CONTIN algorithms of Brookhaven's light scattering software [70] were used.

As Fig. 5 shows, both NNLS and CONTIN do not resolve all the details of the TEM-measured PSD, but CONTIN estimates the width of the PSD better than NNLS. Intensity-averaged diameter $\overline{d_{PCS}}$ and polydispersity Q values of 186.4 nm and 0.5% ($\pm$ 13.0 nm) and 176.4 nm and 6.8% ($\pm$ 46.0 nm) were obtained from the NNLS and CONTIN PSDs, respectively. These results are 29% and 22% greater than the number-averaged TEM-measured diameter of 144.8 $\pm$ 35.9 nm. Conversion of the intensity-averaged NNLS and CONTIN results using the formula of Thomas [72] (discussed in Section II.B.3) yields number-averaged diameters of 181.9 and 146.4 nm, which are 26% and 1% greater than the TEM-measured diameter, respectively.

5. Data Collection Considerations

The average particle diameter estimated from any single measurement can vary significantly from the true value. These variations can be attributed to dust, flare, large particles, or noise in the ACF. Remnants of these artifacts continue to affect an ACF regardless of the duration of an experiment. The methodology suggested by Morrison et al. [83] seems to be a prudent way of overcoming the artifacts introduced by these variations. Morrison et al. recommended collecting multiple data sets, analyzing them independently, and averaging the final results to determine the best fit PSD (or average particle size). This approach assumes that the average of inversions of several different ACF data sets provides a better estimate of the actual PSD than the inverse of the average of several data sets.

This multiple data set analysis approach can be viewed as an external solution constraint in the same way that multi-q analysis approaches [47,49] use additional data as an internal solution constraint. However, there is a point of diminishing returns: regardless of the number of ACFs collected, accurate PSDs predictions require high-quality ACFs that improve as collection time increases. Thus one must make optimal use of the finite amount of time allocated to characterizing a sample. Experimental experience shows that satisfactory sizing results can be consistently achieved by collecting more, rather than fewer, ACFs within an allocated span of time, but the duration of an individual measurement must be "long enough" to ensure a quality ACF with minimal noise.

Multiple data sets can be used to generate more accurate PSD predictions for both Laplace inversion and direct fitting techniques. Various types of additional data sets (besides just collecting multiple data sets at the same conditions) have been explored [49], such as collecting ACFs at multiple sampling times and multiple q values (via multiple angles θ or multiple wavelengths λ). Also, different extrapolation techniques have been employed with direct fitting methods such as dynamic Zimm plots [49] (extrapolation to c and $q = 0$) or R or Γ/q^2 versus q^2 plots [84,61] (extrapolation to $q = 0$).

Additional experimental considerations for PCS include sample preparation (dilution and filtration), choosing the delay time range and spacing among correlator channels, and optical alignment [85]. Sample dilution may be necessary to avoid multiple scattering effects. Depending on the particle and size, sample concentrations on the order of $10^{-3}\%$ particles by volume typically scatter strongly enough for accurate analysis with negligible multiple scattering. Filtration may be necessary to remove dust and other unwanted debris from the sample prior to analysis [85].

Selection of an appropriate delay time range and a scheme for spacing them among correlator channels is necessary to ensure that intensity contributions are sampled from all of the particles in the sample [49,85]. Shaumeyer et al. [62] have explored the effect of the selected value of last delay time on particle size results based on the method of cumulants and single exponential methods. Stepanek [49] discusses the effect of incomplete data sets on Laplace inversion techniques.

Finally, proper alignment to ensure correct positioning of the laser beam, sample, and photodetector are critical for accurate particle size measurements [85]. This is especially important for multiangle measurements because misalignment will produce large systematic errors that are difficult to detect and do not average out. In his review, Stepanek [49] discusses the additional experimental complications of wide correlation functions, square root bias, other sources of systematic error, and resolution of δ function or sharp peaked PSDs.

C. Applications

Although DLS is most often used to size solid colloidal particles, the technique has also been applied to characterize aerosols [78,86,87], emulsion droplets [88,89], amphiphilic systems [90–92], and macromolecular solutions [12,16,93]. Another common application is the study of the fractal structure and kinetics of colloidal aggregation [94–102]. More information about dynamic light scattering and its applications can be found in Refs. 23, 103 (104), and 105, in reviews, Refs. 11, 13, 36, 37, 49, 50, and 106, and in collections of papers Refs. 12, 14, 16, 93 (107), 105, and 108–114.

1. Particle Interactions

As particle concentration increases, particle interactions and multiple scattering invalidate Eq. (33). The cross terms ($j \neq l$) in the static and dynamic structure factors, Eq. (29), no longer cancel out, and thus they lead to more complex relationships [115–119] for $g^{(1)}(q, \tau)$. The diffusive motion of interacting particles also becomes more complex, depending on colloidal and hydrodynamic interactions among the particles and their spatial configurations. DLS measurements of particle motion can provide information about suspension microstructure and particle interactions.

2. Depolarized Dynamic Light Scattering

Up to this point, we have only considered analysis of scattered light with the same polarization as the incident light. Additional dynamic and structural features can be investigated through analysis of scattered light with polarization that differs from that of the incident light. Analysis of depolarized scattered light is known as depolarized dynamic light scattering (DDLS) [120,121]. Because of the relative weakness of the depolarized scattered intensity and the fast decay of its ACF, DDLS experiments are more demanding and require more powerful laser sources and spectrum analyzers rather than correlators. The weak intensity signal makes the noise from glassware, optical components, the suspending solvent, and multiple scattering much more problematic for depolarized scattering than for polarized scattering. Thus the literature contains fewer reports of studies employing DDLS. Nevertheless, DDLS offers a unique way to measure important characteristics of colloidal solutes, especially rodlike particles and macromolecules.

Depolarized scattering occurs because of various forms of particle anisotropy. Distinct classes of depolarizing scatterers include nonspherical particles with uniform isotropic (scalar) polarizabilities (sometimes called form anisotropy), inhomogeneous particles with nonuniform distributions of isotropic polarizability, and particles with anisotropic (tensor) polarizabilities. For each of these classes, the intensity of depolarized light scattered by a particle will change as the particle translates, rotates, or manifests internal rearrangement of its scattering elements. DDLS can provide information on the dynamics of each of these processes.

3. Collective Particle Motions

Dynamic light scattering can also be used to study collective particle motions. However, homodyne techniques cannot be used for this purpose because they cannot sense phase changes in scattered light and thus collective particle motions. Instead, heterodyne techniques and spectral analysis are typically used. Three common DLS collective particle analysis regimes are surface light scattering [122–125], laser Doppler velocimetry (LDV) [50], and electrophoretic light scattering (ELS). Some motions measured by these techniques are surface wave propagations, the natural velocities of bioparticles, and the electrophoretic mobilities of charged particles in the presence of an applied electric field. Surface light scattering and LDV will not be discussed further, but ELS will be described in detail in Section III.

4. Multiple Scattering

As particle concentration increases, particle interactions and multiple scattering effects predominate. These effects are most clearly manifested as deviations of measured correlation functions from exponential decay. Development of more sophisticated models to account for these deviations are motivated by the need to control industrial processes for producing colloidal particles.

Several methods have been proposed to overcome multiple scattering. One simple solution proposes the use of thin samples, but flare effects and wall interactions complicate data interpretation [11]. Alternative solutions involve cross correlation [13,36,126,127] and two color cross correlation techniques [11,13,36] employing simultaneous illumination of the sample by two laser beams with differing wavelengths.

Development of depolarized dynamic light scattering (DDLS) has led to models for double scattering [128]. However, higher levels of multiple scattering are practically impossible to analyze except in the highly multiple scattering limit where the scattering process can be treated as photon diffusion. Application of DLS in this limit is known as diffusing wave spectroscopy (DWS) [129–131].

III. ELECTROPHORETIC LIGHT SCATTERING

Electrophoretic light scattering (ELS) is a subset of DLS (Section II.C.3) that is commonly used to measure the electrophoretic mobility μ of colloidal particles (Section I.A.2). ELS can be used to probe particle surface electrostatic properties via theoretical models of colloidal electrostatics and hydrodynamics that relate μ to particle electrical characteristics.

The key link between ELS experiments and particle electrostatic properties is the theoretical model of colloidal electrohydrodynamics. The required model is considerably more complicated than the one needed in the interpretation of DLS data. DLS relies upon a relatively simple colloidal hydrodynamic model to relate the measured particle diffusivity to particle radius via the Stokes–Einstein Eq. (39). The colloidal electrohydrodynamic model for ELS must account for the complex physical/chemical/electrical structure of the particle surface as well as the distortion of the diffuse part of the electrostatic double layer due to the motion of the particle through the medium.

Models of colloidal electrohydrodynamics relate the electrophoretic mobility μ to the zeta potential ζ, the particle radius R, the composition of the solution via the Debye length $1/\kappa$, and the solvent viscosity and permittivity, η and ε. Dimensional analysis shows that these variables must be related by

$$\zeta = f(\kappa R)\frac{\eta \mu}{\varepsilon} \tag{59}$$

Simplified models may be developed [7,10] for various limiting cases of thin and thick electrostatic double layers ($\kappa R \gg 1$ and $\kappa R \ll 1$, respectively). When the diffuse part of the double layer is thick ($\kappa R \ll 1$), $f \rightarrow 3/2$, a result attributed to Hückel. For thin double layers ($\kappa R \gg 1$), $f \rightarrow 1$ producing a form of Eq. (59)

known as the Helmholtz–Smoluchowski equation. Another approximate model assumes that the double layer retains its equilibrium structure as the particle moves relative to the surroundings. This leads to a closed-form expression for $f(\kappa R)$, known as the Henry equation [7,10], which is valid for any value of κR but is restricted to low values of ζ.

Exact numerical solutions of the full colloidal electrohydrodynamic problem have appeared in recent years. For computational convenience, the numerical schemes treat ζ as an independent variable; the user varies ζ until the predicted and measured values of μ agree. The earliest numerical solutions [132] were hampered by convergence difficulties at relatively low values of ζ. O'Brien and White [133] resolved these numerical problems, and their solution is widely used today. Some recent publications [134,135] document the evolution of analytical and numerical solutions of the colloidal electrohydrodynamics problem and report numerical solutions for particle mobility, suspension conductivity, and suspension dielectric permittivity for both constant and oscillatory applied electric fields.

A. Theory

The discussion of dynamic light scattering (DLS) through Section II.A.1 demonstrates that the field ACF reduces to [Eq. (33)]

$$g^{(1)}(q,\tau) = \frac{1}{N} \sum_{j=1}^{N} \langle \exp\{i\mathbf{q} \cdot [\mathbf{r}_j(0) - \mathbf{r}_j(\tau)]\} \rangle = \langle \exp\{i\mathbf{q} \cdot [\Delta\mathbf{r}(\tau)]\} \rangle \quad (60)$$

for identical particles in dilute solutions. Here $\Delta\mathbf{r} = \mathbf{r}(\tau) - \mathbf{r}(0)$ denotes the change of position of a particle over the time interval τ. Evaluation of the ensemble average in Eq. (60) requires a theoretical description of the motion of the particles. The theory for DLS assumes that the particles move solely due to random Brownian diffusion. We require a more sophisticated model to analyze cases in which the particles move collectively (i.e., nonrandomly) due to applied external forces such as gravity, electromagnetic fields, or flow.

In this case, the motion of a single noninteracting particle can be described in terms of a Brownian random walk characterized by D_0 superimposed on a constant velocity $\mathbf{v}$ arising due to the applied field. The probability $p(\Delta\mathbf{r}, \tau)$ of a particle displacement $\Delta\mathbf{r}$ in time τ satisfies the diffusion equation [Eq. (34)]

$$\frac{\partial p(\Delta\mathbf{r}, \tau)}{\partial \tau} = D_0 \nabla^2 p(\Delta\mathbf{r}, \tau) + \mathbf{v} \cdot \nabla p(\Delta\mathbf{r}, \tau) \quad (61)$$

which now includes a convective term that depends on the collective particle velocity $\mathbf{v}$.

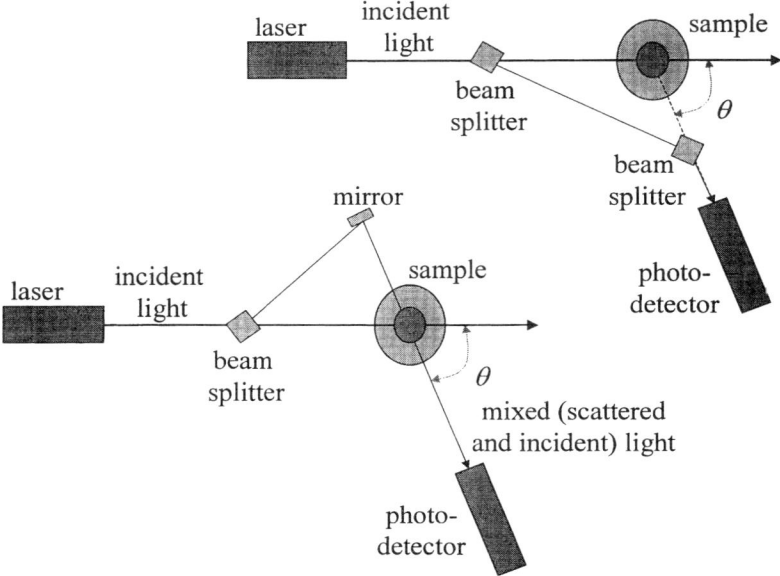

FIG. 6 Typical reference beam detection schemes. (Adapted from Ref. 50.)

as the "detected" field. Considerable mathematical simplification results if we assume that the incident light has linear polarization perpendicular to the scattering plane. In this case, we have $\mathbf{k} \times (\mathbf{k} \times \mathbf{E}_0) = k^2 \mathbf{E}_0$, allowing us to write

$$E_D = E_I + E_S \tag{72}$$

$$= E_0 \exp[i(\mathbf{k} \cdot \mathbf{L} - \omega t)] \left\{ 1 + \sum_{j=1}^{N} b_j(q,t) \exp[i\mathbf{q} \cdot \mathbf{r}_j(t)] \right\}$$

in terms of electric field magnitudes using Eq. (16). The scattered intensity is

$$I_S(t) = \frac{1}{2}\left(\frac{\varepsilon}{\mu}\right)^{1/2} \text{Re}[E_D(t) E_D^*(t)] \tag{73}$$

$$= I_I \, \text{Re} \left(\begin{array}{l} 1 + \sum_{j=1}^{N} b_j \exp[i\mathbf{q} \cdot \mathbf{r}_j(t)] + \sum_{j=1}^{N} b_j \exp[-i\mathbf{q} \cdot \mathbf{r}_j(t)] \\ + \sum_{j=1}^{N} \sum_{l=1}^{N} b_j b_l \exp[-i\mathbf{q} \cdot [\mathbf{r}_l(t) - \mathbf{r}_j(t)]] \end{array} \right)$$

Characterizing Colloidal Materials Using DLS

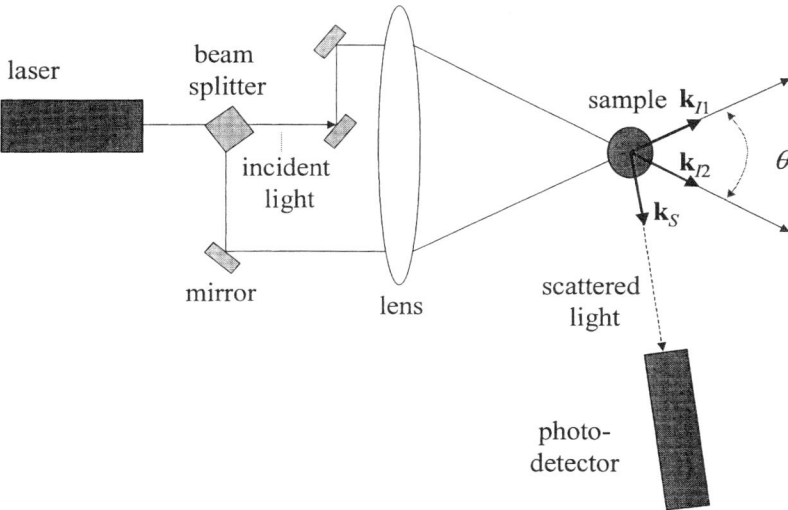

FIG. 7 Typical symmetric real fringe experimental arrangement.

In the limit of dilute suspensions, this reduces to

$$I_S(t) = I_I \, \text{Re}\left(1 + \sum_{j=1}^{N} b_j \{b_j + \exp[i\mathbf{q}\cdot\mathbf{r}_j(t)] + \exp[-i\mathbf{q}\cdot\mathbf{r}_j(t)]\}\right) \quad (74)$$

$$\approx I_I \left\{1 + 2\sum_{j=1}^{N} b_j \cos[\mathbf{q}\cdot\mathbf{r}_j(t)]\right\}$$

assuming that the b_j^2 term in the first line can be neglected. We see clearly that the scattered intensity in the dual-beam arrangement carries phase information.

The intensity ACF $G^{(2)}$ can be defined in terms of the detected field as

$$G^{(2)}(\tau) = \langle I_D(0) I_D(\tau) \rangle = \langle E_D(0) E_D^*(0) E_D(\tau) E_D^*(\tau) \rangle \quad (75)$$

If we assume that the scattered field E_S is a Gaussian random variable, then the detected field E_D is a Gaussian random variable as well. Based on the properties of Gaussian random variables [41], most of the terms in the product of the fields in Eq. (75) vanish, leaving

$$G^{(2)}(\tau) = |\langle E_D(0) E_D^*(0) \rangle|^2 + |\langle E_D(0) E_D^*(\tau) \rangle|^2 \quad (76)$$

Using Eq. (72), we can show that

$$\langle E_D(0)E_D^*(0)\rangle \qquad (77)$$

$$= E_0^2 \left\langle \begin{array}{c} 1 + \sum_{j=1}^{N} b_j \exp[i\mathbf{q}\cdot\mathbf{r}_j(0)] + \sum_{j=1}^{N} b_j \exp[-i\mathbf{q}\cdot\mathbf{r}_j(0)] \\ + \sum_{j=1}^{N}\sum_{l=1}^{N} b_j b_i \exp[-i\mathbf{q}\cdot[\mathbf{r}_l(0)-\mathbf{r}_j(0)]] \end{array} \right\rangle$$

$$\approx C_1 N \overline{b^2} E_0^2$$

The second equality results through recognition that all the terms in the brackets are constants. The term $N\overline{b^2}$ has been factored out for later convenience. We also have

$$\langle E_D(0)E_D^*(\tau)\rangle \qquad (78)$$

$$= E_0^2 \left\langle \begin{array}{c} 1 + \sum_{j=1}^{N} b_j \exp[i\mathbf{q}\cdot\mathbf{r}_j(0)] + \sum_{j=1}^{N} b_j \exp[-i\mathbf{q}\cdot\mathbf{r}_j(\tau)] \\ + \sum_{j=1}^{N}\sum_{l=1}^{N} b_j b_l \exp[i\mathbf{q}\cdot[\mathbf{r}_j(0)-\mathbf{r}_l(\tau)]] \end{array} \right\rangle$$

$$= E_0^2 \left[N\overline{b^2} C_2 + \sum_{j=1}^{N}\sum_{l=1}^{N} \langle b_j b_l \exp[i\mathbf{q}\cdot[\mathbf{r}_j(0)-\mathbf{r}_l(\tau)]]\rangle \right]$$

Upon averaging the terms in brackets, the first and second are constants (again, with $N\overline{b^2}$ factored out for convenience), and the third (a Gaussian random variable similar to E_S) vanishes.

For dilute suspensions of identical particles, the last term in Eq. (78) simplifies (see Section II.A.1), giving

$$\langle E_D(0)E_D^*(\tau)\rangle = N\overline{b^2}E_0^2[C_2 + \langle \exp[i\mathbf{q}\cdot[\mathbf{r}_j(0)-\mathbf{r}_l(\tau)]]\rangle] \qquad (79)$$

$$= N\overline{b^2}E_0^2[C_2 + g^{(1)}(\tau)]$$

with $g^{(1)}$ identified by comparison with Eq. (60). Equation (76) becomes

$$G^{(2)}(\tau) = |C_1 N\overline{b^2}E_0^2|^2 + N\overline{b^2}E_0^2[C_2 + g^{(1)}(\tau)]|^2 \qquad (80)$$

or, upon normalization of the intensity ACF,

$$g^{(2)}(\tau) = C_1^2 + |[C_2 + g^{(1)}(\tau)]|^2 \qquad (81)$$

Using Eq. (68), it is not difficult to show that

$$g^{(2)}(\tau) = C_1^2 + [C_2 + \exp(-\Gamma\tau)\cos(\mathbf{q}\cdot\mathbf{v}\tau)]^2 \tag{82}$$
$$= C_1^2 + C_2^2 + 2C_2\exp(-\Gamma\tau)\cos(\mathbf{q}\cdot\mathbf{v}\tau) + \exp(-2\Gamma\tau)\cos^2(\mathbf{q}\cdot\mathbf{v}\tau)$$
$$\approx C_3 + C_4\exp(-\Gamma\tau)\cos(\mathbf{q}\cdot\mathbf{v}\tau)$$

The last line neglects the term containing the higher order exponential. This result demonstrates that the intensity ACF measured using the reference beam heterodyne arrangement can yield information on collective particle motions represented by $\mathbf{q}\cdot\mathbf{v}\tau$.

(b) *Real Fringe Arrangement.* The reference beam arrangement (Section III.A.2.a) uses a portion of the incident light to interfere with the scattered light, producing a detected intensity signal that contains phase information. An alternative method creates an interferometer by splitting the incident light into two beams and crossing them in the sample cell (Fig. 7). The two crossing beams have different incident wave vectors ($\mathbf{k}_{I1}$ and $\mathbf{k}_{I2}$) due to their redirection by various mirrors and lenses. Assuming both beams retain vertical polarization, the combined incident electric field $\mathbf{E}_{RF}$ in the crossing region is

$$\mathbf{E}_{RF} = \mathbf{E}_{I1} + \mathbf{E}_{I2} \tag{83}$$
$$= \mathbf{E}_0\{\exp[i(\mathbf{k}_{I1}\cdot\mathbf{z} - \omega t)] + \exp[i(\mathbf{k})_{I2}\cdot\mathbf{z} - \omega t)]\}$$

with intensity

$$I_{RF} = \frac{1}{2}\left(\frac{\varepsilon}{\mu}\right)^{1/2}\mathrm{Re}[E_{RF}(t)E_{RF}^*(t)] \tag{84}$$
$$= 2I_1\{1 + \cos[(\mathbf{k}_{I1} - \mathbf{k}_{I2})\cdot\mathbf{z}]\}$$
$$= 2I_1[1 + \cos(\mathbf{q}_{12}\cdot\mathbf{z})]$$

where $\mathbf{q}_{12} \equiv \mathbf{k}_{I1} - \mathbf{k}_{I2}$. This indicates that the intersecting beams establish an interference fringe pattern (Fig. 8) with a sinusoidal variation of intensity that manifests alternating regions of constructive and destructive interference. The fringes are aligned parallel to the bisector of the beams and separated by 2π radians or, in terms of distance s,

$$s = \frac{2\pi}{|\mathbf{k}_{I1} - \mathbf{k}_{I2}|} = \frac{2\pi}{q_{12}} = \frac{\lambda_0}{2n\sin(\theta/2)} \tag{85}$$

where $|\mathbf{k}_{I1}| = |\mathbf{k}_{I2}| = 2\pi n/\lambda_0$ and θ is the angle between the incident beams.

Since the incident field in the fringe region can be represented as the vector sum of the fields of the component beams, the scattered field magnitude can also be represented as a sum,

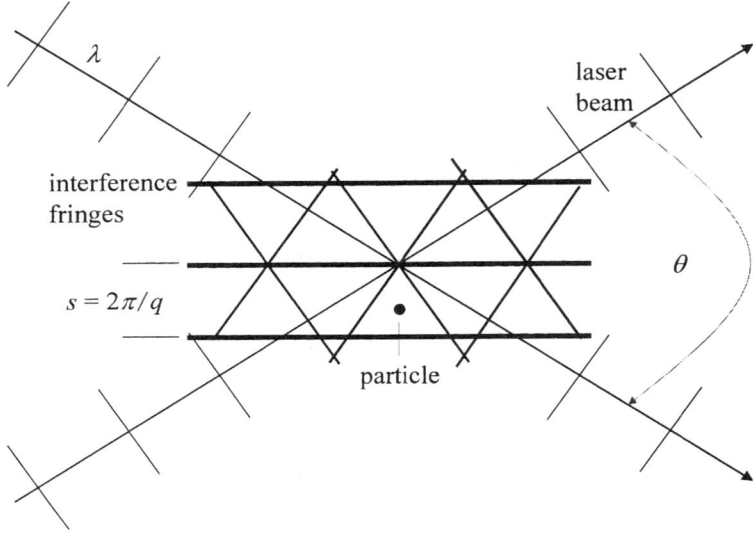

FIG. 8 Interference fringe pattern established by intersecting laser beams. (Adapted from Ref. 137.)

$$E_D = E_{S1} + E_{S2} \tag{86}$$

$$= E_0 \exp[i(\mathbf{k}_S \cdot \mathbf{L} - \omega t)] \sum_{j=1}^{N} b_j \{\exp[i\mathbf{q}_1 \cdot \mathbf{r}(t)] + \exp[i\mathbf{q}_2 \cdot \mathbf{r}(t)]\}$$

where the scattering vectors are defined as $\mathbf{q}_1 \equiv \mathbf{k}_{I1} - \mathbf{k}_S$ and $\mathbf{q}_2 \equiv \mathbf{k}_{I2} - \mathbf{k}_S$, and N denotes the number of particles illuminated in the crossing region. The scattered intensity becomes

$$I_S(t) = I_I \sum_{j=1}^{N} \sum_{l=1}^{N} b_j b_i \left\{ \begin{array}{l} \exp[i\mathbf{q}_1 \cdot (\mathbf{r}_j - \mathbf{r}_l)] + \exp[i\mathbf{q}_2 \cdot (\mathbf{r}_j - \mathbf{r}_l)] \\ + \exp[i(\mathbf{q}_1 \cdot \mathbf{r}_j - \mathbf{q}_2 \cdot \mathbf{r}_l)] \\ + \exp[i(\mathbf{q}_2 \cdot \mathbf{r}_j - \mathbf{q}_1 \cdot \mathbf{r}_l)] \end{array} \right\} \tag{87}$$

where all the particle positions are understood to be functions of time t. For dilute suspensions of identical particles, this expression simplifies considerably, giving

$$I_S(t) = I_I \sum_{j=1}^{N} b_j^2 \{2 + \exp[i\mathbf{q}_{12} \cdot \mathbf{r}_j(t)] + \exp[-i\mathbf{q}_{12} \cdot \mathbf{r}_j(t)]\} \tag{88}$$

Characterizing Colloidal Materials Using DLS

in which we recognize $\mathbf{q}_{12} \equiv \mathbf{q}_1 - \mathbf{q}_2 = \mathbf{k}_{I1} - \mathbf{k}_{I2}$. Furthermore, if we assume that all particles are identical, and consider only the real part of the intensity, we have

$$I_S(t) = 2I_1 N \overline{b^2} \, \text{Re}\{1 + \exp[i\mathbf{q}_{12} \cdot \mathbf{r}(t)]\} \tag{89}$$
$$= 2I_1 N \overline{b^2} \{1 + \cos[\mathbf{q}_{12} \cdot \mathbf{r}(t)]\}$$

As in the reference beam arrangement [Section III.A.2.a, Eq. (74)], the scattered intensity carries phase information encoded in $\mathbf{q}_{12} \cdot \mathbf{r}(t)$.

More importantly, the scattered intensity is independent of $\mathbf{k}_S$ and thus the direction of the photodetector [50]. Single-beam (homodyne) and dual-reference-beam arrangements require coherent detection, necessitating the use of a pinhole to select scattered light with essentially unique values of $\mathbf{q}$ and $\mathbf{k}_S$. This reduces the detection area and the strength of the scattered light signal, so experiments must employ more intense incident light and longer durations. The lack of dependence of scattered intensity on $\mathbf{k}_S$ in the real fringe arrangement permits the use of incoherent detection. The absence of a pinhole leads to a greater signal-to-noise ratio and shorter experimental runs using a less powerful laser. For this reason, the real fringe arrangements are preferred over reference beam arrangements in most electrophoretic light scattering systems.

Still assuming dilute suspensions of identical particles, the intensity ACF can be calculated from Eq. (89), yielding

$$G^{(2)}(\tau) = \langle I_S(0) I_S^*(\tau) \rangle \tag{90}$$
$$= (2I_1 N \overline{b^2})^2 \left\langle \begin{array}{l} 1 + \exp[i\mathbf{q}_{12} \cdot \mathbf{r}(0)] + \exp[-i\mathbf{q}_{12} \cdot \mathbf{r}(\tau)] \\ + \exp\{i\mathbf{q}_{12} \cdot [\mathbf{r}(0) - \mathbf{r}(\tau)]\} \end{array} \right\rangle$$

The ensemble average of the second term is a constant, the third term (a Gaussian random variable) vanishes, and the fourth term can be identified as $g^{(1)}$ from Eq. (60) with $\mathbf{q}$ replaced by $\mathbf{q}_{12}$. Thus we have

$$G^{(2)}(\tau) = (2I_1 N \overline{b^2})^2 [C_1 + g^{(1)}(\tau)] \tag{91}$$
$$= (2I_1 N \overline{b^2})^2 [C_1 + \exp(-\Gamma \tau) \exp(i\mathbf{q}_{12} \cdot \mathbf{v}\tau)]$$
$$= (2I_1 N \overline{b^2})^2 [C_1 + \exp(-\Gamma \tau) \cos(\mathbf{q}_{12} \cdot \mathbf{v}\tau)]$$

where Eq. (68) provides the expression for $g^{(1)}$. The last line recognizes that our interest lies only in the real part of the intensity ACF.

B. Experimental Methods

There are two commonly applied ELS experimental methods: laser Doppler electrophoresis (LDE) [11], a subset of laser Doppler velocimetry (LDV) [37,50,138], and phase analysis light scattering (PALS) [11,137,139–143]. The primary difference between the techniques is that LDV (and thus LDE) employs intensity or spectral analysis of the scattered intensity signal to extract **v** (and thus μ), while PALS employs phase analysis. Both reference beam [136] and real fringe arrangements [50] have been used for LDV experiments. PALS is designed around the real fringe arrangement [137].

1. Laser Doppler Electrophoresis

The application of laser Doppler velocimetry (LDV) to measure the electrophoretic mobility μ of charged colloidal particles is known as laser Doppler electrophoresis (LDE). In a typical LDE experiment, an applied electric field drives the collective motion of charged colloidal particles. The particles pass through an interference pattern created by a dual-beam experimental setup (Section III.A.2). The collective electrophoretic velocity of the particles is then determined via intensity- or spectrum-based analysis of the scattered light, and the electrophoretic mobility μ is calculated by dividing the velocity by the applied electric field strength.

Spectral analysis provides a convenient means for analyzing the scattered intensity signal in LDE. The Fourier transform of the intensity ACF, Eq. (91), gives

$$S_I(\omega) = \frac{(2I_1 N \overline{b^2})^2}{2\pi} \left[C_1 \delta(\omega) + \frac{\Gamma}{\Gamma^2 + (\omega - \mathbf{q}_{12} \cdot \mathbf{v})^2} + \frac{\Gamma}{\Gamma^2 + (\omega + \mathbf{q}_{12} \cdot \mathbf{v})^2} \right]$$

(92)

for the power spectrum of the scattered intensity. The slow variation in the incident intensity I_I across the width of the scattering volume can be neglected (and I_I treated as a constant) because the fringe spacing s is much smaller than the width of the laser beams and the scattering volume. Thus the spectrum consists of two Lorentzian peaks separated by a frequency difference of $2\mathbf{q}_{12} \cdot \mathbf{v}$ plus a spike at zero frequency due to the baseline (DC) component of the signal. Of the two Lorentzian peaks, the spectrum analyzer provides only the one at positive frequency ($\omega > 0$). The electrophoretic velocity v can be calculated from the location of this peak at $\omega = \mathbf{q}_{12} \cdot \mathbf{v}$. Particle size can also be extracted from the half-width of the peak at half-height (Γ). However, optical misalignment and other artifacts tend to broaden the peak without shifting its location. Consequently, minor optical misalignment does not impair real fringe LDE measurements of mobility, but particle size measurements are rendered less accurate.

Complications with intensity or spectral analysis sometimes hamper LDE experiments. For small particles with large diffusion coefficients, or for particles with low electrophoretic velocities, the decay constant Γ becomes large compared to $q_{12}v$ [Eq. (92)]. As a result, the power spectrum broadens, making it difficult to resolve the exact location of the peak and thus the electrophoretic velocity. This problem, known as diffusion broadening, is exacerbated by the presence of the low-frequency spike. In practice, the low-frequency spike appears as a broadened peak (or "pedestal") due to fluctuations in the laser power, the finite size of the particles relative to the fringe spacing, any difference in the intensities of the two crossing incident beams, and minor misalignment of optical components. The pedestal makes signal filtering impractical: if the constant C_1 has a significant magnitude, high-pass analog or digital filtering can jeopardize the integrity of the rest of the data [140]. However, this does not present a problem for data analysis as long as $\mathbf{q}_{12} \cdot \mathbf{v}$ is large so that the Lorentzian peak is shifted away from the pedestal.

LDE experiments generally require large electric fields to produce sufficient particle velocities, and large fields can produce thermal convection and electrolysis. Square wave or other alternating current (AC) fields, which produce periodic collective velocities, are often used to avoid electrolysis effects [137]. To measure collective velocities using LDV and AC fields, the amplitude of the velocities must be larger than the average Brownian motion over one field period and larger than the effective fringe spacing $s = 2\pi/q$ (Section III.A.2.b). Alternatively, phase analysis, which will be discussed in Section III.B.2, can be used to determine $\mathbf{v}$. Phase analysis has been shown to have enhanced sensitivity to small collective particle motions such as electrophoretic mobilities relative to LDV because of the nature of the signal analysis methods.

Figure 9 shows the LDE power spectrum for aqueous CdS particles [144] measured at 24°C and $\theta = 24°$ using a moving real fringe setup [137] (Section III.A.2.b) and a 1 Hz, 1.37 V/mm square wave electric field. For comparison, the inset of Fig. 9 shows the LDE power spectrum obtained with no applied field and all the other conditions held constant. The mobility of the particles can be determined from the separation distance between the peaks and the origin in Fig. 9. This distance should equal $\mathbf{q} \cdot \mathbf{v} = qv = q\mu E$ for the real fringe arrangement. For reference beam heterodyne arrangements, the additional $\cos(\theta/2)$ term would appear in this expression, $\mathbf{q} \cdot \mathbf{v} = qv\cos(\theta/2) = q\mu E \cos(\theta/2)$. Analysis of the data in Fig. 9 yielded a mobility of 52×10^{-10} m^2/Vs (i.e., $v = 7.6$ μm/s). The average diffusivity D_0 of the particles can also be estimated from the width of the peaks in Fig. 9 at half height, which should equal $\Gamma = q^2 D_0$. The average radius of the particles could then be calculated using the Stokes–Einstein Eq. (39). However, as mentioned earlier, this analysis is generally inaccurate due to experimental complications and therefore was not performed.

2. Phase Analysis Light Scattering

LDE is limited by the experimental difficulty of distinguishing between collective and diffusive motion of colloidal particles [137]. Phase analysis light scattering (PALS) was invented to overcome this difficulty. Instead of correlat-

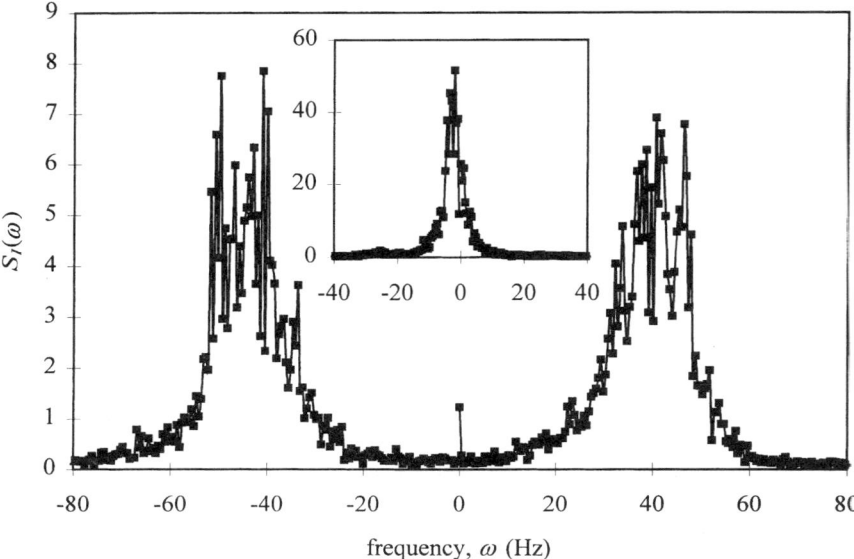

FIG. 9 Power spectrum of CdS particles [144] at 24°C and $\theta = 24°$ subject to a 1 Hz, 1.37 V/mm square wave field using a real fringe setup. The inset shows the power spectrum with no applied field under the same conditions.

ing the intensity of the scattered light as in LDE, the PALS method analyzes the time-dependent phase of the scattered light. PALS uses a real fringe optical setup with interference fringes of separation s (Section III.A.2.b) as well as an electronic setup that creates moving interference fringes to demodulate the phase of the detected intensity signal. This is accomplished by introducing a frequency shift of ω_s in one of the incident laser beams using Bragg cells [137,142]. Because of the theoretical development required, we will not discuss PALS further here, but Refs. 50, 137, 140, and 142 provide a detailed discussion of the theoretical and experimental considerations of PALS relative to LDE.

Figure 10 shows the amplitude-weighted phase difference (AWPD) and phase structure (AWPS) functions [137] measured using PALS for the aqueous CdS particles [144] discussed in Fig. 9. The conditions are identical (24°C and $\theta = 24°$) with the experiment employing a moving real fringe setup [137] and a 30 Hz, 1.37 V/mm sine wave electric field. The autotrack results utilize a feature of the experimental analysis software [140] that corrects for other convective effects such as sedimentation. The mobility of the particles can be determined by analyzing the data in Fig. 10 with appropriate models for AWPD and AWPS functions [137]. Analysis of the data in Fig. 10 yielded AWPD and AWPS

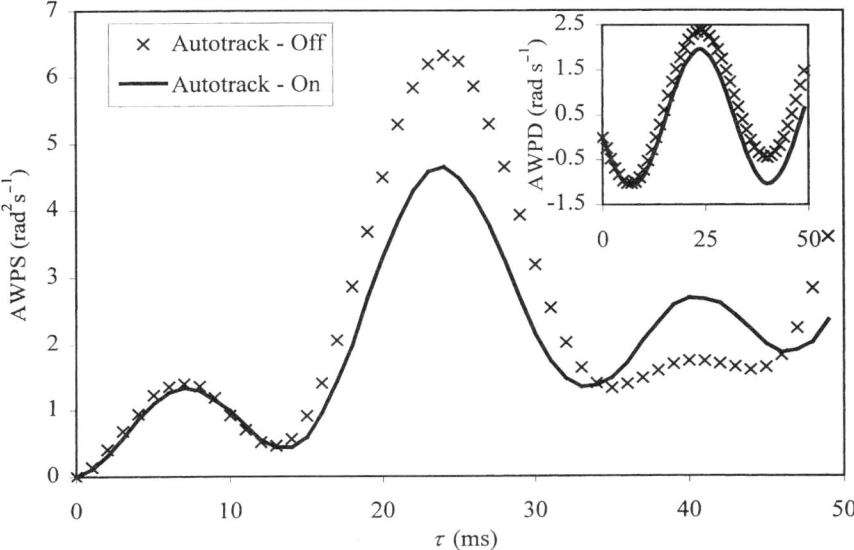

FIG. 10 Amplitude-weighted phase difference (AWPD) and phase structure (AWPS) functions measured using PALS with autotrack on and off for aqueous CdS particles [144] at 24°C and $\theta = 24°$ subject to 30 Hz, 1.37 V/mm sine wave electric field.

mobilities of 289 ± 6 and $331 \pm 7 \times 10^{-10}$ m^2/Vs (i.e., velocities of 5.3 ± 0.7 and 5.7 ± 0.8 μm/s), respectively. The average radius of the particles can also be calculated via analysis of the AWPS data; however, like LDE particle size analysis, experimental complications generally prevent accurate particle size determination from AWPS data.

C. Applications

Particle electrostatic properties play a crucial role in controlling suspension stability and rheology (at least for aqueous suspensions). Consequently, characterization of particle electrostatic properties, especially surface charge density and surface electric potential, is of considerable interest. Electrophoretic light scattering has been used extensively for this purpose. A random sampling of the literature reveals the utility of ELS in a variety of applications, including characterization of colloidal particles [145–149] used in adsorption studies, calibration, chromatography [150], and paint formulation [162]; development of colloidal semiconductors [151,152]; and environmental waste characterization [153–155] and separation research [156,157]. Biological applications include the development of particles for clinical diagnostic tests [158] as well as the design and characterization of particles for bioanalysis [159–161] and drug delivery systems [162,163].

ACKNOWLEDGMENTS

We gratefully acknowledge the financial support of the U.S. National Science Foundation through grants CTS-9258137 and EPS-9630167.

REFERENCES

1. Hunter, R. J., *Foundations of Colloid Science*, Vol. 1, Oxford University Press, Oxford, 1987, pp. 3, 157.
2. Hunter, R. J., *Introduction to Modern Colloid Science*, Oxford University Press, Oxford, 1993, p. 3.
3. Barth, H. G., and Flippen, R. B. Anal. Chem. 1995, *67*(12):257R–272R.
4. Black, D. L., McQuay, M. Q., and Bonin, M. P. Prog. Energy Combust. Sci. 1996, *22*:267–306.
5. Provder, T. Prog. Org. Coat. 1997, *32*:143–153.
6. Haskell, R. J. J. Pharm. Sci. 1998, *87*(2):125–129.
7. Hiemenz, P. C., and Rajagopalan, R., *Principles of Colloid and Surface Chemistry*, 3d ed., Marcel Dekker, New York, 1997.
8. Israelachvili, J. N., *Intermolecular and Surface Forces*, 2d ed., Academic Press, London, 1991, p. 73.
9. Atkins, P. W., *Physical Chemistry*, 4th ed., W. H. Freeman, New York, 1990.
10. Russel, W. B., Saville, D. A., and Schowalter, W. R., *Colloidal Dispersions*, Cambridge University Press, New York, 1989.
11. Schätzel, K., in *NASA Laser Light Scattering Advanced Technology Development Workshop—1988* (Meyer, W. V., ed.), NASA Conference Publication 10033, NASA, Cleveland, 1989.
12. *Laser Light Scattering in Biochemistry* (Harding, S. E., Sattelle, D. B., and Bloomfield, V. A., eds.), Royal Society of Chemistry, Cambridge, 1992, Parts II-III.
13. Schätzel, K. Adv. Col. Int. Sci. 1993, *46*:309–332.
14. *Static and Dynamic Light Scattering in Medicine and Biology* (Nossal, R. J., and Pecora, R., eds.), SPIE Proceedings Series 1884, SPIE, Bellingham, WA, 1993.
15. *Light Scattering: Principles and Developments* (Brown, W., ed.), Clarendon Press, Oxford, 1995.
16. *Dynamic Light Scattering: Applications of Photon Correlation Spectroscopy* (Pecora, R., ed.), Plenum, New York, 1985.
17. Bohren, C. F., and Huffman, D. R., *Adsorption and Scattering of Light by Small Particles*, John Wiley, New York, 1983, pp. 25–28.
18. Landau, L. D., Lifshitz, E. M., and Pitaevskiĭ, L. P., *Electrodynamics of Continuous Media*, 2d ed., Pergamon Press, Oxford, 1984.
19. Becker, R., *Electromagnetic Fields and Interactions*, Dover, New York, 1964.

20. Liebert, R. B., and Prieve, D. C. Ind. Engr. Chemistry Res. 1995, *34*:3542.
21. Walz, J. Y., and Prieve, D. C. Langmuir, 1992, *8*:3043.
22. Feynman, R. P., Leighton, R. B., and Sands, M., *The Feynman Lectures on Physics*, Vol. II, Addison-Wesley, Reading, MA, 1964, Ch. 21.
23. Chu, B., *Laser Light Scattering*, 2d ed., Academic Press, San Diego, 1991.
24. Appl. Opt. 1997, *36*(30):7493–7507.
25. Flammer, I., and Ricka, J. Appl. Opt. 1997, *36*(30):7508–7517.
26. Rovati, L., Fankhauser, F., II, and Ricka, J. Rev. Sci. Instrum. 1996, *67*(7):2615–2620.
27. Gisler, T., Rüger, H., Egelhaaf, S.U., Tschumi, J., Schurtenberger, P., and Ricka, J. Appl. Opt. 1995, *34*(18):3546–3553.
28. Könz, F., Ricka, J., Frenz, M., and Fankhauser, F., II. Opt. Eng. 1995, *34*(8):2390–2395.
29. Suparno, Deurloo, K., Stamatelopolous, P., Srivastva, R., and Thomas, J. C. Appl. Opt. 1994, *33*(20):7200–7205.
30. Dhadwal, H. S., Khan, R. R., and Suh, K. Appl. Opt. 1993, *32*(21):3901–3904.
31. McClymer, J. P. Rev. Sci. Instrum. 1990, *61*(7):2001–2002.
32. Dautet, H., Deschamps, P., Dion, B., MacGregor, A., MacSween, D., McIntyre, R., Trottier, C., and Webb, P. Appl. Opt. 1993, *32*(21): 3894–3900.
33. Podoleanu, A. G., Harding, R. K., and Jackson, D. A. Appl. Opt. 1997, *36*(30): 7523–7530.
34. Phillies, G. D. J. Rev. Sci. Instrum. 1996, *67*(10):3423–3427.
35. Sampson, D. D., Dove, W. T., and Jackson, D. A. Appl. Opt. 1993, *32*(21):3905–3916.
36. Finsy, R. Adv. Col. Int. Sci. 1994, *52*:79–143.
37. Brown, G. W., and Smart, A. E. Appl. Opt. 1997, *36*(30):7480–7492.
38. Henry, C. Anal. Chem. 1998, *70*(1):59A–63A.
39. Finsy, R., and Jaeger, N. D. Part. Part. Syst. Charact. 1991, *8*:187–193.
40. Schätzel, K., in *Dynamic Light Scattering: The Method and Some Applications* (Brown, W., ed.), Clarendon Press, Oxford, 1993.
41. McQuarrie, D. A., *Statistical Mechanics*, Harper and Row, New York, 1975, Chs., 13, 21, and 22.
42. Thomas, J. C., in *Photon Correlation Spectroscopy: Multicomponent Systems* (Schmitz, K. S., ed.), SPIE Proceedings Series 1430, SPIE, Bellingham, WA, 1991.
43. Kerker, M., *The Scattering of Light, and Other Electromagnetic Radiation*, Academic Press, New York, 1969.
44. van de Hulst, H. C., *Light Scattering by Small Particles*, John Wiley, New York, 1957.
45. Cantu, L., Corti, M., Lago, P., Musolino, M., in *Photon Correlation Spectroscopy: Multicomponent Systems* (Schmitz, K. S., ed.), SPIE Proceedings Series 1430, SPIE, Bellingham, WA, 1991.
46. Pike, E. R., and McNally, B. Appl. Opt. 1997, *36*(30):7531–7538.
47. Bryant, G., Abeynayake, C., and Thomas, J. C. Langmuir 1996, *12*(26):6224–6228.

48. Johnsen, R. M., and Brown, W., in *Laser Light Scattering in Biochemistry* (Harding, S. E., Sattelle, D. B., and Bloomfield, V. A., eds.), Royal Society of Chemistry, Cambridge, 1992.
49. Stepanek, P., in *Dynamic Light Scattering: The Method and Some Applications* (Brown, W., ed.), Clarendon Press, Oxford, 1993.
50. Schätzel, K. Appl. Phys. B 1987, *42*:193–213.
51. Ruf, H., Grell, E., and Stelzer, E. H. K. Eur. Biophys. J. 1992, *21*:21–28.
52. Press, W. H., Flannery, B. P., Teukolsky, S. A., and Vetterling, W. T. *Numerical Recipes*, Cambridge University Press, New York, 1988.
53. Barrodale, I., and Zala, C., in *Numerical Algorithms* (Mohamed, J. L., Walsh, J., eds.), Oxford University Press, Oxford, 1986.
54. Fletcher, G. C., and Ramsay, D. J. Optica Acta, 1983, *30*(8):1183–1196.
55. Bargeron, C. B. J. Chem. Phys. 1974, *60*(6):2516–2519.
56. Brookhaven Instruments Corporation's Data Analysis Software (BI-ISDA), version 6.5, Brookhaven Instruments Corporation (www.bic.com), Holtsville, New York, 1989.
57. Bargeron, C. B. J. Chem. Phys. 1974, *61*(5):2134–2138.
58. Livesey, A. K., Licinio, P., and Delaye, M. J. Chem. Phys. 1986, *84*(9):5102–5107.
59. Provencher, S. W. Computer Phys. Comm. 1982, *27*:213–227.
60. Kopple, D. E. J. Chem. Phys. 1972, *57*(11):4814–4820.
61. Pusey, P. N., Koppel, D. E., Schaefer, D. W., Camerini-Otero, R. D., and Koenig, S. H. Biochemistry 1974, *13*(5):952–960.
62. Shaumeyer, J. N., Briggs, M. E., and Gammon, R. W. Appl. Opt. 1993, *32*(21):3871–3878.
63. Peters, R., in *Dynamic Light Scattering: The Method and Some Applications* (Brown, W., ed.), Clarendon Press, Oxford, 1993.
64. Lorusso, G. F., Minafra, A., and Capozzi, V. Appl. Opt. 1993, *32*(21):3867–3870.
65. Honerkamp, J., Maier, D., and Weese, J. J. Chem. Phys. 1993, *98*(2):865–872.
66. Will, S., and Leipertz, A. Appl. Opt. 1993, *32*(21):3813–3821.
67. Beretta, S., Lunelli, L., Chirico, G., and Baldini, G. Appl. Opt. 1996, *35*(19):3763–3770.
68. Stöber, W., Fink, A. E., and Bohn, E. J. Col. Int. Sci. 1968, *26*(1):62–69.
69. van Blaaderen, A., and Vrij, A., in *The Colloid Chemistry of Silica* (Bergna, H. E., ed.), ACS Symposium Series 234, American Chemical Society, Washington, DC, 1994.
70. Brookhaven Instruments Corporation's Data Analysis Software (BI-ISDA); version 6.5, Brookhaven Instruments Corporation (www.bic.com), Holtsville, New York, 1989.
71. A Hitachi model H-8000 electron microscope, an AMT 1024 × 1024 digital Kodak camera system with version 1.55 software, and NIH's image analysis software were used for the TEM analysis. The TEM images were calibrated at each magnification using a 0.5 μm diffraction grating coated with 261 nm polystyrene particles (Electron Microscopy Sciences, #80055). The particles were used as the primary size standard and the lines of the grating as a secondary

standard. TEM analysis of 336 particles yielded a number-weighted diameter of 144.8 ± 35.9 based on ellipsoid axes measurements. Area measurements based on a circular cross section yielded equivalent results.

72. Thomas, J. C. J. Col. Int. Sci. 1987, *117*(1):187–192.
73. Hanus, L. H., and Ploehn, H. J. submitted for publication in Langmuir.
74. Hallet, F. R., Craig, T., Marsh, J., and Nickle, B. Can. J. Spectrosc. 1989, *34*(3):63–70.
75. Ostrowsky, N., Sornette, D., Parker, P., and Pike, E. R. Optica Acta 1981, *28*(8):1059–1070.
76. Chu, B., and DiNapoli, A., in *Measurement of Suspended Particles by Quasi-Elastic Light Scattering* (Dahneke, B. E., ed.), John Wiley, New York, 1983.
77. Morrison, I. D., and Grabowski, E. F., in *Measurement of Suspended Particles by Quasi-Elastic Light Scattering* (Dahneke, B. E., ed.), John Wiley, New York, 1983.
78. Itoh, M., and Takahashi, K. J. Aerosol Sci. 1991, *22*(7):815–822.
79. Schätzel, K. Appl. Opt. 1993, *32*(21):3880–3885.
80. Provencher, S. W., and Vogel, R. H., in *Numerical Treatment of Inverse Problems in Differential and Integral Equations* (Deuflhard, P., and Hairer, E., eds.), Birkhäuser, Boston, 1983, pp. 304–319.
81. Provencher, S. W., in *Laser Light Scattering in Biochemistry* (Harding, S. E., Sattelle, D. B., and Bloomfield, V. A., eds.), Royal Society of Chemistry, Cambridge, 1992.
82. Provencher, S. W., and Stepanek, P. Part. Part. Syst. Charact. 1996, *13*:291.
83. Morrison, I. D., Grabowski, E. F., and Herb, C. A. Langmuir 1985, *1*(4):496–501.
84. Egelhaaf, S. U., and Schurtenberger, P. Rev. Sci. Instrum. 1996, *67*(2):540–545.
85. Weiner, B. B., and Tscharnuter, W. W., in *Particle Size Distribution: Assessment and Characterization* (Provder, T. ed.), ACS Symposium Series 332, American Chemical Society, Washington, DC, 1987.
86. Weber, R., and Schweiger, G. J. Aerosol Sci. 1996, *26*:S29–S30.
87. Weber, R., and Schweiger, G. J. Aerosol Sci. 1996, *27*:S533-S534.
88. Caldwell, K. D., and Li, J. J. Col. Int. Sci. 1989, *132*(1):256–268.
89. Ansari, R. R., Harbans, D. S., Cheung, H. M., and Meyer, W. V. Appl. Opt. 1993, *32*(21):3822–3827.
90. Mazer, N. A., in *Dynamic Light Scattering: Applications of Photon Correlation Spectroscopy* (Pecora, R., ed.), Plenum, New York, 1985.
91. Rigsbee, D. R., and Dubin, P. L., in *Photon Correlation Spectroscopy: Multicomponent Systems* (Schmitz, K. S., ed.), SPIE Proceedings Series 1430, SPIE, Bellingham, WA, 1991.
92. Magid, L., in *Dynamic Light Scattering: The Method and Some Applications* (Brown, W., ed.), Clarendon Press, Oxford, 1993.
93. *Dynamic Light Scattering: The Method and Some Applications* (Brown, W., ed.), Clarendon Press, Oxford, 1993.
94. Lin, M. Y., Lindsay, H. M., Weitz, D. A., Ball, R. C., Klein, R., and Meakin, P. Nature 1989, *339*:360–362.

95. Lin, M. Y., Lindsay, H. M., Weitz, D. A., Klein, R., Ball, R. C., and Meakin, P. J. Phys.: Condens. Matter 1990, 2:3093–3113.
96. Lin, M. Y., Lindsay, H. M., Weitz, D. A., Ball, R. C., Klein, R., and Meakin, P. Phys. Rev. A 1990, 41(4):2005–2020.
97. Carpineti, M., and Giglio, M. Adv. Col. Int. Sci. 1993, 46:73–90.
98. Hidalgo-Alvarez, R., Martin, A., Fernandez, A., Bastos, D., Martinez, F., and de las Nieves, F. J. Adv. Col. Int. Sci. 1996, 67:1–118.
99. Holthoff, H., Egelhaaf, S. U., Borkovec, M., Schurtenberger, P., and Sticher, H. Langmuir 1996, 12(23):5541–5549.
100. Kyriakidis, A. S., Yiatsios, S. G., and Karabelas, A. J. J. Col. Int. Sci. 1997, 195:299–306.
101. Holthoff, H., Schmitt, A., Fernandez-Barbero, A., Borkovec, M., Cabrerizo-Vilchez, M. A., Schurtenberger, P., and Hidalgo-Alvarez, R. J. Col. Int. Sci. 1997, 192(2):463–470.
102. Holthoff, H., Borkovec, M., and Schurtenberger, P. Phys. Rev. E 1997, 56(6):6945–6953.
103. Schmitz, K. N., *Dynamic Light Scattering by Macromolecules*, Academic Press, New York, 1990.
104. Phillies, G. D. J. Phys. Today 1991, 44:66.
105. Refs. in Sect. 7 of Brown, G. W., and Smart, A. E. Appl. Opt. 1997, 36(30):7480–7492.
106. Phillies, G. D. J. Anal. Chem. 1990, 62(20):1049A–1057A.
107. Chu, B. Anal. Chem. 1995, 67(5):182A.
108. Appl. Opt. 1997, 36(30):7477–7677.
109. Appl. Opt. 1993, 32(21):3811–3916.
110. *Photon Correlation Spectroscopy: Multicomponent Systems* (Schmitz, K. S., ed.), SPIE Proceedings Series 1430, SPIE: Bellingham, WA, 1991.
111. *Particle Size Distribution: Assessment and Characterization* (Provder, T., ed.), ACS Symposium Series 332, American Chemical Society: Washington, DC, 1987.
112. *Measurement of Suspended Particles by Quasi-Elastic Light Scattering* (Dahneke, B. E., ed.), John Wiley, New York, 1983.
113. *Scattering Techniques Applied to Supramolecular and Non-equilibrium Systems* (Chen, S. H., Nossal, R., and Chu, B. eds.), Plenum, New York, 1981.
114. Refs. of Meyer, W. V., Smart, A. E., Brown, R. G. W., and Anisimov, M. A. Appl. Opt. 1997, 36(30):7477–7479.
115. Pusey, P. N. J. Phys. A 1975, 8:1433.
116. Brown, J. C., Pusey, P. N., Goodwin, J. W., and Ottewill, R. H. J. Phys. A 1975, 8(5):664–682.
117. Pusey, P. N., and Tough, R. J. A., in *Dynamic Light Scattering: Applications of Photon Correlation Spectroscopy* (Pecora, R., ed.), Plenum, New York, 1985.
118. van Megan, W., Underwood, S. M., Ottewill, R. H., Williams, N., and Pusey, P. N. Faraday Discuss. Chem. Soc. 1987, 83:47–57.
119. van Megan, W., and Underwood, S. M. J. Chem. Phys. 1989, 91(1):552–559.
120. Fytas, G., and Patkowski, A., in *Dynamic Light Scattering: The Method and Some Applications* (Brown, W., ed.), Clarendon Press, Oxford, 1993.

121. Zero, K., and Pecora, R., in *Dynamic Light Scattering: Applications of Photon Correlation Spectroscopy* (Pecora, R., ed.), Plenum, New York, 1985.
122. Earnshaw, J. C., and McCoo, E. Langmuir 1995, *11*(4):1087–1100.
123. Earnshaw, J. C. Appl. Opt. 1997, *36*(30):7583–7592.
124. Appl. Opt. 1997, *36*(30):7583–7628.
125. Earnshaw, J. C. Adv. Col. Int. Sci. 1996, *68*:1–29.
126. Appl. Opt. 1997, *36*(30):7551–7577.
127. Phillies, G. D. J. Chem. Phys. 1981, *74*:260–262.
128. Sorensen, C. M., Mockler, R. C., and O'Sullivan, W. J. Phys. Rev. A 1976, *14*(4):1520–1532.
129. Weitz, D. A., and Pine, D. J., in *Dynamic Light Scattering: The Method and Some Applications* (Brown, W., ed.), Clarendon Press, Oxford, 1993.
130. J. Opt. Soc. Am. A 1997, *14*(1):136–342.
131. Appl. Opt. 1997, *36*:9–231.
132. Wiersma, P. H., Loeb, A. L., and Overbeek, J. Th. G. J. Colloid Interface Sci. 1966, *22*:78.
133. O'Brien, R. W., and White, L. R. J. Chem. Soc., Faraday Trans. 2 1978, *74*:1607.
134. Ohshima, H. J. Col. Int. Sci. 1996, *179*:431–438.
135. Mangelsdorf, C. S., and White, L. R. J. Chem. Soc., Faraday Trans., 1997, *93*(17):3145–3154.
136. Ware, B. R., and Flygare, W. H. Chem. Phys. Lett. 1971, *12*(1):81–85.
137. Miller, J. F., Schätzel, K., and Vincent, B. J. Colloid Interface Sci. 1991, *143*(2):532–554.
138. Durst, F., Melling, A., and Whitelaw, J. H., *Principles and Practice of Laser-Doppler Anemometry*, 2d ed., Academic Press, London, 1981.
139. Schätzel, K., and Merz, J. J. Chem. Phys. 1984, *81*(5):2482–2488.
140. Miller, J. F., Ph.D. dissertation, University of Bristol, 1990.
141. Gimsa, J., Eppmann, P., and Prüger, B. Biophys. J. 1997, *73*:3309–3316.
142. Miller, J., Velev, O., Wu, S. C. C., and Ploehn, H. J. J. Col. Int. Sci. 1995, *174*:490–499.
143. Larsson, A., and Rasmusson, M. Carbohydrate Research, 1997, *304*:315–323.
144. Synthesized using the methods described in Mahtab, R., Rogers, J. P., Singleton, C. P., and Murphy, C. J. J. Am. Chem. Soc. 1996, *118*(30):7028–7032.
145. Bastos, D., and de las Nieves, F. J. Coll. Polym. Sci. 1996, *274*:1081–1088.
146. Gittings, M. R., and Saville, D. A. Langmuir 1995, *11*(3):798–800.
147. Krabi, A., Allan, G., Donath, E., and Vincent, B. Coll. Surf. A 1997, *122*:33–42.
148. Donath, E., Walther, D., Shilov, V. N., Knippel, E., Budde, A., Lowack, K., Helm, C. A., and Möhwald, H. Langmuir 1997, *13*(20):5294–5305.
149. Tuin, G., Senders, J. H. J. E., and Stein, H. N. J. Col. Int. Sci. 1996, *179*:522–531.
150. Dietrich, P. G., Lerche, K. H., Reusch, J., and Nitzsche, R. Chromatographia 1997 *44*(7/8):362–366.
151. Guindo, M. C., Zurita, L., Duran, J. D. G., and Delgado, A. V. Mater. Chem. Phys. 1996, *44*:51–58.
152. Boxall, C., and Kelsall, G. H. J. Chem. Soc., Faraday Trans. 1991, *87*(12):3537–3545.

153. Seaman, J. C., Bertsch, P. M., and Strom, R. N. Environ. Sci. Technol. 1997, *31*(10):2782–2790.
154. Boxall, C., Kelsall, G., and Zhang, Z. J. Chem. Soc., Faraday Trans. 1996, *92*(5):791–802.
155. Malyschew, A., Schmidt, H. J., Weil, K. G., and Hoffmann, P. Atmospheric Environment 1994, *28*(9):1575–1581.
156. Chandrakanth, M. S., and Amy, G. L. Environ. Sci. Technol. 1996, *30*(2):431–443.
157. Ongerth, J. E., and Pecoraro, J. P. J. Environ. Eng. 1996, *122*(3):228–231.
158. Peula-Garcia, J. M., Hidalgo-Alvarez, R., and de las Nieves, F. J. Coll. Surf. A 1997, *127*:19–24.
159. Xia, J., Dubin, P. L., Kim, Y., Muhoberac, B. B., and Klimkowski, V. J. J. Chem. Phys. 1993, *97*(17):4528–4534.
160. McNeil-Watson, F. K., and Parker, A., in Laser Light Scattering in Biochemistry (Harding, S. E., Sattelle, D. B., and Bloomfield, V. A., eds.), Royal Society of Chemistry, Cambridge, 1992.
161. Langley, K. H., in Laser Light Scattering in Biochemistry (Harding, S. E., Sattelle, D. B., and Bloomfield, V. A., eds.), Royal Society of Chemistry, Cambridge, 1992.
162. Brindley, A., Davis, S. S., Davies, M. C., Watts, J. F. J. Col. Int. Sci. 1995, *171*:150–161.
163. Fritz, H., Maier, M., and Bayer, E. J. Col. Int. Sci. 1997, *195*:272–288.

8
Light Scattering Studies of Microcapsules in Suspension

TOSHIAKI DOBASHI Department of Biological and Chemical Engineering, Gunma University, Kiryu, Gunma, Japan

BENJAMIN CHU Department of Chemistry, State University of New York at Stony Brook, Stony Brook, New York

I.	Introduction	250
II.	Theoretical Background	251
	A. Static light scattering	251
	B. Dynamic light scattering	253
III.	Experimental Applications	257
	A. Membrane thickness determination by combination of light and x-ray scattering	257
	B. Adsorbed layer thickness determination by combination of dynamic light scattering and adsorption isotherm	258
	C. Micellar and microemulsion shell thickness determination by combination of light scattering and neutron scattering	261
	D. Gel-to-liquid-crystalline phase transition of liposomes by dynamic light scattering and anisotropy measurements	262
	E. Osmotic swelling of living cells by combination of static and dynamic light scattering	262
	F. Scattering from a single microcapsule	263
	G. Scanning near field microscopy	265
	References	266

I. INTRODUCTION

The term microcapsule is used for a variety of complex materials including not only synthetic microcapsules but also microemulsions, emulsions, micelles, coated latices, vesicles (such as liposomes) and even living biological cells [1–4]. The key feature of these colloidal entities is that they are spheres having an inner core and an outer shell. A microemulsion is sometimes described as a dispersed microphase having a flexible interfacial film separating it from the dispersing medium [5]. Micelles can often be considered to be composed of a liquid spherical core and a heavily solvated corona [6]. Latex particles are frequently coated by bioactive agents, in which the core can be regarded as an amorphous polymer solid. Biomembranes are fluidlike lipid bilayers into which proteins are embedded with differentiated lateral domains and that are supported by cytoskeletons [7]. The size of most microcapsules is in the range from 0.01 μm to 100 μm, for which light scattering is an appropriate characterization technique, complementary to optical and electron microscopy.

From static and dynamic light scattering experiments, the size and size distribution, shape, and anisotropy of microcapsules, the thickness of the membrane, and the aggregation behavior of microcapsules can be detected in a noninvasive way. Important applications of scattering methods on each material have been surveyed in reviews and articles [8–13]. It is advantageous to use a combination of different techniques, such as static and dynamic light scattering, small-angle x-ray scattering, and small-angle neutron scattering, to characterize the complex structure of microcapsules.

Microcapsules are classified according to several aspects. When the outer shell is charged, the suspension that consists of microcapsules and the dispersing medium is stabilized mainly by Coulombic interactions. On the other hand, when the outer shell is not charged, the suspension is unstable unless some protective colloids are adsorbed on the microcapsule surface. The surface roughness of the charged shell depends on the distribution of charges on the surface, whereas the protective polymer for the noncharged shell can take various conformations according to the nature of the polymer, including molecular weight, concentration, and even temperature. In either case, the fine structure of the surface between the shell and the dispersing medium can be related to the stability of the suspension.

The microcapsule membrane can also be divided into two categories depending on the constituent molecules, e.g., cross-linked polymer networks or associated low-molecular-weight components held together by weaker forces such as hydrophobic interactions. The membrane having a cross-linked network is much more stable than that consisting of low-molecular-weight components when thermodynamic conditions are changed. For example, liposomes, emulsions/microemulsions, and micelles can easily change their size and structure depending on temperature and concentration. Microcapsules are usually polydisperse. They tend to aggregate easily and to grow in size. To determine the structure of microcapsules, it is advisable to be aware of all these intrinsic properties of microcapsules.

In Section II of this review, we describe the theoretical background and in Section III we give examples of practical applications to determine the microcapsule structure by using a combination of different scattering techniques. We also touch on recent topics, such as scanning near field microscopy.

II. THEORETICAL BACKGROUND

A. Static Light Scattering

The form factors of particles with different sizes and shapes were theoretically derived by scientists in a variety of fields, such as biophysics, astronomy, and physical chemistry. The books by Kerker [14] and van de Hulst [15] are among the recommended texts for studying light scattering of small particles. A table for the form factor of particles with various shapes was given by Burchard and Patterson [16].

The simplest case is for an optically isotropic spherical shell [17,18]. Then the particle scattering (or form) factor $P(q)$ has the form

$$P(q) = \left[\frac{3}{(r_o q)^3 (1-\gamma^3)}\right]^2 \qquad (1)$$

$$\times \left[\sin(r_o q) - \sin(\gamma r_o q) - r_o q \cos(r_o q) + r_o q \gamma \cos(r_o q \gamma)\right]^2$$

where r_o and γ are the outer radius of the shell and the inner-to-outer radius ratio; q is the magnitude of the scattering wave vector [$= (4\pi n_s/\lambda) \sin(\theta/2)$] with n_s, λ and θ being the refractive index of the dispersing medium, the wavelength in vacuum and the scattering angle, respectively. In case the membrane is infinitely thin, Eq. (1) is replaced by [19,20]

$$P(q) = \left[\frac{\sin(rq)}{rq}\right]^2 \qquad (2)$$

and in case the inner diameter of the shell is infinitely small (solid sphere), Eq. (1) is reduced to [21]

$$P(q) = \left[\frac{3}{(r_o q)^3}\right]^2 [\sin(r_o q) - r_o q \cos(r_o q)]^2 \qquad (3)$$

In a double logarithmic plot of $P(q)$ vs. q, the asymptotic slope at high q of the upper envelope of the scattering curve depends on the ratio of the wall membrane thickness to the outer diameter, e.g., -2 for an infinitely thin shell from Eq. (2) and -4 for a solid sphere from Eq. (3). By fitting an experimental scattering curve to the theoretical functional form of Eq. (1), two parameters, outer radius and inner-to-outer radius ratio, can be determined simultaneously by using a least squares method.

When the interfacial thickness between the membrane and the dispersing medium is finite, a term proportional to $\exp(-\sigma^2 q^2)$ is multiplied by $P(q)$ in Eqs. (1–3) at large q, where σ denotes an index of the surface thickness [22,23]. Thus σ can be determined from the asymptotic slope of the semilogarithmic plot of $P(q)q^2$ (infinitely thin shell) or $P(q)q^4$ (solid sphere) versus q^2.

The form of microcapsules is frequently expressed by an ellipsoidal hollow shell of revolution rather than a spherical shell. In the case of a thin oblate ellipsoidal shell, the form factor is represented by

$$P(x) = \int_0^1 \frac{\sin^2[x(1-Qt^2)^{1/2}]}{x^2(1-Qt^2)} dt \tag{4}$$

where $Q = 1 - \rho^{-2}$, $x = qa$, and ρ is the axial ratio defined as a/b with a the major semiaxis and b the minor semiaxis of the ellipsoidal shell. This equation also holds for the thin prolate ellipsoidal shell if Q and x are replaced by $-Q$ and qb, respectively [24]. Light scattering from particles of general shape is a very difficult topic. However, we note that the T-matrix method or the extended-boundary-condition method is often efficient in calculating the light scattering properties of fairly complex systems [25].

The form factor for anisotropic spherical shells is related not only to the size and the shape but to the anisotropy of the spherical shells [18,26–28]. The form factor for an unpolarized incident beam and without an analyzer is related to the partial scattering factors by the relation

$$P_T(q) = \frac{P_{Vv}(q) + P_{Vh}(q) + P_{Hv}(q) + P_{Hh}(q)}{1 + \cos^2\theta} \tag{5}$$

where H and V denote the horizontal and vertical components of the scattered light, respectively, and h and v the horizontal and vertical components of the incident light, respectively. In case of cylindrical symmetry of the radially oriented scattering elements in a particle,

$$P_{Vh}(q) = P_{Hv}(q) = 0 \tag{6}$$

$$P_{Hh} = \left[d_0 \cos\theta - \left(1 + \frac{\cos\theta}{3}\right) \frac{\delta_h d_2'}{2} \right]^2 \tag{7}$$

and

$$P_{Vv} = \left[d_0 + \frac{\delta_h d_2'}{3} \right]^2 \tag{8}$$

where d_0 and d_2' are the isotropic and anisotropic form factors of a single-layered shell, δ_h the anisotropy ratio defined by $(\alpha_\parallel - \alpha_\perp)/\langle\alpha\rangle$ with $\alpha_\parallel$, $\alpha_\perp$ and $\langle\alpha\rangle$ being the component of the polarizability along the long axis, the component perpen-

LS Studies of Microcapsules in Suspension

dicular to it, and the average polarizability given by $(\alpha_\parallel + 2\alpha_\perp)/3$, respectively. The isotropic form factor d_0 is given by the square root of $P(q)$ in Eqs. (1–3), and the anisotropic form factor d_2' is given by [18,27]

$$d_2' = \frac{3}{(r_o q)^3 (1 - \gamma^3)} [3Si(r_o q) - 3Si(r_o q\gamma) + r_o q \cos(r_o q) \qquad (9)$$
$$- r_o q\gamma \cos(r_o q\gamma) - 4\sin(r_o q) + 4\sin(r_o q\gamma)]$$

for hollow and solid spheres and

$$d_2' = \frac{3}{(rq)^3} \left[\sin(rq) - rq \cos(rq) - \frac{(rq)^2}{3} \sin(rq) \right] \qquad (10)$$

for an infinitely thin sphere.

B. Dynamic Light Scattering

Dynamic light scattering is a powerful tool to determine the size, the size distribution, and the shape of the particles in suspension. Theoretical and experimental aspects of this technique are available in standard monographs [29–31].

The intensity autocorrelation function $G^{(2)}(t)$ has the form

$$G^{(2)}(t) = A(1 + \beta g^{(1)}(t)^2) \qquad (11)$$

where t is the delay time, A the base line, β a spacial coherence factor, and $g^{(1)}(t)$ the first-order normalized electric field time correlation function, which is related to the normalized characteristic line width (Γ) distribution $G(\Gamma)$ by the Laplace transformation

$$g^{(1)}(t) = \int G(\Gamma) \exp(-\Gamma t) \, d\Gamma \qquad (12)$$

The z-average translational diffusion coefficient D_T is calculated from the equation $D_T = \Gamma/q^2$. For a collection of identical spheres undergoing ordinary Brownian motion in solution,

$$|g^{(1)}(t)| = \exp(-D_T q^2 t) \qquad (13)$$

According to the Einstein relation, the diffusion coefficient is inversely proportional to the translational friction coefficient f at infinite dilution by the expression

$$D_{T0} = \frac{k_B T}{f} \qquad (14)$$

Here, k_B is the Boltzmann constant, T the absolute temperature, and D_{T0} the translational diffusion coefficient extrapolated to zero concentration. The friction coefficient for a sphere of radius r is given by the Stokes law:

$$f = 6\pi \eta r \qquad (15)$$

where η is the solvent viscosity. The size distribution for a collection of spheres with heterogeneous diameters can be calculated through the inversion of the field correlation function by an inverse Laplace transform or by using the cumulants method.

The friction coefficient of an ellipsoidal shell should be hydrodynamically equivalent to that of an ellipsoid of revolution [24] and is given by [32,33]

$$f = \frac{6\pi \eta b (a^2/b^2 - 1)^{1/2}}{\arctan(a^2/b^2 - 1)^{1/2}} \tag{16}$$

for an oblate with semiaxes a, a, and b, and

$$f = \frac{6\pi \eta a (1 - b^2/a^2)^{1/2}}{\ln[\{1 + (1 - b^2/a^2)^{1/2}\}a/b]} \tag{17}$$

for a prolate with semiaxes a, b, and b, where a is the major semiaxis and b is the minor semiaxis of the ellipsoid.

The equivalent friction coefficient of the assembly for a sphere surrounded by a loosely packed shell of adsorbed molecules is expressed as [34,35]

$$f = 6\pi \eta r_o \left\{ 1 - \lambda_2^{-1} \frac{\lambda_3^{-1} \tanh(\lambda_2 - \lambda_3) + \lambda_1^{-1} \tanh(\lambda_1)}{\lambda_3^{-1} + \lambda_1^{-1} \tanh(\lambda_1) \tanh(\lambda_2 - \lambda_3)} \right\} \tag{18}$$

where $\lambda_1 = [(2\xi\rho_1)/(3\eta)]^{1/2} r_i$, $\lambda_2 = [(2\xi\rho_2)/(3\eta)]^{1/2} r_o$, and $\lambda_3 = [(2\xi\rho_2)/(3\eta)]^{1/2} r_i$, with ξ the friction coefficient of a polymer subunit, r_i and r_o the radii of the sphere and the shell, and ρ_1 and ρ_2 the number densities of the particles packing the sphere and the shell. For an impenetrable sphere, Eq. (18) is reduced to

$$f = 6\pi \eta r_o \left[1 - \frac{\tanh(\lambda_2 - \lambda_3)}{\lambda_2} \right] \tag{19}$$

For nonspherical particles, $|g^{(1)}(t)|$ in Eq. (13) is replaced by

$$|g^{(1)}(t)| = \exp(-D_T q^2 t)(1 + B_2 \exp(-6D_R t) + \cdots) \tag{20}$$

where D_R is the rotational diffusion coefficient and B_2 is a coefficient that depends on the size and shape of the particles.

Though we cannot determine the fine internal structure of the particle with the current technique by using dynamic light scattering, the q^2 dependence of the diffusion coefficient obtained from dynamic light scattering strongly depends on the shell structure of vesicles [36].

Particle scattering factors and translational friction coefficients of microcapsules with different shapes are summarized in Table 1.

TABLE 1 Particle Scattering Factor and Translational Friction Coefficient of Microcapsule

A. Particle scattering factor
1. Isotropic

Model	Particle scattering factor $P(q)$	Defining parameters	Ref.
Hollow sphere			
(Finite shell thickness)	$[3/\{(r_o q)^3(1-\gamma^3)\}]^2[\sin(r_o q) - \sin(\gamma r_o q) - r_o q\cos(r_o q) + r_o q\gamma\cos(r_o q\gamma)]^2$	r_o; outer radius, γ; inner-to-outer radius ratio $\gamma = r_i/r_o$	17,18
(Infinitely thin)	$[\sin(rq)/(rq)]^2$	r; radius	19,20
Solid sphere	$[3/(rq)^3]^2[\sin(rq) - rq\cos(rq)]^2$	r; radius	21
Thin ellipsoidal shell			
(Oblate, semiaxes a, a, b)	$\int_0^1 \sin^2[x(1-Qt^2)^{1/2}]/[x^2(1-Qt^2)]\,dt$	$Q = 1 - \rho^{-2}$, $x = qa$,	24
(Prolate, semiaxes a, b, b)	$\int_0^1 \sin^2[x(1+Qt^2)^{1/2}]/[x^2(1+Qt^2)]\,dt$	$Q = 1 - \rho^{-2}$, $x = qb$ a; major semiaxis, b; minor semiaxis ρ; axial ratio a/b	

2. Anisotropic

Model	Particle scattering factor	Defining parameters	Ref.
Hollow and solid sphere (radially oriented scattering elements with cylindrical symmetry)	$(P_{Vv}(q) + P_{Vh}(q) + P_{Hv}(q) + P_{Hh}(q))$ $/(1+\cos^2(\theta))$ $P_{Vh}(q) = P_{Hv}(q) = 0,$ $P_{Hh} = [d_0\cos\theta - (1+(\cos\theta)/3)\delta_h d_2'/2]^2$ $P_{Vv} = [d_0 + (\delta_h d_2')/3]^2$	H, V; horizontal and vertical components of scattered light h, v; horizontal and vertical components of incident light δ_h; the anisotropy ratio $(\alpha_\parallel - \alpha_\perp)/\langle\alpha\rangle$, θ; scattering angle $\alpha_\parallel, \alpha_\perp, \langle\alpha\rangle$; components of polarizability along the long axis, perpendicular to it and the average polarizability $(\alpha_\parallel + 2\alpha_\perp)/3$ d_0; isotropic form factor $P(q)^{1/2}$, d_2'; anisotropic form factor	18,26–28

(continued)

TABLE 1 (*Continued*)

A. Particle scattering factor
2. Anisotropic

Model	d'_2
Hollow sphere	
(Finite thickness)	$[3/\{r_o q\}^3(1-\gamma^3)]\{[3\text{Si}(r_o q) - 3\text{Si}(r_o q\gamma) + r_o q \cos(r_o q) - r_o q\gamma \cos(r_o q\gamma) - 4\sin(r_o q) + 4\sin(r_o q\gamma)]$
(Infinitely thin)	$[3/(rq)^3][\sin(rq) - rq\cos(rq) - \{(rq)^2/3\}\sin(rq)]$
Solid sphere	$[3/(rq)^3][3\text{Si}(rq) + rq\cos(rq) - 4\sin(rq)]$

B. Translational friction coefficient

Model	Translational friction coefficient	Defining parameters	Ref.
Sphere	$6\pi\eta r$	η; solvent viscosity, r; radius	
Ellipsoid of revolution (Oblate, semiaxes a, a, b) (Prolate, semiaxes a, b, b)	$6\pi\eta b(a^2/b^2 - 1)^{1/2}/\arctan(a^2/b^2 - 1)^{1/2}$ $6\pi\eta a(1 - b^2/a^2)^{1/2}/\ln[\{1 + (1 - b^2/a^2)^{1/2}\}a/b]$	a; major semiaxis, b; minor semiaxis	32,33
Sphere surrounded by loosely packed shell of adsorbed polymer	$6\pi\eta r_o\{1 - \lambda_2^{-1}[\lambda_3^{-1}\tanh(\lambda_2 - \lambda_3) + \lambda_1^{-1}\tanh(\lambda_1)]$ $/[\lambda_3^{-1} + \lambda_1^{-1}\tanh(\lambda_1)\tanh(\lambda_2 - \lambda_3)]\}$ $\lambda_1 = [(2\xi\rho_1)/(3\eta)]^{1/2}r_i, \lambda_2 = [(2\xi\rho_2)/(3\eta)]^{1/2}r_o,$ $\lambda_3 = [(2\xi\rho_2)/(3\eta)]^{1/2}r_i$	ξ; friction coefficient of polymer subunit r_i, r_o; radii of sphere and shell ρ_1, ρ_2; number densities of the particles packing the sphere and the shell	34,35

III. EXPERIMENTAL APPLICATIONS

A. Membrane Thickness Determination by Combination of Light and X-Ray Scattering

Several approaches have been developed to simplify the study of the complex structure of microcapsules. A typical example can be given for a microcapsule with no charge on the surface. If we prepare a model microcapsule whose membrane and core have the same refractive index but different electron densities, we can obtain the core size from small-angle x-ray scattering (SAXS) and the microcapsule shell size from a combination of static light scattering (SLS) and dynamic light scattering (DLS).

One of the usual methods of preparing microcapsules is by interfacial polymerization, which is typically used for microcapsules as recording agents on facsimile papers. The interior molecules penetrate through the membrane matrix depending on its solubility during the interfacial polymerization. Thus the structure can be core-shell type or homogeneous spheres. A demonstration of the latter case has been made for poly(urea-urethane) microcapsules with sizes of the order of 0.1 μm using a combination of static and dynamic laser light scattering, synchrotron small-angle x-ray scattering, viscosimetry, and scanning electron microscopy [37]. These microcapsules contain phosphoric organic solvent and are suspended in water with copoly(vinyl alcohol-vinylacetate) as the protective colloid. The radius of gyration determined by SLS and synchrotron SAXS yielded the same value, indicating that most of the core solvent entered into the shell wall. This conclusion is consistent with an observation by transmission electron microscopy based on the freeze-fracture method for microcapsules having a high weight ratio of core/wall materials [38]. By taking the microcapsule as a solid sphere, the radius of gyration, estimated from the hydrodynamic radius (the observed correlation function is shown in Fig. 1) and from the viscosity measurements, also agreed with the values obtained by SLS and SAXS. This coincidence suggested that the protective colloid had stuck tightly onto the surface of the microcapsule in wet form. On the other hand, the radius of gyration calculated from the size distribution based on electron microscopy was slightly smaller than those obtained in the wet state.

Glatter demonstrated a characterization of amphiphilic systems using a combination of DLS and SAXS for lipid IV_A (a bioactive precursor of lipid A of the outer membrane of Gram-negative bacteria) vesicles [39]. It was not easy to determine the overall size and shape of such large lipid aggregates of 250 nm from SAXS and to determine the thickness of 4.8 nm with a hydrocarbon chain length of 1.2 nm from DLS. Glatter determined the size and the size distribution of the vesicles by DLS and the internal structure by fitting a theoretical function to a desmeared SAXS scattering curve. The agreement of the measured SAXS curve and its approximate function is shown in Fig. 2.

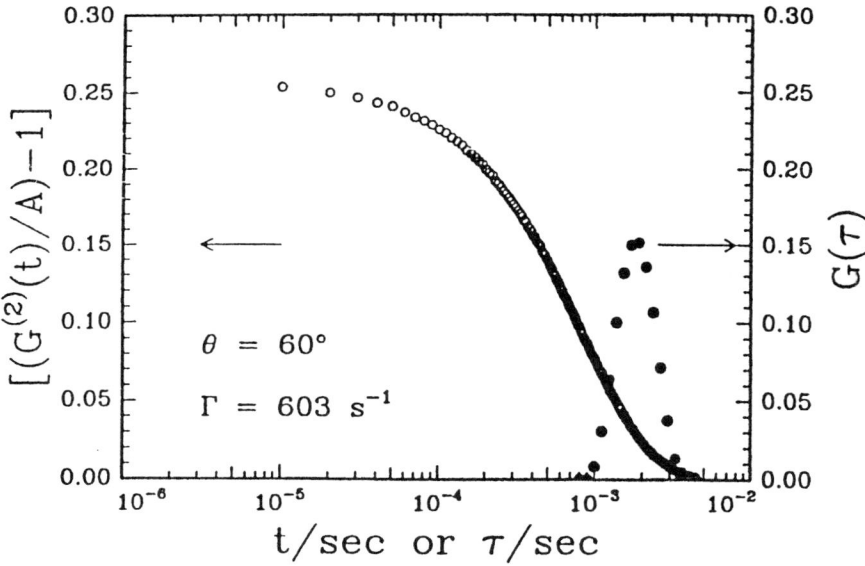

FIG. 1 Normalized intensity–intensity time correlation function of poly(urethane-urea) microcapsule suspension at $c = 5 \times 10^{-5}$ g/cm^3 measured at 60° and at 30°C. $\tau(1/\Gamma)$ is the characteristic decay time. (Reprinted with permission from the paper entitled An experimental investigation on the structure of microcapsules, by T. Dobashi, F. Yeh, Q. Ying, K. Ichikawa, and B. Chu. Langmuir *11*:4278. Copyright 1995 American Chemical Society.)

A combination of laser light scattering, SAXS, and transient electric birefringence has revealed the water-induced micellization of triblock copolymer Pluronic L64 (EO$_{13}$PO$_{30}$EO$_{13}$) [40]. Translational diffusion coefficients and rotational diffusion coefficients of the micelles were determined by dynamic light scattering and transient electric birefringence, respectively. Form factor fitting for SAXS and the hydrodynamic radius obtained from the DLS measurements with an assumption of an ellipsoid of revolution were consistent with an oblate form for the micelles, as shown in Fig. 3.

B. Adsorbed Layer Thickness Determination by Combination of Dynamic Light Scattering and Adsorption Isotherm

Hydration and conformation of protective polymers generally depend on the thermodynamic parameters. The thickness of an adsorbed layer of polymer has been studied using latex as the substrate. An apparent hydrodynamic thickness

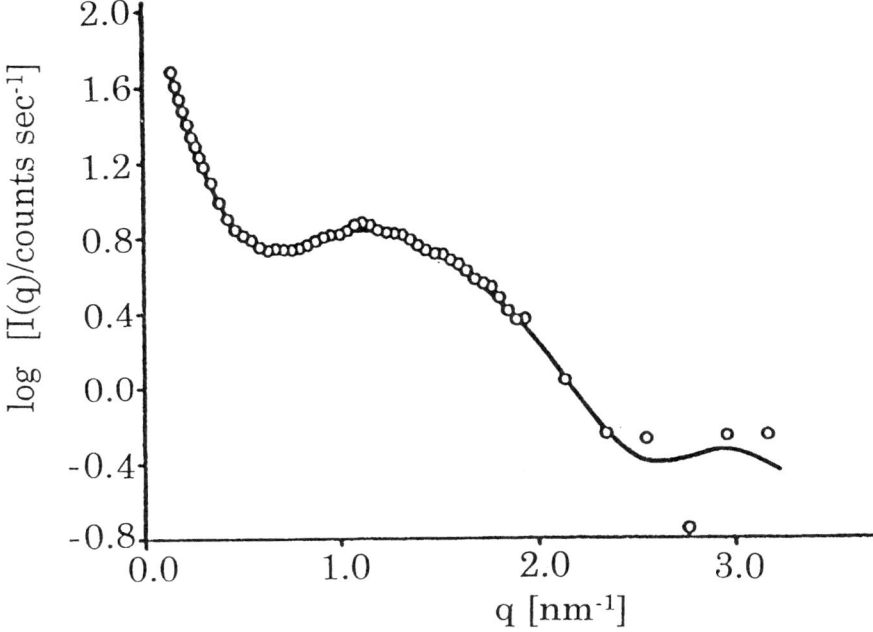

FIG. 2 Observed SAXS curve for lipid IV$_A$ vesicles and its approximation function. (Reprinted with permission from the paper entitled Scattering studies on colloids of biological interest (amphiphilic systems), by O. Glatter, Progr. Colloid Polym. Sci. *84*:52. Copyright 1991 Steinkopff Publishers, Darmstadt, FRG.)

of the adsorbed layer could be estimated from the difference between the Stokes radius of the bare sphere in the solvent alone and the value for the particle in the dilute polymer or surfactant solution using DLS. Morrissey and Han carried out dynamic light scattering measurements for aqueous suspensions of monodisperse polystyrene latex spheres (with a diameter of 91 nm) coated with γ-globulin, which was employed serologically in diagnostic tests for macroglobulin factors associated with disease conditions [41]. They determined the hydrodynamic radius of the coated latices from the half-height half-width of the Lorentzian power spectrum (shown in Fig. 4) using Eq. (19) for an impenetrable sphere surrounded by a loosely packed shell of adsorbed molecules. Here, they used the friction coefficient ξ of adsorbed γ-globulin estimated by assuming a maximum number density for hexagonally close packed cylindrical γ-globulin molecules on the latex surface. The number of adsorbed γ-globulin molecules per latex particle was estimated from the hydrodynamic radius and the adsorption isotherm determined by measuring the γ-globulin content in the supernatant solution after centrifuga-

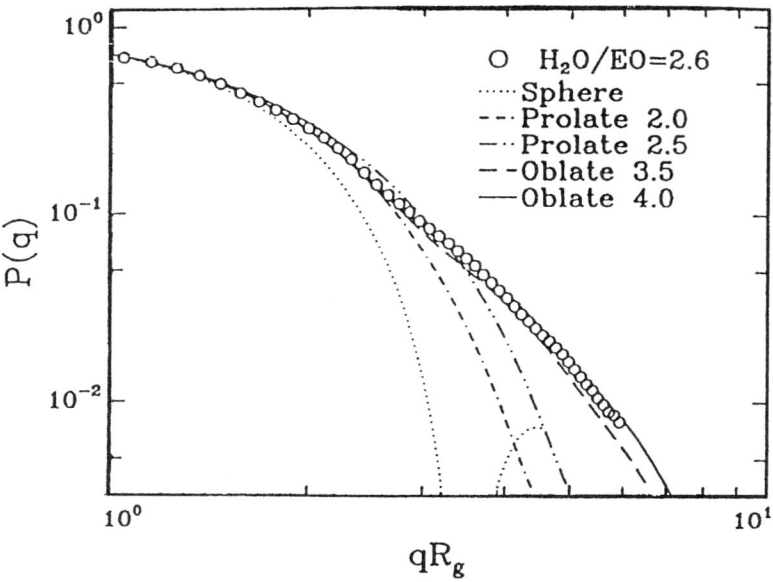

FIG. 3 Comparison of theoretical form factors with experimental SAXS data at a water-to-EO molar ratio of micelles at 2.6. R_g is the radius of gyration. (Reprinted with permission from the paper entitled Water-induced micellar structures of block copoly(oxyethylene-oxypropylene-oxyethylene) in o-xylene, by G. Wu, Z. Zhou, and B. Chu. J. Polym. Sci.: Part B: Polymer Physics *31*:2045. Copyright 1993 John Wiley & Sons, Inc.)

tion. The results indicated that the molecular extension of adsorbed γ-globulin at high coverage was consistent with end-on adsorbed molecules and a molecular flattening at very low surface concentration. Kato et al. also used the combination of DLS and adsorption measurements to determine the conformation of poly(oxyethylene) adsorbed on "monodisperse" polystyrene latex spheres (ranging in diameter 0.1–1.0 μm) over a wide molecular weight range from 2×10^4 to 130×10^4 gmol^{-1} [42]. They showed that the adsorbed polymers overlapped one another considerably. The molecular weight dependence of the adsorbed layer thickness was related to the excluded volume effect of the polymer chain in bulk solution. Brown and Rymden made dynamic light scattering measurements for suspensions of methacrylic acid modified butyl acrylate/styrene latex spheres with a hydrodynamic radius of 72 nm and high surface charge density (-40 μC/cm^2) in aqueous solutions of various cellulose derivatives [43]. They used a multiexponential analysis of the time correlation function and showed a bimodal distribution: the fast mode and the slow mode were consistent with the

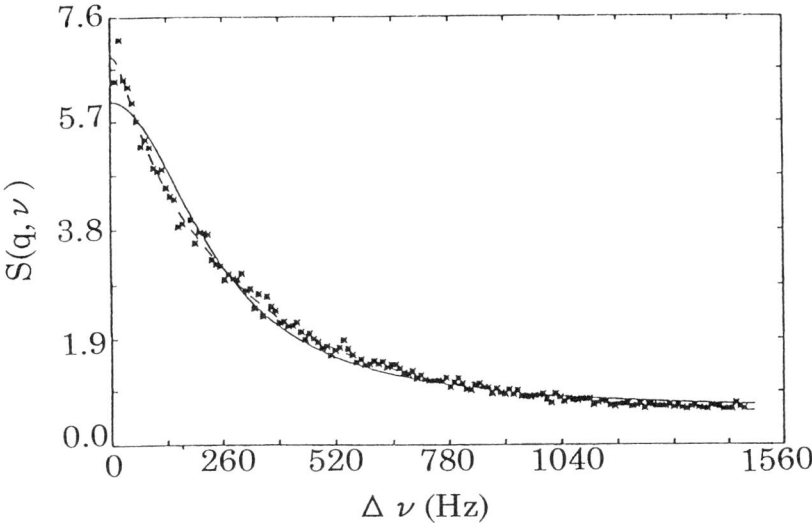

FIG. 4 Power spectrum of aqueous suspension of polystyrene latices coated with γ-globulin at $c = 3 \times 10^{-3}$/mg mL^{-1}. The solid and dashed curves represent single and bimodal Lorentzian fit to the data, respectively. (Reprinted with permission from the paper entitled The conformation of γ-globulin adsorbed on polystyrene latices determined by quasielastic light scattering, by B. W. Morrissey and C. C. Han. J. Colloid Int. Sci. 65:428. Copyright 1978 Academic Press.) It should be noted that present-day digital photon correlation measurements are much more precise than analog power spectral analysis.

diffusion of isolated latex particles and dimer or trimer bridged latex aggregates, respectively.

C. Micellar and Microemulsion Shell Thickness Determination by Combination of Light Scattering and Neutron Scattering

Diblock and triblock copolymers in selective solvents also show a micellization behavior that is similar to that of nonionic surfactants. Small-angle neutron scattering (SANS) and SAXS are appropriate for studying the scattering behavior by such small particles (<100 nm). To obtain a reasonable signal-to-noise ratio, the concentration frequently needs to be sufficiently high for SAXS and SANS measurements. Then it may become necessary to correct the effects due to intermicellar interactions for the analysis of the scattered intensity profile. Based on laser light scattering and SANS of Pluronic L64 in xylene/D$_2$O over a wide concentration range (<0.2 g/ml), it was shown that the hard sphere model could be used to correct intermicellar interactions by using an equivalent hard sphere size even when the micellar shape was slightly asymmetric [6,44]. For this solution the equivalent hard sphere radius was determined as the micellar core radius plus one-half the micellar shell thickness.

Microemulsions consist of oil, water and an oil–water interfacial film. DLS and SLS have been used to determine the translational diffusion coefficient and the interaction potential of microemulsions [45–47]. The thickness of the interfacial film and its curvature were measured by the contrast variation method in neutron scattering [48,49]. This method is based on changing the scattering strength by changing the relative amount of light and heavy water in the microemulsion.

D. Gel-to-Liquid-Crystalline Phase Transition of Liposomes by Dynamic Light Scattering and Anisotropy Measurements

Unilamellar vesicles are usually formed from lipid dispersions with sonication. Thus the suspension is metastable, and vesicles aggregate to form a multilamellar structure. The decay time of sonicated phosphatidylcholine vesicles, as measured by DLS, showed a bimodal distribution [50]. The hydrodynamic radius estimated from the smaller decay time was consistent with the values obtained by other experimental techniques such as ultracentrifugation.

The time evolution of the size distribution of sonicated dimyristoyl phosphatidylcholine vesicles was measured by Sornette and Ostrowsky [51]. They showed that as time developed, the size distribution of the vesicles changed more at a temperature lower than the gel-to-liquid-crystalline phase transition temperature T_g than at a higher temperature. Milon et al. used the ether injection method to prepare dimyristoylphosphatidylcholine vesicles in buffered solution containing EDTA, NaN_3 and showed a drastic and reversible change in the size of the vesicles near the transition temperature [52]. They showed that a 5–6.5% increase in the hydrodynamic radius appeared at the transition temperature, and the extent of the change did not depend on the size of the vesicles. The gel-to-liquid-crystalline phase transition of vesicles can also be detected by a change in the depolarization ratio, and the orientation of lipid molecules can be clarified from the change in the anisotropy ratio of lipid molecules. An upper bound for the anisotropy ratio and its temperature dependence of sonicated dipalmitoylphosphatidylcholine have been reported by several researchers [27,28].

E. Osmotic Swelling of Living Cells by Combination of Static and Dynamic Light Scattering

Light scattering has also been used to study living cells. Osmotic swelling of rod outer segment disk membranes and brush border membrane vesicles was one of the typical successful examples. The form factor [Eq. (4)] and the translational diffusion coefficient [Eqs. (14) and (16)] for an oblate ellipsoidal shell are functions of semiaxes a and b. Norisuye and Yu determined the major semiaxis a and the axial ratio $\rho = a/b$ of bovine rod outer segment disk membranes as a function of sucrose concentration of the dispersing medium, i.e., osmotic pressure, for each unbleached and bleached state, using a combination of SLS and DLS [24]. The scattering curves are shown in Fig. 5. They showed the difference in the axial ratio between two photochemical states; axial ratio steeply increased with increasing concentration for the bleached membrane, whereas that for the unbleached membrane increased only gradually. They also analyzed the sucrose concentration dependence of the translational diffusion coefficient by a simple model based on the calcium ion efflux from the intravesicular space to the bathing medium upon bleaching.

The cell membranes having a transport system such as a brush border membrane and a secretory cell membrane are inferred to change their shape flexibly to activate their biological functions. The elastic modulus of the lipid bilayer was measured by an osmotic swelling experiment because the change in the diameter of the vesicle could be related to the osmotic pressure through the elastic modulus assuming van't Hoff's law. Fujime et al. determined the diameter of the rat brush border membrane vesicles using DLS both in the absence and in the presence of a glucose-transport inhibitor [36,53]. The results showed that the vesicle membrane became flexible when its transport function was activated.

F. Scattering from a Single Microcapsule

Most microcapsules have a polydisperse size distribution. For a collection of heterogeneous microcapsules, the characteristic features become smeared because of the superposition of scattering curves of particles with different diameters. To measure the scattering function of each microcapsule independently, it is necessary to design a light scattering apparatus whose sample cell contains one single microcapsule in the scattering volume. A light scattering apparatus using a capillary tubing sample cell was developed and used to determine the outer diameter d and the membrane thickness δ of poly(L-lysine-alt-terephthalic acid) microcapsules by fitting the observed form factor to Eq. (1), as shown in Fig. 6 [54]. A volumetric phase transition of the hydrogel membrane was demonstrated by the proportional change in d and δ.

To obtain information on a single particle is important for studying the dynamic properties of living cells. Tishler and Carlson used a DLS microscope spectrometer to study the dynamics of the plasma membrane of individual human red blood cells. They combined an inverted microscope with the photon correlation technique [55,56]. The scattering volume of an 8 μm diameter cross section and the spread in scattering angle of 5.5° enabled them to study the correlation time due to the spontaneous fluctuations of cells. The detected fluctuation

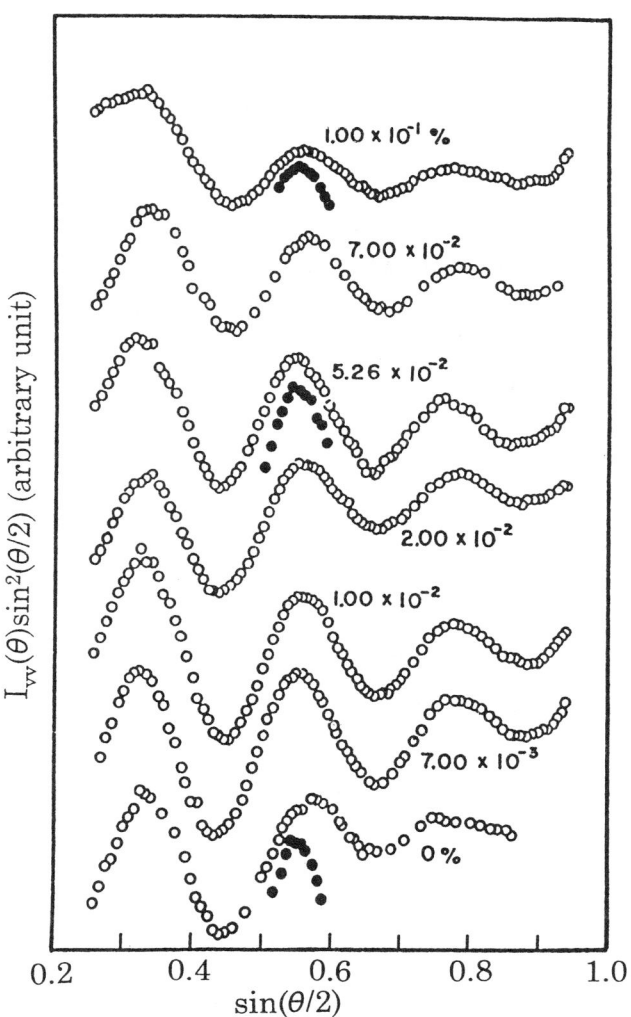

FIG. 5 Light scattering curves for isolated bovine rod outer segment disk membrane in aqueous sucrose of the indicated concentrations. (●) unbleached and (○) bleached membranes. (Reprinted from T. Norisuye and H. Yu. Osmotically-induced and photo-induced deformations of disk membranes. Biochim. Biophys. Acta *471*:441. Copyright 1977. With kind permission from Elsevier Science NL, Sara Burgerhartstraat 25, 1055 KV Amsterdam, The Netherlands.)

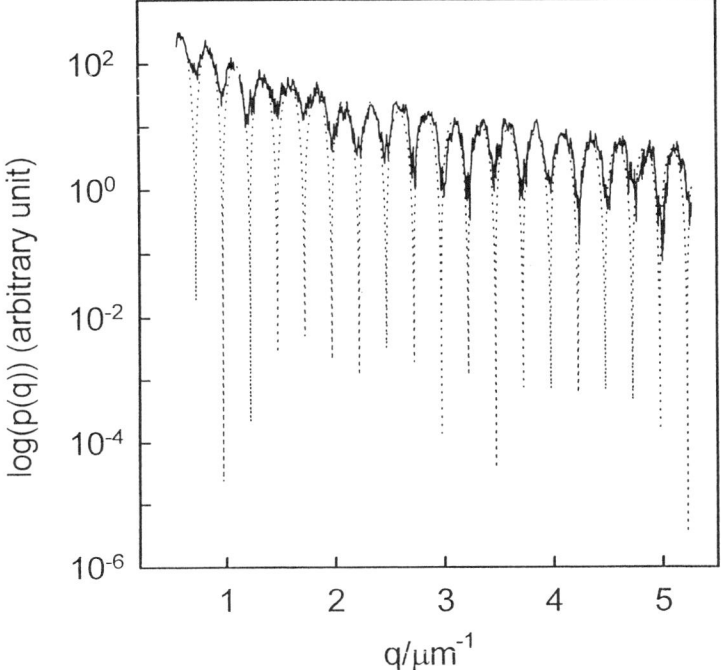

FIG. 6 Form factor for a single poly(L-lysine-alt-terephthalic acid) microcapsule at pH 4. Solid and dotted curves are observed and calculated ones, respectively. (Reprinted with permission from the paper entitled Light scattering of single microcapsule with hydrogel membrane, by T. Dobashi, T. Narita, J. Masuda, K. Makino, T. Mogi, H. Ohshima, M. Takenaka, and B. Chu. Langmuir *14*:748. Copyright 1998 American Chemical Society.)

signal was identified as the motion of red cell membrane/cytoskeleton. Physiological changes were measured in studies of age-separated red cells using the value of the correlation time as an index.

G. Scanning Near Field Microscopy

Scanning probe microscopies (SPM) such as STM and AFM are powerful tools for analyzing solid surfaces. A combination of these microscopic methods and the scattering techniques could give us a new way of determining the fine structures of microcapsule surfaces. Recently, another SPM, scanning near field microscopy, has been developed [57]. The extreme limit of the resolving power of the optical microscope based on the Abbe diffraction theorem can be raised to the seminanometer scale by using a scanning probe beam with a width narrower than the wavelength in the near field of the sample [58]. Initially it required a special technique to operate the apparatus because very thin capillary tubings were used as probes to obtain an aperture whose size was less than the incident light wavelength; but recently probe tips of metal or ceramic materials and the photocantilever approach make the measurements much easier [59]. It is hoped that these new technologies can be used routinely as complementary methods to the scattering methods in both the dry and wet states.

REFERENCES

1. J. R. Nixon, *Microencapsulation*, Marcel Dekker, New York, 1976.
2. T. Kondo, in *Surface and Colloid Science* (E. Matijevic, ed.), Plenum, New York, 1978, pp. 1–43.
3. *Microencapsulation and Related Drug Processes* (P. B. Deasy, ed.), Marcel Dekker, New York, 1984.
4. *Microcapsules and Nanoparticles in Medicine and Pharmacy* (M. Donbrow, ed.), CRC Press, Boca Raton, 1992.
5. P. G. de Gennes and C. Taupin. J. Phys. Chem. 86:2294 (1982).
6. B. Chu, G. Wu, and D. K. Schneider. J. Polym. Sci. B. Polym. Phys. 32:2605 (1994).
7. R. B. Gennis, *Biomembrane*, Springer Verlag, New York, 1989.
8. *Micellization, Solubilization, and Microemulsions* (K. L. Mittal, ed.), Plenum, New York, 1977.
9. *Scattering Techniques Applied to Supramolecular and Nonequilibrium Systems* (S. H. Chen, B. Chu, and R. Nossal, eds.), Plenum, New York, 1981.
10. H. Z. Cummins, in *NATO Adv. Sci. Inst. Ser., Ser. A. 59*, 1983, pp. 171–207.
11. M. A. Cohen Stuart, T. Congrove, and B. Vincent, Adv. Colloid Interface Sci. 24:143 (1986).
12. *Micelles, Membranes, Microemulsions, and Monolayers* (W. M. Gelbert, A. Bem-Shaul, and D. Roux, eds.), Springer Verlag, New York, 1994.
13. J. H. Van Zanten, in *Surfactant Sci. Ser. 62*, 1996, pp. 239–294.
14. H. C. van de Hulst, *Light Scattering by Small Particles*, John Wiley, New York, 1957.
15. M. Kerker, *The Scattering of Light and Other Electromagnetic Radiation*, Academic Press, New York, 1969.
16. W. Burchard and G. D. Patterson, *Light Scattering from Polymers, Adv. Polym. Sci. 48*, Springer Verlag, Berlin, 1983.
17. M. Kerker, J. P. Kratohvil, and E. Matijevic. J. Opt. Soc. Am. 52:551 (1962).
18. S. R. Aragon and R. Pecora. J. Chem. Phys. 66:2506 (1977).
19. D. O. Tinker. Chem. Phys. Lipids 8:230 (1972).
20. G. Oster and D. P. Riley. Acta Cryst. 5:1 (1952).
21. J. W. Rayleigh. Proc. Roy. Soc. A 90:219 (1914).
22. H. Hashimoto, M. Fujimura, T. Hashimoto, and H. Kawai. Macromolecules 14:844 (1981).

23. M. Shibayama and T. Hashimoto. Macromolecules *19*:740 (1986).
24. T. Norisuye and H. Yu. Biochim. Biophys. Acta *471*:436 (1977).
25. P. W. Barber and S. C. Hill, *Light Scattering by Particles: Computational Methods*, World Scientific, Singapore, 1990.
26. K. Mishima. Colloid Interface Sci. *73*:448 (1980).
27. S. R. Aragon and R. Pecora. J. Colloid Interface Sci. *89*:170 (1982).
28. K. Mishima. Chem. Phys. Lett. *96*:552 (1983).
29. B. Chu, *Laser Light Scattering*, Academic Press, New York, 1974; *Laser Light Scattering: Basic Principles and Practice*, 2d. ed., Academic Press, Boston, 1991.
30. B. Berne and R. Pecora, *Dynamic Light Scattering*, Interscience, New York, 1975.
31. H. Z. Cummins, in *Photon Correlation and Light Beating Spectroscopyh* (H. Z. Cummins and E. R. Pike, eds.), Plenum, New York, 1973.
32. F. Perrin. J. Phys. Radium *7*:1 (1936).
33. R. Pecora, in *Dynamic Light Scattering: Applications of Photon Correlation Spectroscopy* (R. Pecora, ed.), Plenum, New York, 1985.
34. J. A. McCammon, J. M. Deutch, and B. U. Felderhof. Biopolymers *14*:2613 (1975).
35. V. Bloomfield, K. E. Van Holde, and W. O. Dalton. Biopolymers *5*:149 (1967).
36. S. Fujime, M. Takasaki-Ohsita, and S. Miyamoto. Biophys. J. *53*:497 (1988).
37. T. Dobashi, F.-J. Yeh, Q. Ying, K. Ichikawa, and B. Chu. Langmuir *11*:4278 (1995); T. Dobashi, M. Takenaka, F.-J. Yeh, G. Wu, K. Ichikawa, and B. Chu. J. Colloid Int. Sci. *179*:640 (1996).
38. K. Ichikawa. J. Appl. Polym. Sci. *54*:1321 (1994).
39. O. Glatter. Progr. Colloid Polym. Sci. *84*:46–54 (1991). Trends Colloid Interface Sci. 5.
40. G. Wu, Z. Zhou, and B. Chu. J. Polym. Sci. Part B, Polym. Phys. Ed. *31*:2035 (1993).
41. B. W. Morrissey and C. C. Han. J. Colloid Int. Sci. *65*:423 (1978).
42. T. Kato, K. Nakamura, M. Kawaguchi, and A. Takahashi. Polym. J. *13*:1037 (1981).
43. W. Brown and R. Rymden. Macromolecules *19*:2942 (1986); *20*:2867 (1987).
44. G. Wu, Q. Ying, and B. Chu. Macromolecules *27*:5758 (1994).
45. A. M. Cazabat and D. Langevin. J. Chem. Phys. *74*:3148 (1981).
46. A. M. Cazabat, D. Langevin, and A. Pouchelon. J. Colloid Interface Sci. *73*:1 (1980).
47. A. A. Calje, W. G. M. Agterof, and A. Vrij, in *Micellization, Solubilization, and Microemulsions*, Vol. 2, Plenum, New York, 1977.
48. M. Lagues, R. Ober, and C. Taupin. J. Phys. Lett. (Paris) *39*:487 (1978).
49. M. Dvolaitzky, M. Guyot, M. Lagues, J. P. Lepesant, R. Ober, C. Sauterey, and C. Taupin. J. Chem. Phys. *69*:3279 (1978).
50. F. C. Chen, A. Chrzeszczyk, and B. Chu. J. Chem. Phys. *64*:3403 (1976); *66*:2237 (1977).
51. D. Sornette and N. Ostrowsky, in *Scattering Techniques Applied to Supramolecular and Nonequilibrium Systems* (S. H. Chen, B. Chu, and R. Nossal, eds.), Plenum, New York, 1981, pp. 351–362.

52. A. Milon, J. Ricka, S. T. Sun, T. Tanaka, Y. Nakatani, and G. Ourisson. Biochim. Biophys. Acta 777:331 (1984).
53. S. Miyamoto, T. Maeda, and S. Fujime. Biophys. J. 53:505 (1988).
54. T. Dobashi, T. Narita, J. Masuda, K. Makino, T. Mogi, H. Ohshima, M. Takenaka, and B. Chu. Langmuir 14:745 (1998).
55. P. S. Blank, R. B. Tishler and F. D. Carlson. Appl. Optics 26:351 (1987); R. B. Tishler and F. D. Carlson. Biophys. J. 51:993 (1987).
56. R. B. Tishler and F. D. Carlson. Biophys. J. 65:2586 (1993).
57. *Near Field Optics* (D. W. Pohl and D. Courjon, eds.), Kluwer, Dordrecht, 1993; D. W. Pohl, *Scanning Near-field Optical Microscopy*, Academic Press, London, 1990.
58. E. H. Synge. Phil. Mag. 6:356 (1928).
59. For example, R. Bachelot, P. Gleyzes, and A. C. Boccara. Ultramicroscopy 61:111 (1995); K. Fukuzawa, Y. Tanaka, S. Akamine, H. Kuwano, and H. Yamada. J. Appl. Phys. 78:7376 (1995).

9
Three-Dimensional Particle Tracking of Micronic Colloidal Particles

Y. GRASSELLI and GEORGES BOSSIS Department of Physics, CNRS–University of Nice, Nice, France

I.	Introduction	269
II.	Tracking Device with a Conventional Microscope	270
III.	Some Examples of Applications	274
	A. Diffusion coefficient	274
	B. Sedimentation velocity	276
	C. Wall–particle interaction and the roughness of particles	276
	D. Electrophoresis	280
IV.	Other Single-Particle Tracking Techniques	282
V.	Conclusion	284
	References	284

I. INTRODUCTION

Single particle tracking is a promising technique that allows us to obtain information on the physical properties of a particle such as hydrodynamic radius, density, refractive index, zeta potential, and also the surface roughness and interparticle forces. This technique is useful when we need knowledge of the variance of such properties on a given population of particles. Although it is in principle more powerful than conventional methods like dynamic or static light scattering, this method can suffer from the need to do statistical analysis on a large number of tracked particles. The time needed to get all the information on a single particle is related to the number of independent samplings on a Brownian trajectory whose characteristic time Δt for the loss of memory of its previous position is $\Delta t = m/\xi = (2/9)\rho a^2/\eta$ with m the particle mass, ρ the density of the particle, η the fluid viscosity, and a the particle radius. For a particle of diameter 1 μm in water this time is smaller than a microsecond, and the distance it has moved during this time $(6D\Delta t)^{1/2}$ is less than a nanometer. In practice the minimum sampling time of the position will be fixed by the spatial resolution of the device or by the speed of the image analysis program. With a usual CCD camera, and if each image can be analyzed in real time by the computer, the sampling time will be about 0.05 second. A 1% uncertainty on the diffusion coefficient will necessitate about 10^4 samplings (the error decreases as the square root of the number of samples), which will take 200 seconds for one particle. During this time the particle will usually have sedimented enough to be completely out of focus. Tracking in three dimensions is then a necessary condition if we want to obtain a good accuracy on the physical properties of the particles.

We shall describe a simple and low-cost method of tracking particles in three dimensions that is efficient for a particle diameter of about 0.5 μm to 1 mm. After a section devoted to the experimental device we shall give different examples of utilization: size and density determination, measurement of zeta potential, measurement of roughness. In the last section we shall discuss some other methods and compare their respective advantages.

II. TRACKING DEVICE WITH A CONVENTIONAL MICROSCOPE

This tracking technique is based on a direct observation of colloidal particles with an optical microscope [1]. The experimental cell containing the particles is placed under the microscope objective and the images are recorded and digitized with the help of a CCD camera connected to a computer equipped with an image analysis device (Fig. 1). The recorded images, generally retrieved in 256 colors or grey levels, have to be thresholded in two (black and white) or three (black, grey and white) levels in order to separate the observed particles from the liquid background (Fig. 2). This whole tracking technique is based on the color analysis of video pixels in real time.

Let us first describe the tracking in the focal plane of the microscope objective. First, the chosen particle has to be selected by, for example, moving a cross on the screen. Its position is given by the barycenter of the white or black zone. To do so, after having found a first white pixel (three-level situation), a recursive algorithm must be applied to find the coordinates of all the other white pixels of the central zone. One thus obtains the position of the particle at a time t. To determine its new position at time $t + dt$, we have to turn around on a spiral from the last position known until we find a new white pixel, and the analysis of the new white zone found will give the new position of the particle. Depending

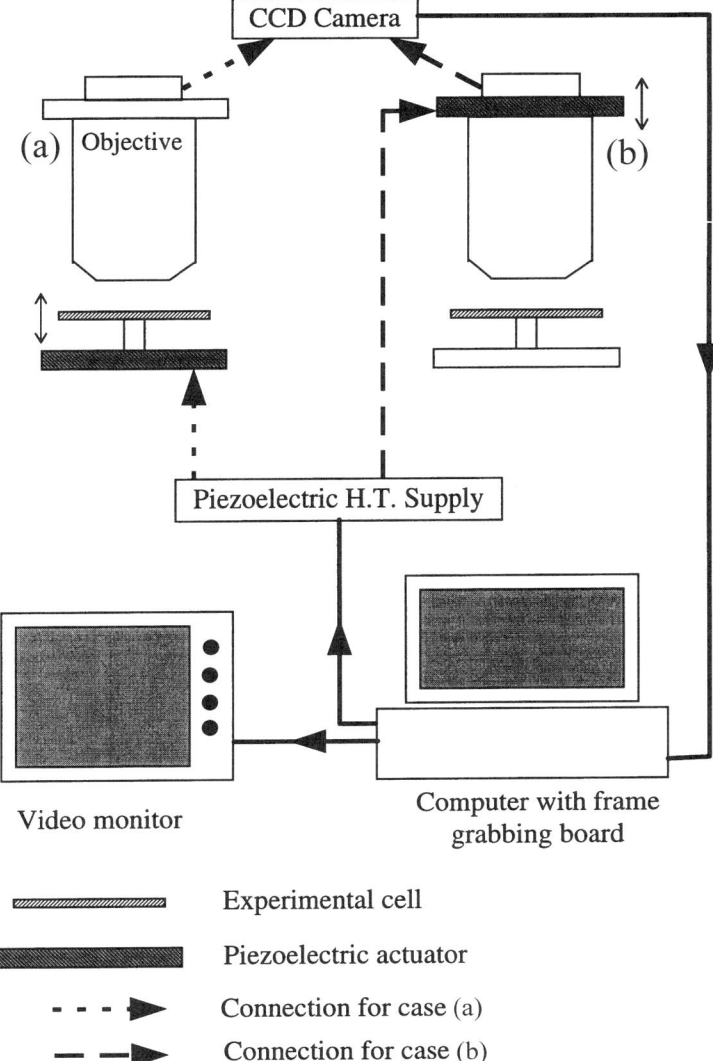

FIG. 1 Schematic representation of the tracking device. The piezoelectric actuator, which keeps the focalization constant and allows one to access the third dimension of the motion, can be mounted either on the cell (a) or on the objective (b).

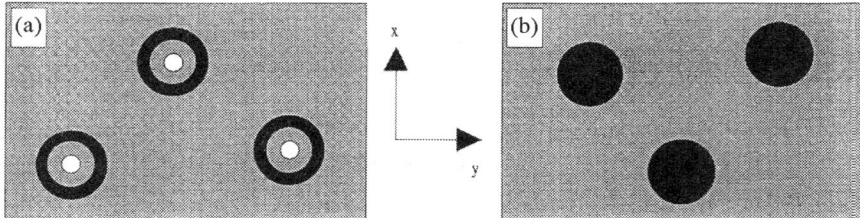

FIG. 2 Colloidal particle representation of a thresholded image. (a) Three levels—the white zone corresponds to the center of the particle. (b) Two levels—the whole particle is represented by a single color.

on the velocity of the particles, the experimental process, the video equipment performance, etc. one can reduce or increase the size of the search spiral, which generally takes the larger part of the time treatment of an image. In the two-level case, the whole black zone has to be treated; this situation will be more useful for large particles (size larger than 20–30 μm) and/or small magnification.

In the general case, the density of the particle is different from that of the suspending liquid, and the particle will also move perpendicularly to the focal plane. The idea of accessing the third dimension of the displacement is to move either the cell or the objective in order to keep the particle focused during its motion. Several solutions can be used, but we think that piezoelectric transducers are best because of their fast response time and also because they do not introduce external vibrations into the experiments. The displacement of the moving part is controlled by the size of the white zone at the center of the particle (three-level threshold). If one can maintain the particle at a given altitude in the cell (a sedimented particle, for example), it is possible to record the number of significant white pixels n versus the relative position of the particle from the focal plane: n is a maximum when the particle is close to the focal plane and decreases rather sharply when it goes out of focus. A set of typical curves giving the number of white pixels versus the vertical positions of the centers of the particle relative to the focal plane is represented in Fig. 3 for silica particles of nominal diameter $d = 1$ μm. These curves have been recorded for particles sedimented on the bottom glass plate by moving the cell up and down with the piezoelectric actuator, the applied voltage being also controlled by the computer. We notice that, for this kind of particle, we have a linear region with a sharp slope between 8 μm and 5 μm below the focal plane. It is recommended that one calculate the slope in pixels per microns for each new kind of particle, since it can change with the index of refraction or with the absorption. On the other hand, it is quite insensitive to the size of the particles (at least if the size does not vary too much). If we begin to record the vertical motion with the center of

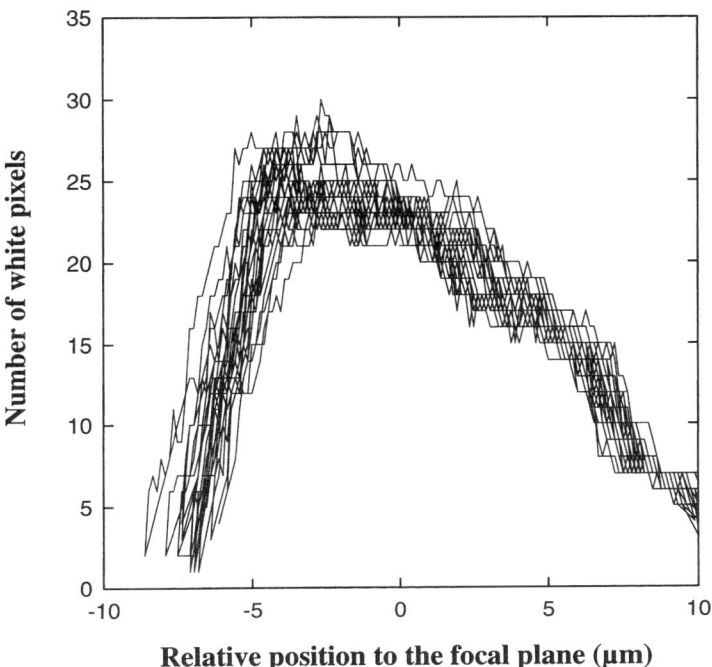

FIG. 3 Plot of the number of white pixels of the central zone (three-level threshold, cf. text) relative to the position of the particle from the focal plane of the objective. Curves retrieved for a silica particle of diameter $d = 1$ μm. (From Ref. 1.)

the particle in the middle of this zone (typically we look for a 15 pixel white zone) then an increase of the white zone will correspond to an upward motion of the particle and a decrease to a backward motion. When we know the slope of the curve (here 7 pixels per μm), a voltage is sent by the computer to the piezoelectric transducer, which moves the whole cell or the objective so that the white zone comes back to 15 pixels. The new position corresponding to this displacement is recorded by the computer as a function of time, and a new image is analyzed and the same process is repeated. The maximum possible resolution in time on an IBM 486 microcomputer is 0.1 s, which corresponds to the loss of more than one image in two on an ordinary CCD camera (25 images per second). Actually, the performance of the tracking system is mainly governed by the speed of the digitization device. On the other hand, the particle can be lost if during 0.1 s it has moved vertically over a distance that puts it completely out of focus; in this example, this distance is about 2 μm ($\cong$ 15 pixels), so the maximum vertical velocity we can follow is about 20 μm/s for a particle of diameter close to 1 μm.

In general, diluted suspensions are easy to study, but particles can also be tracked in concentrated suspensions if we use particles with the same refractive index as that of the suspending liquid (invisible to optical observations) and some "colored" particles that play the role of tracers. Nonspherical particles could also be tracked by this technique; the only thing needed is the relation between the number of white pixels and the vertical displacement. Nevertheless, for strongly elongated particles, this relation will depend on the orientation of the particle relative to the optical axis and will complicate the tracking algorithm. Finally, with this method we have also succeeded in tracking several particles simultaneously when their relative vertical motion is small. The knowledge of the relative motion of two particles is a direct way of obtaining information on interparticle forces.

III. SOME EXAMPLES OF APPLICATIONS

A. Diffusion Coefficient

We have used a suspension made of commercial monodisperse latex particles (ref. Rhône Poulenc Estapor L. 300) dispersed in water with a radius $a = 1.5$ μm and a density $\rho = 1.045$. The volume fraction ($\phi = 10^{-4}$) is low in order to reduce particle interactions and collisions. The diffusion coefficient is determined by analyzing the trajectory of a Brownian particle in a plane perpendicular to the optical axis. A typical trajectory consisting of 22,000 different positions with a time interval of 0.1 s projected on the x–y plane is represented in Fig. 4. The diffusion coefficients along the x and y axes can be calculated independently from the records of the $x(t)$ and $y(t)$ positions:

$$\frac{1}{N}\sum_{i=1}^{N}(x(t_i + \tau) - x(t_i))^2 = 2D_x\tau$$

$$\frac{1}{N}\sum_{i=1}^{N}(y(t_i + \tau) - y(t_i))^2 = 2D_y\tau$$
(1)

The plots of the mean square displacements along the x axis versus the time interval τ are also represented in Fig. 4. The experimental points are well fitted by a straight line with a slope corresponding to a diffusion coefficient $D_x = 1.438 \ 10^{-13}$ m^2/s. A similar treatment along the y axis gives $D_y = 1.412 \ 10^{-13}$ m^2/s. This difference gives an idea of the uncertainty, which is of order 2% for nearly 20,000 independent positions. Note that the correlation time of the velocities, $\tau = m/6\pi\eta a$, with η the viscosity of the suspending fluid, is less than

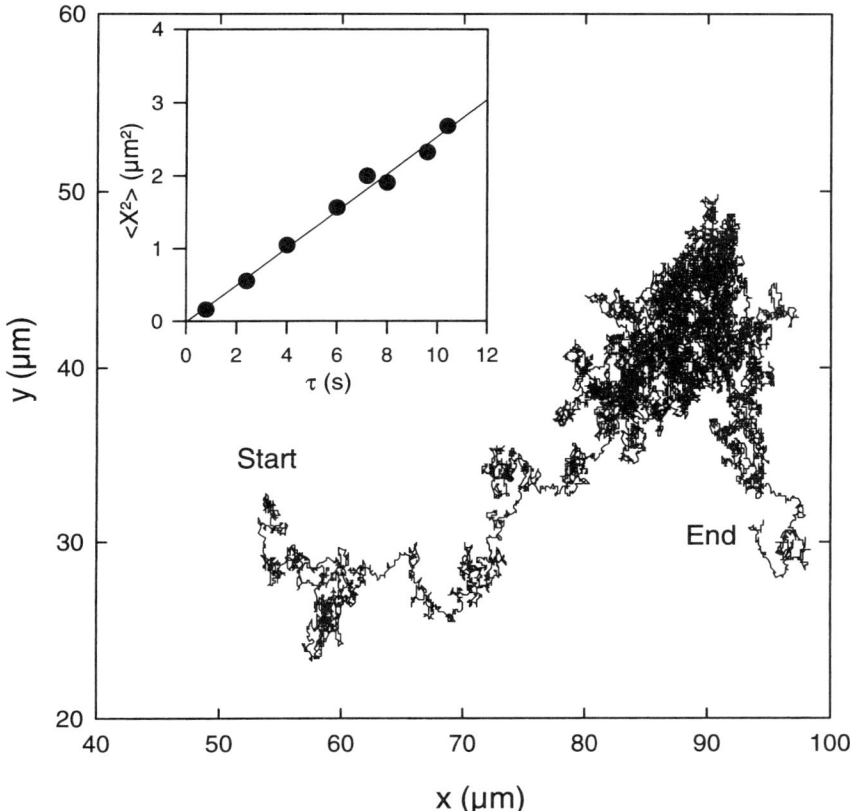

FIG. 4 Diffusion of a Brownian latex particle (radius $a = 1.5$ μm) in the plane perpendicular to the optical axis. 22,000 different positions of the particles have been recorded in about 40 mn.

10^{-6} s, whereas the time step between two consecutive records is 0.1 s; then each new position is uncorrelated with the preceding one. Taking the average of these two values, we can deduce the hydrodynamic radius from the Stokes equation:

$$a = \frac{kT}{3\pi\eta(D_x + D_y)}$$

We find $a = (1.45 \pm 0.01)$ μm, which well agrees with the value of 1.5 μm given by the manufacturer.

B. Sedimentation Velocity

We can use the same record of the trajectory of a particle to determine at the same time its hydrodynamic radius and its density. For sedimentation at low Reynolds number, the average vertical velocity is given by the balance between the body force and the friction force:

$$6\pi\eta a\lambda\left(\frac{h}{a}\right)\frac{dz}{dt} = \frac{4}{3}\pi a^3(\rho_p - \rho_f)g \tag{2}$$

The coefficient $\lambda(h/a)$ has been introduced by H. Brenner [2] and represents the modification of the Stokes friction due to the bottom wall located at a distance h from the particle. If $h/a \to \infty$, then $\lambda(h/a) = 1$ and we recover the usual Stokes law for an isolated particle. On the other hand, if the particle comes very close to the plane ($h/a \to 1$), then $\lambda(h/a) \to \infty$ and the particle will never reach the plane, owing to the presence of lubrication forces. The sedimentation velocity follows from Eq. (2):

$$V_{\text{sed}} = \frac{2a^2(\rho_p - \rho_l)g}{9\eta\lambda(h/a)}$$

The experimental velocity is obtained from the average slope of the curve $z(t)$ representing the vertical displacement as function of time when the particle is far enough from the lower wall. In this case $\lambda(h/a) = 1$, and from the experimental slope (cf. Fig. 5) we obtain $V_{\text{sed}} = 0.23 \pm 0.02$ μm/s, which gives us a density $\rho_p = 1.046 \pm 0.01$ g/cm^3. This density is in good agreement with the one given by the manufacturer: $\rho_p = 1.045$ g/cm^3. In Fig. 5 one can also observe the hydrodynamic slowing down as the particle comes closer to the wall; the solid line is the whole solution, which fits the experimental data very well.

C. Wall–Particle Interaction and the Roughness of Particles

The determination of the effective roughness of the surface of a particle can be quite important if we are interested in the rheology of concentrated suspensions, since it will determine the short-range repulsive force when the particles come into contact. It is also important in adhesion and more generally for any phenomena involving surfaces. A simple and nondestructive way to determine this effective roughness consists in measuring the time needed for a particle to fall from a flat glass wall [3,4]. Actually, the minimum distance ε between the particle and the glass wall is the effective roughness. If we assume that the presence of the asperities does not modify the hydrodynamic interaction between a sphere and a plane, the velocity of the particle will, as discussed in the preceeding section, be determined by the hydrodynamic interactions between a sphere and a wall separated by a distance ε. If this distance is small, the lubrication

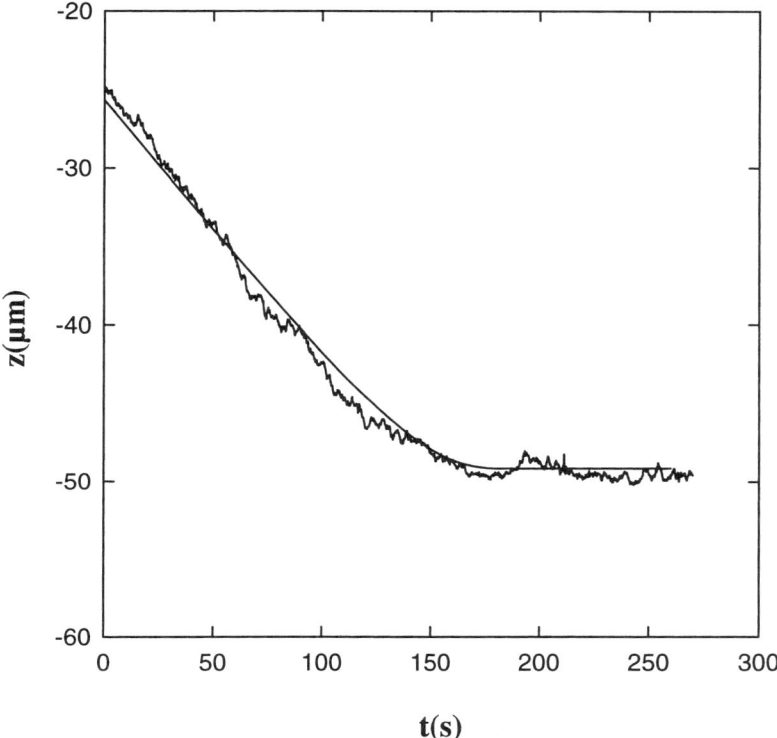

FIG. 5 Sedimentation curve recorded for a latex particle (radius $a = 1.5$ μm). z represents the particle height in the cell. One can observe the hydrodynamic slowing down as the particle comes closer to the lower wall. (——— experiments; ——— theory.) (From Ref. 1.)

force is very strong, and the sedimentation process will take a long time. On the other hand, for larger initial separation, the particle will start to sediment more quickly. The problem of a particle moving between two parallel walls is much more complicated, since one must take into account the hydrodynamic reflections between three bodies: one sphere and two walls. This problem is relevant for many experiments involving observation of colloidal particles with a microscope, because the experimental cell is made of two glass plates enclosing a thin layer of suspension. A simple and efficient approach in this case is obtained by combining the single-wall solution with the first terms obtained by the reflection method in the presence of two walls [5]. Although the determination of the surface roughness mainly depends on the interaction of the particle with a

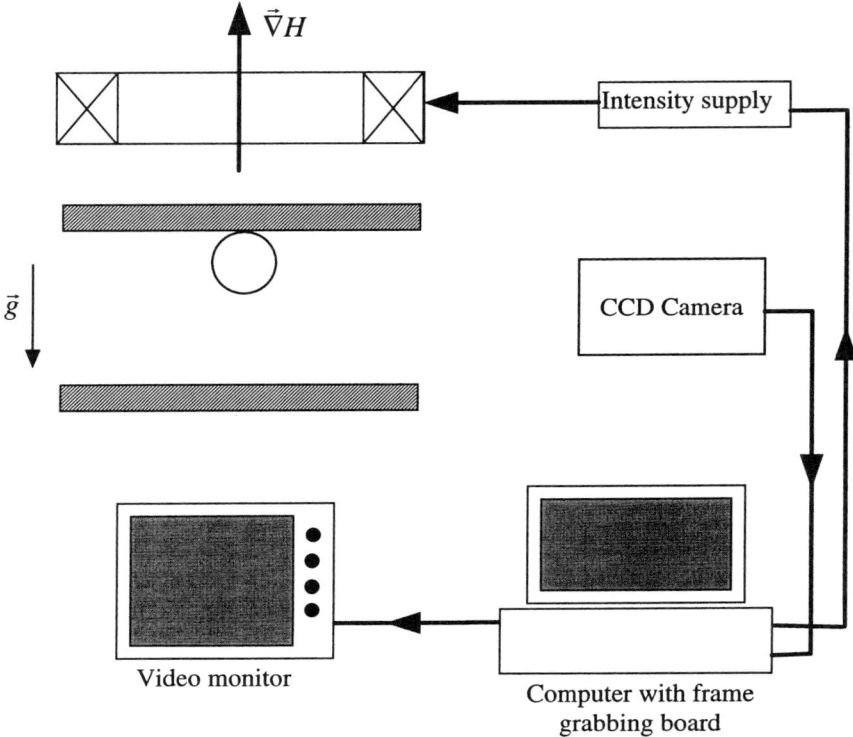

FIG. 6 Schematic representation of the experimental tracking device for the investigation of wall–particle interactions.

single plane, we have tested the whole theory for a millimetric particle starting from the upper glass plate and falling by gravity onto the bottom glass plate. In this experiment, we have used an iron particle (diameter $d = 1$ mm) in silicone oil (viscosity $\eta = 5000$ cp) initially stuck to the upper wall by a magnetic field gradient (cf. Fig. 6). In this way we are able to master exactly the initial time of sedimentation by turning off the magnetic field. Owing to the large size of the particle, the range of motion given by the piezoelectric transducer is too small, so we have placed the camera perpendicular to the trajectory and have used a two-level thresholding. In this experiment the magnification is chosen so that the whole trajectory is recorded in a single frame, but the accuracy could be greatly improved with a larger magnification and a camera mounted on a translation device. Without this additional device, the maximum precision achieved in the particle position is about 20 μm, which is sufficient in regard to the size of the

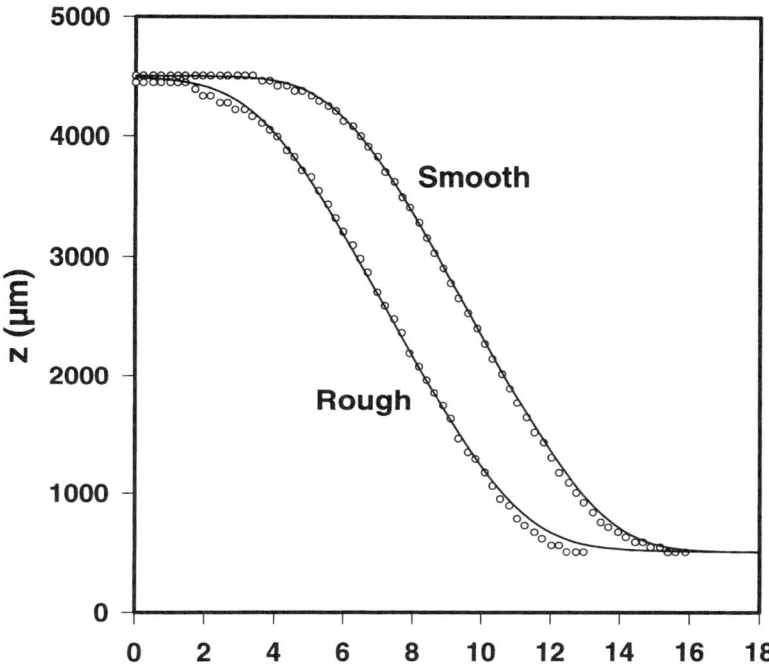

FIG. 7 Wall–particle interaction. Sedimentation curves $z = f(t)$ for identical iron particles with a smooth and rough surface. The distance between the two horizontal walls is 5 mm. The particles are held to the upper wall by a magnetic field gradient. The initial time corresponds to the beginning of the sedimentation (i.e., to the removing of the field). Owing to surface roughness, which increases the initial particle-to-upper-wall distance, the rough particle sediments away faster from the upper wall. (From Ref. 4.)

particle. In Fig. 7, we have plotted the sedimentation curves recorded for two identical iron particles having either a smooth or a rough surface (the distance between the two horizontal walls is $d = 5$ mm). The initial time corresponds to the instant where we turn off the magnetic field, i.e., sedimentation starts. We can see that the rough particle sediments faster because of the larger distance between its surface and the wall. From a fit of these curves with the theory taking into account the presence of the two walls, we can determine exactly the initial distance ε. In the present experiment, we find $\varepsilon = 200$ nm for the smooth particle and $\varepsilon = 1.9$ μm for the rough one. Of course the surface roughness just influences the particle motion at the beginning of the sedimentation, but it is worth noting that, using the general theory with the hydrodynamic reflections between the two planes, we have an excellent agreement with the experimental data for the whole sedimentation. This is not the case if we simply add separately the interaction of the sphere with each plane [4].

D. Electrophoresis

In general, colloidal particles exhibit surface charges that induce the formation of the well-known ionic double layer. The determination of this surface charge, or of the surface potential, can be achieved from the measurement of the electrophoretic velocity of the particle in the presence of an electric field. The conventional technique is based on electrophoretic light scattering (Doppler shift or interference fringes) in a configuration where the light beam is perpendicular to the electirc field [6]. In this configuration the electric field is parallel to the walls of the cell, and the motion of the double layer on the walls of the cell gives rise to an electro-osmotic flow. In order to measure the true electrophoretic velocity the scattering volume must be chosen in a precise region where this electro-osmotic flow cancels.

With the tracking along the optical axis of the microscope it is possible to use a cell where the electric field is parallel to the optical axis and so perpendicular to the larger walls of the cell, which prevents to a large extent the onset of electrohydrodynamic instabilities. In our device the cell is made of two glass plates covered with a conducting and transparent tin–indium oxide film and separated by a few hundred microns. We have recorded the motion of silica particles (diameter $d = 1$ μm) in a silicone oil (viscosity $\eta = 20$ cp). These suspensions of highly polarizable particles in an organic medium are known as electrorheological fluids, because their viscosity can change by several orders of magnitude in the presence of an electric field [7,8]. The electrophoretic mobility of these particles usually comes from ionization due to aqueous solvent contained inside or at the surface of the particles.

We apply an alternative squared electric field in order to prevent the particles from reaching the electrodes, and the frequency is low enough to allow for a good tracking. The equation of motion along the vertical axis, neglecting the hydrodynamic interactions with the walls, can then be written as

$$\frac{4}{3}\pi a^3 (\rho_p - \rho_l)g = 6\pi \eta a \frac{dz}{dt} + qE(t) \qquad (3)$$

with q being the particle charge and $E(t)$ the applied electric field. Equation (3) tells us that the expected displacement of the particle is a triangular path superposed on an average drift coming from the pure sedimentation process. The recorded path of a particle is presented in Fig. 8. The observed motion is well described by Eq. (3), and in particular we can see that inertial forces do not play any role: the direction of the motion changes as soon as the field is reversed. An

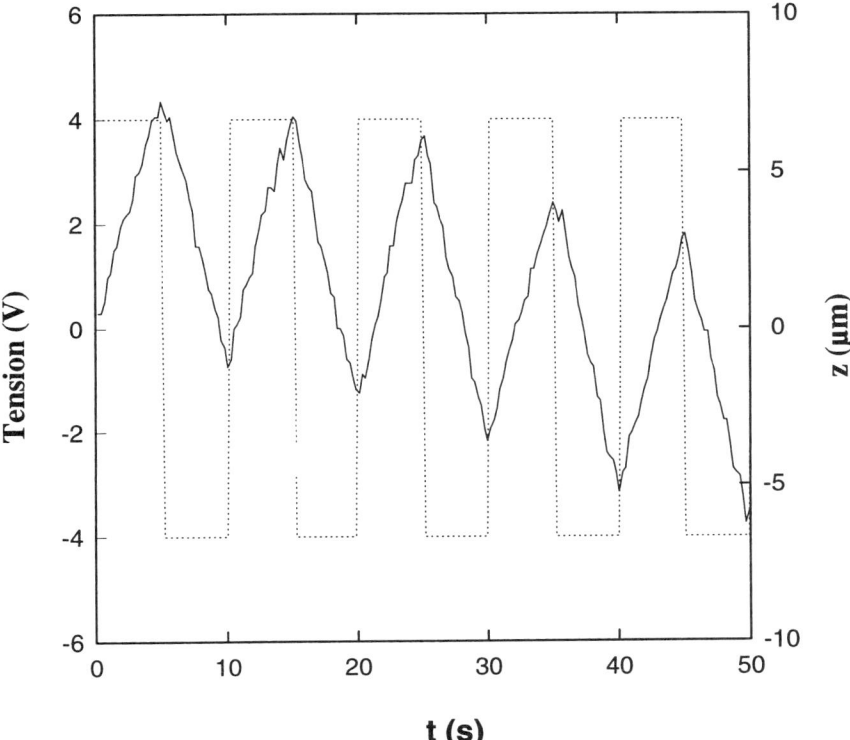

FIG. 8 Electrophoretic motion of a silica particle (diameter $d = 1$ μm) in silicone oil (viscosity $\eta = 20$ cp). The applied electric field is a square function with a frequency of 0.1 Hz and an amplitude of 555 V/cm. (— vertical position of the particle; --- electric field.) (From Ref. 1.)

average over a few periods allows us to determine the electrophoretic mobility with good accuracy. We obtain $\langle v \rangle = 3 \pm 0.1$ μm/s with a field $E = 555$ V/cm, which gives a mobility $\langle \mu_E \rangle = 5.3 \pm 0.15 \times 10^{-11}$ m^{2}s^{-1}V^{-1}. If we assume a thin double layer, the zeta potential is given by $\zeta = \eta \mu_E / \varepsilon$. With the permittivity of silicone oil $\varepsilon = 2.8$ we obtain $\zeta = 43$ mV.

It is worth noting that, due to the rather high viscosity and low permittivity of the suspending fluids used in electrorheology, the electrophoretic mobility is much smaller than in an aqueous medium (typical values in aqueous media are 1 μm/s for 1 V/cm, and we have almost three orders of magnitude less). In this case our tracking system is more efficient and far less expensive than conventional devices based on Doppler velocimetry.

IV. OTHER SINGLE-PARTICLE TRACKING TECHNIQUES

The technique we have described is cheap and quite easy to install. It essentially rests on a software able to determine quickly the size of a zone of a given grey level in a thresholded image. The size of this zone determines the altitude of the particle relative to the focus plane. With an objective of 50× and white light we were able to follow particles of diameter 0.8 μm. Smaller particles can be tracked as well if we use for instance fluorescence microscopy. Furthermore, fluorescence can be used to study the diffusion at quite high volume fraction: Schaertl and Sillescu [9] have studied this dependence for particles of diameter 0.5 μm up to volume fraction of 22% but with 2-D tracking. Kao and Verkman [10] have succeeded in tracking fluorescent particles of diameter 0.26 μm in 3-D. The z position was obtained by inserting a cylindrical lens between the objective lens of the microscope and the detector plane of the camera. The resulting image of a spherical particle is spherical when the particle is in the focus plane but becomes ellipsoidal and larger when it goes away from the focal plane. By analyzing the shape of the image, they were able, after a calibration, to deduce the displacement along the z axis with a precision of 12 nm. The time needed to analyze the shape of the image and deduce the position was about 0.5 s on a 33 MHz 486DX PC, which is substantially longer than the 0.1 s we need, with the same computer, just to calculate the size of a white zone.

Although not using a 3-D tracking, we have to mention the works related to single-particle tracking devices for such special purposes as motion in a flow or in an electric field.

The tracking of micronic colloidal particles in a tube flow was realized by Mason et al. [11]. In their device, the microscope and camera stage were fixed, whereas a platform supporting the tube and the syringe was moved in the direction opposite to the flow. The velocity of the platform was selected in order to give an apparent zero velocity for the particles located on the axis of the tube. By moving the platform with an hydraulic system (to eliminate vibrations), the particle can be kept in the focus plane over distances up to 2.5 cm at velocities from 0.001 to 0.6 cm s^{-1}. The motion of the particles was filmed with a high-speed camera and then analyzed frame by frame by a computer. This system was successfully used for studies of aggregation and microrheology. A similar system was also used by Cornell et al. [12] to study the kinetics of floculation of micronic aggregates.

For the motion of particles submitted to an electric field we have described in the last section a single particle tracking method that allows one to determine the zeta potential. Nevertheless, if we want to have some statistics on this measurement, the selection and data recording for one particle can take typically between one and two minutes, which makes a determination on one hundred particles a quite long and boring process for statistics which still will remain poor. An attempt to record several trajectories at the same time was made by Noordmans et al. [13]. A given number of video images (sixteen) are digitized at a constant time interval. Then they are binarized and combined into a single image. A particle trajectory then appears as a single track, the slope of which gives the velocity and so the zeta potential. Because of the intersection of different tracks or of particles moving out of focus, a statistical analysis must be done in order to obtain an average velocity. The advantage of the method is that it can function automatically without the need to select particles one after the other. Nevertheless, calibration procedures must be used, and the range of application and the accuracy of this method are difficult to evaluate.

A common drawback of tracking with a video camera is the limitation of the number of images per second (typically a few hundred for high-speed cameras) and also the time needed for calculating the position of the particle from image analysis. In the following method there is no camera, and the signal relative to a single particle is recorded on a photodiode.

In this technique, developed by Schätzel et al. [14,15], a single particle is illuminated by a strongly focused laser beam. The particle is submitted to an axial force exerted by the light pressure, which pushes it away in the direction of the incident beam, and also by ponderomotive forces, which are generated by the spatial gradients of optical intensity. The Gaussian profile of the intensity of the beam produces radial forces, perpendicular to the beam and directed towards its center, which are strong enough to trap the particle inside the beam. On the other hand, owing to the focusing lens, there is an axial gradient that will drive the particle toward the beam focus, but unlike the case of the optical tweezer this axial force is smaller than the pressure radiation force, and the particle is just driven forward axially in the beam with a constant velocity given by the Stokes law. In their experimental device, the particle is tracked by recording the backscattered intensity on several photodiodes placed after different beam splitters. The axial position along the laser beam is given by recording the difference of signal between two small-aperture photodiodes located respectively before and behind the image plane so that the difference of signal is zero when the particle is in focus. Then, when the particle moves forward axially, the objective lens is moved at the same time with the help of a piezoelectric translation stage. Transverse tracking is realized by tilting a mirror in front of the focusing lens with the help of three small piezoelectric transducers, and the resulting backscattered intensity is recorded by a high-sensitivity silicon quadrant diode. The resolution of this system is respectively 20 nm for transverse motion and 50 nm for axial motion. The maximum axial velocity that can be recorded is 90 μm s^{-1}; it gives an upper limitation for the diameter of particles around 0.5 μm (the axial velocity grows monotonically with the size of the particles) and the lower limit—about 0.3 µm—is given by the optical contrast (resulting from residual internal reflections of the microscope objective).

V. CONCLUSION

Single-particle tracking of colloidal particles is a direct way to obtain the size and the surface properties (roughness, zeta potential) of particles.

The determination of the surface roughness, but also of the interaction forces between two particles or between a particle and a wall, can be realized if the particle can be first located close to a glass wall or close to another particle. We have used a magnetic field gradient to bring a magnetic particle onto the upper glass plate; for a nonmagnetic particle we could use an electric field gradient as well, but the optical tweezer realized by a focused laser beam is certainly the best way to move a particle. It is possible first to locate the particle either close to a wall or close to another particle (using two laser beams). Then the particle can be simply released by turning off the light and the tracking realized by a video camera and an image analysis as described above. The combination of a laser beam trap and fast video recording is certainly a promising technique for the measurement of individual particle properties.

REFERENCES

1. Y. Grasselli and G. Bossis. J. Coll. Int. Sci. *170*:269 (1995).
2. H. Brenner. Chem. Eng. Sci. *16*:242 (1961).
3. Jeffrey R. Smart and David T. Leighton. Phys. Fluids A. *1*:52 (1989).
4. Y. Grasselli and L. Lobry. Phys. Fluids *12*:3929 (1997).
5. L. Lobry and N. Ostrowsky. Phys. Rev. B. *53*:18 (1996).
6. A. K. Gaigalas, S. Woo, and J. B. Hubbard. J. Coll. Int. Sci. *136*:213 (1990).
7. A. P. Gast and C. F. Zukoski. Advances in Colloid and Interface Science *30*:153–202 (1989).
8. H. Block and J. P. Kelly. J. Phys. A. Appl. Phys. *21*:1661–1677 (1988).
9. W. Schaertl and H. Sillescu. J. Coll. Inteface. Sci. *155*:313 (1993).
10. H. P. Kao and A. S. Verkman. Biophysical Journal *67*:1291 (1994).
11. E. B. Vadas, H. L. Goldsmith, and S. G. Mason. J. Coll. Interface. Sci. *43*:630 (1973).
12. R. M. Cornell, J. W. Goodwin, and R. H. Ottewill. J. Coll. Interface Sci. *71*:254 (1979).
13. J. Noordmans, J. Kempen, and H. J. Busscher. J. Coll. Interface Sci. *156*:394 (1993).
14. K. Schätzel, W. G. Neumann, J. Müller, and B. Materzok. Applied Optics *31*:770–778 (1992).
15. N. Garbow, J. Muller, K. Schätzel, and T. Palberg. Physica A *235*:291–305 (1997).

10
Low-Mass Luminescent Organogels

PIERRE TERECH Département de Recherche Fondamentale sur la Matière Condensée, UMR 5819, CEA–CNRS–Université J. Fourier, Grenoble, France

RICHARD G. WEISS Department of Chemistry, Georgetown University, Washington, D.C.

I.	Introduction	286
II.	Physical Gelation of Organic Liquids by Low-Mass Molecules	286
III.	Electronic Absorption and Luminescence Properties	292
IV.	Some Useful Techniques and Concepts for Elucidating the Structures of Organogels	295
	A. Rheology	295
	B. Phase diagrams	296
	C. Electron microscopies	298
	D. Small-angle scattering techniques	300
	E. Luminescence in aggregated systems	303
V.	Classes of Luminescent Organogelators	308
	A. Anthryl derivatives	309
	B. Anthryl and anthraquinone appended steroid-based gelators	314
	C. Azobenzene steroid-based gelators	329
	D. Sorbitol and polyol derivatives	338
VI.	Applications	339
	References	341

I. INTRODUCTION

This chapter describes the microscopic and macroscopic properties of the aggregate states of a class of relatively small molecules (MW ≈ 1000) which both luminesce and can combine in small amounts (typically < 2 wt%) with selected organic liquids to form thermally reversible gels. Under specific conditions—defined principally by the liquid type, concentration, and temperature—these gelators form colloid aggregates. Luminescence (fluorescence or phosphorescence) in many cases is a convenient and sensitive indicator of molecular aggregation and the process during which the system changes rheologically from a solution to a gel. All gelators form linear aggregates (fibers and chains), which raises the possibility of one-dimensional energy transfer through the molecular networks after the absorption of a photon.

Here, rheology is used to characterize the gel state, whose stability, as measured thermodynamically or kinetically, can be described by temperature–concentration phase diagrams or simply time. The structural features of gelator aggregates at nanoscopic scales are described via data from the complementary techniques of electron microscopy and scattering techniques. Finally, the optical properties, including absorption and luminescence, are detailed.

II. PHYSICAL GELATION OF ORGANIC LIQUIDS BY LOW-MASS MOLECULES

During the last several years, the number of published investigations involving low-mass physical organogelators has increased enormously. The growth of interest in the field is motivated by (1) the emergence of powerful investigation tools (e.g., scattering, diffraction techniques, electron microscopy), (2) a clarification of the pertinent physical concepts associated with these systems, (3) identification of challenging ''chemical'' or ''physical'' questions, and (4) the development of applications using the reversibility or the consistency of the systems in actuators, sensors, medical drug transporters, etc. A recent review [1] describes the state of the art in the field and illustrates the great variety of the systems. Examples mentioned in this chapter focus on published works and certainly constitute only the embryo of a developing field.

The described gels are binary mixtures of gelator molecules in organic liquids. As a consequence of the preparation procedure (vide infra), a continuous structure with macroscopic dimensions is formed and is permanent on a time scale of at least thousands of seconds at a given temperature within the gel domain of the phase diagram. A pertinent analytical technique to qualify a system as a gel is rheology. The solidlike behavior observed in the gels should be classified as ''soft,'' given the low concentrations of gelator molecules. A more accurate definition will be provided. However, we note that systems exhibiting viscoelasticity only for short times are better described as "jellies" [2].

Based upon the many chemical compositions and physical properties of gels that are known, several fundamental issues concerning their most common features, and bizarre behavior, need to be addressed. For instance, the structural requirements for a molecule to gel an organic liquid are still not well understood; most discoveries of gelation have occurred (at the onset) serendipitously. To determine the dependence of structures of the aggregates at different length scales on the mechanical, optical, and other properties of the gels will require a formidable effort. An invaluable aid in this regard is the determination of the relationship between the structural packing of the gelators in their bulk crystalline states and in their gel aggregates; many gelators are polymorphous. To understand the variety of physical processes involved in this type of physical gelation—a mix of the features of spinodal decomposition [3], unidirectional crystallization [4,5], mesophase formation, micellar aggregation, swelling/shrinking behaviors, and ordered microstructures in random nanoscopic 3-D networks—requires the efforts of both chemists and physicists.

The phenomenology of physical organogels and jellies is extremely rich, and their comportments are similar in some aspects to those of both surfactants in solution (e.g., lyotropism and crystallization) and polymer solutions [6] (e.g., swelling/shrinking behaviors and microscopic mass motion). Gels can be considered as being at the interface between "complex fluids" (i.e., micellar systems) and phase-separated states of matter. The main properties and concepts appropriate to describe the gels and the basic principles of techniques for their study will be reviewed here.

The formation of an organogel is a spontaneous phenomenon from the moment that the gelator molecules are put into an unstable state (usually at low temperature and relatively high concentration). Under such conditions, which correspond to the gel domain of the phase diagram, autoassociation proceeds, leading to the formation of the physical gels. Typically, the gels or jellies are prepared by warming a gelator in an organic liquid until the solid dissolves and then cooling the solution (or sol) to below the aggregation transition temperature T_{ag}. The equilibrated system does not flow over long periods for gels or shorter ones (< 1 s) for jellies. This rheological behavior is rather surprising, since it involves in some cases < 0.5 wt% of gelator. The viscoelasticity of the gels reveals that the primary aggregates (whose diameters typically are 20–2000 Å) are linked in complex three-dimensional networks leading to colloids that immobilize the liquid component principally by surface tension forces. This phenomenon cannot reasonably be explained using the macroscopic laws for complete wetting describing the capillary rise, in which the Young–Laplace capillary pressure is balanced by the hydrostatic pressure. Since the pertinent length scale for a gel network is its statistical mesh size (i.e., microscopic scale), forces such as van der Waals (for organic liquids) must be important contributors [7].

$$\frac{2\gamma}{R_1}\cos\theta_e = \rho g h \tag{1}$$

ρ is the density difference between liquid and vapor, R_1 is the radius of the capillary, h is the height in the capillary, γ is the interfacial energy, and θ_e is the contact angle ($\theta_e = 0$ for complete wetting).

The immobilization of a large volume of liquid by a small quantity of gelator is achieved efficiently if the elementary assemblies are rodlike and have large aspect ratios. Such linear structures are determined by specific binding forces associated with the chemical constitution of the gelators. In nonaqueous liquids, the attractive forces are mainly the van der Waals type and can be supplemented by dipolar interactions, intermolecular hydrogen bonds, metal-coordination bonds, or electron transfers, etc.

Aggregation of surfactants in organic liquids has been the subject of different theoretical and quantitative treatments [8]. Beyond a concentration threshold C_{mic} that is normally much less pronounced than the analogous critical micellar concentration (CMC) of aqueous systems, finite spherical micelles are formed that may undergo a sphere-to-rod transition when the concentration is increased or a small amount of an appropriate cosurfactant is added. Gelators in the gel domain form very long linear aggregates at a concentration C_{ag} that corresponds to the onset of viscoelastic properties. The kinetic behavior is similar to that of the so-called supersaturation gels [9]. The warmed homogeneous gelator/hydrocarbon solutions are unstable at temperatures in the gel domain of the phase diagram. Since the gelator solubility C_{sol}^T at temperature T is low in most gelled liquids, the supersaturation degree ($\sigma_T = C/C_{\mathrm{sol}}^T$) is usually high. Most of the gelator molecules are involved in the aggregation/gelation process in which unidirectional crystallization and fiber entanglement phenomena occur. The variation of the induction time (as short as σ_T is high), as well as the kinetics of aggregation (as fast as σ_T is high), are consistent with the features of nucleation phenomena in other supersaturated systems (see Refs. 10 and 11). A sigmoid kinetic curve that exhibits the characteristics—induction delay, rapid initiation (i.e., autocatalytic), terminal first-order reaction rate—of a collective kinetic process is usually observed. The corresponding theoretical variation can be obtained [12] by solving the differential equation associated with a two-step model with two successive reactions (an autocatalytic one followed by a first-order step):

$$\frac{d^2\alpha}{dt^2} + \frac{d\alpha}{dt}\left[\frac{1/k_0 - k_0 k_1 (1-\alpha)^2 + k_2(1-\alpha)}{1-\alpha}\right] + k_0 k_2 (1-\alpha) = 0$$

α is the reaction rate, and k_0, k_1, and k_2 are the related kinetic constants.

Low-Mass Luminescent Organogels

The difference in the cohesive energies between the gelators in the aggregate and dispersed states is responsible for the aggregation phenomenon. The free energy μ_N^0 of an aggregate of N molecules decreases during the aggregation towards an asymptotic limit μ_∞^0 ($N \to \infty$) [13]:

$$\mu_N^0 = \mu_\infty^0 + \frac{\beta kT}{N} \qquad (2)$$

βkT represents the energy of the gelator–gelator interaction in the aggregate. The rodlike aggregates are in thermal equilibrium with the surrounding bulk liquid and engaged in dynamic exchange with dispersed gelator molecules. At equilibrium, corresponding to minimization of the free energy of the system, the optimal geometry for a rodlike aggregate in suspension in a liquid is obtained by considering the influences of the ratio of the cross-sectional areas of the various parts of the gelator and the polarity of the solvent. Owing to the balance of attractive and repulsive forces, an optimal geometric arrangement prevails. Hydrophobic attraction and ionic repulsion components are clearly identified with aqueous systems, while the situation with aprotic organic liquids is more complex [8]: dipolar interactions (as well as specific H-bonds and/or coordinations with metals) constitute the main attractive force and are supplemented by van der Waals interactions between aliphatic groups; the primary positive contribution to free energy may involve the reduction in translational and rotational freedom of motion that occurs during aggregation. Thus the additional contributors and opposing forces make the transitions from unaggregated to aggregated molecules less sharply defined than in the case of aqueous micelles. For instance, gelator molecules in organic liquids can coexist in small and large aggregate states under many conditions of temperature and concentration, but in aqueous micellar systems (especially above the well-defined CMC), the concentration of smaller aggregates is vanishingly low. The one-dimensional aggregates are a very special situation compared to the formation of finite discs or spherical micelles.

The extremities of the chains or fibers are in the form of caps, forcing the molecules involved to adopt higher energy packing arrangements. Depending on the amount of additional free energy involved at the caps, either "infinitely long" giant micelles or numerous shorter ones (the entropically favored case) can be obtained. An abrupt transition leads to long fibers with a mean *concentration dependent* aggregation number $\langle n \rangle$. The density distribution of gelator molecules in fibers of N molecules is broad and peaks at $n_{\max} = \sqrt{(Ce^\beta)}$. The distribution also appears to be sensitive to any modification of the interaction parameter β (upon addition of "poisons," etc.). Even low concentrations of a cosurfactant "poison" with a strong affinity for gelator molecules can function as an effective "chain terminator" or "chain-growth inhibitor," affect the length distribution by interrupting gelator–gelator interactions, influence the morphology of the aggregates, and/or shift the threshold C_{ag}. Thus the composition and the method of preparation of the gelling mixtures must be carefully controlled.

The energy of scission of a fibrillar aggregate ($E_{scis} \propto \beta kT$) depends on the type of intermolecular interaction, whether it be H-bonding or electrostatic, and the number of gelator molecules within a cross section. The simplest case is for molecular threads with only one gelator molecule per cross-sectional repetition unit. The gelator–gelator interaction energy determines the aggregation threshold $C_{ag} \propto e^{-\Psi E_{scis}}$ and the size distribution of the chainlike aggregates, which is in turn correlated with the contour length L ($\langle L \rangle \propto \phi^\delta \exp(E_{scis}/kT)$; δ is 1/2 in a mean-field approach) and flexibility $\langle l_p \rangle$. Fiberlike aggregates are formed while phase separation into macroscopically separated crystalline and liquid layers is avoided by the interconnection of the fibers in a solidlike three-dimensional network stabilized by a balance of opposing forces. The extreme sensitivity of gelation ability to minor modifications of the chemical structure of a gelator (vide infra) shows how delicately balanced are the various energetic factors determining the type of molecular aggregate in a nucleating system and why there have been relatively few well-documented examples of gelation of organic liquids by low-mass gelators.

In spite of broad distributions of aggregate lengths, cross-sectional dimensions are frequently monodisperse. The magnitudes of the dimensions are controlled by the physical mechanisms of molecular connections: multiples of bimolecular lengths are observed when head-to-head associations prevail. Beyond a critical concentration C_{ag}, the linear aggregates are no longer separated (as in the "dilute" concentration regime), their rotation becomes hindered, and the equivalent spherical volumes of the chains overlap. A porous network with an average mesh size ξ is formed. By contrast to polymeric networks in good solvents, which can present homogeneously swollen structures, surfactant-made networks are less ideal, and a random distribution of heterogeneities always adorns the mesh of fibers. The origin of the heterogeneities is twofold: (1) the domains where fibers are entangled constitute the "nodes" or, more precisely, "junction zones" of the network; (2) the equilibrium state of most systems is the phase-separated bulk solid and liquid, although some systems are stable for decades! Nevertheless, localized crystallization sites can develop within a gel. The molecular packing with selected gels and corresponding xerogels have been shown to be close, but not identical, to that of the bulk crystal [14]. Packing restrictions and different modes of nucleation lead to unidirectional crystallization (and fiber formation) instead of the usual 3-D bulk crystal growth. At a given temperature in the phase diagram defining the aggregation domain corresponding to the gel, the networks are either permanent or transient depending on the lifetime of their fiber associations. This distinction relates to the difference between gels and jellies, and numerous experiments have correlated the mechanical properties of the aggregates and either their rigidity (in gels) or semiflexibility (in jellies).

This chapter deals primarily with examples in the semidilute concentration range, corresponding to overlapped chainlike aggregates of "isotropic gels." As mentioned, the heterogeneity of the networks makes possible the existence of more concentrated microdomains, where chains are somewhat ordered in crystalline or lyotropic structures (corresponding to anisotropic gels).

The onset of viscoelasticity, which reveals the gelation phenomenon, is affected by the same factors as those responsible for the aggregation process. Besides concentration and temperature, factors include the solvent nature (including its polarity), conformational lability, and molecular shape of the gelator, and the possible participation of cosurfactants. The phase diagram defines the boundaries between the sol state (including microgels or nonconnected "lumps") and the gel or aggregated and interconnected states (containing an infinite aggregate). Some structural and kinetic aspects of the aggregation/gelation process are also consistent with its being a phase separation involving nucleation. This emphasizes the role of the interfacial free energy of the gelators (in contrast to spinodal decomposition phenomena; vide infra).

Rheological measurements are a better means to identify the gel state than various classifications based on morphology, consistency, or chemical and mechanistic considerations [2,9,15]. The sol-to-gel and gel-to-sol (melting) phase transitions can be studied by rheology in the same manner as with macromolecular physical gels (e.g., gelatin in water). Gelation can be described by statistical physics and, in particular, percolation models in the vicinity of the critical gelation threshold, which is accompanied by a divergence of the connectivity correlation length (as for second-order phase transitions) [16].

Gels exhibit a sharp sol-to-gel phase transition at a specific temperature T_{SG}, while jellies have a less abrupt threshold that is somewhat analogous to the glass transition of polymers. Both the permanent and transient networks are thermally reversible—the supramolecular architecture is melted and individual molecules are redispersed in the bulk solution—but exhibit very different viscoelasticity and relaxation properties. As an example, gels are soft solids with a high rigidity, while jellies are viscoelastic liquids whose static and dynamic quantities may follow power laws obtained by incorporating the concentration and temperature dependences of the length of the strand assemblies in the scaling laws for semidilute polymers in good solvents. This is exemplified with the "living polymers" [17,18], which are "giant micelles" undergoing scission/recombination reactions. Figure 1 illustrates schematically four structural possibilities encountered with the heterogeneous networks of low-mass organogelators in organogels. The microdomains where the fiberlike aggregates merge can be long-lived, as with crystalline or lyotropic zones, or shorter lived for long-range interactions or transient entanglements and associations. The long-lived solidlike gels exhibit high yield stress values σ^*, thixotropic suspensions of rods have moderate values, and viscoelastic liquids lack significant σ^* values.

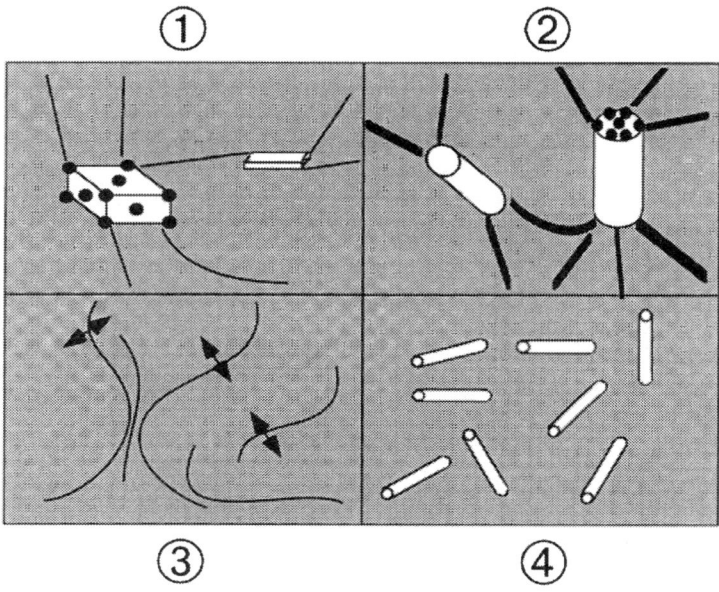

FIG. 1 Cartoon representation of the four types of junction zones in the networks of low-mass organogels [1]. (1) Crystalline; (2) lyotropic; (3) transient entanglements and scission/recombination processes as in "living giant micelles"; (4) long-range interactions. (From Ref. 120.)

The systems presented here all are "true" gels, which develop crystalline or lyotropic ordered "junction zones."

III. ELECTRONIC ABSORPTION AND LUMINESCENCE PROPERTIES [19]

The electronic absorption characteristics of chromophores within potential gelators can provide an important experimental monitor of the microscopic environment in which they reside. This is especially true when the information includes optical rotatory dispersion (ORD) and circular dichroism (CD) data for potential gelators that are chiral. Dichroism relates to the absorptivity difference between the two components of circularly polarized light, which constitutes the incident plane of linearly polarized light as described by the Kronig–Kramers transform. The intensity of UV/vis absorption depends on corresponding quantum transition. The wavelengths at which nonzero circular dichroism may be observable in the CD spectrum can be discerned from the shape of the absorption bands. The intensity of CD absorption is a function of the scalar product of the rotational strength of the electronic transition and its related magnetic transition moment. The angle between the two transition moments accounts for the bisignet nature of CD bands. The signals arise when there is a perceivable environmental asymmetry felt by the chromophore; helically arranged arrays represent an important nonmolecular example of how such environments can be created. Calculations using a classical model with complex electronic polarizabilities, exciton theory, and other quantum approaches have been made [20,21] to predict the absorption and optical activity of dilute solutions of molecular aggregates. With optically active organogelators, aggregation frequently involves either local or long-range helical arrangements of molecules. The macrostructures can be characterized by analyses of the signatures of the dichroic curves or, sometimes, by the selective reflection component (which depends on the sense of the pitch of the helices).

Excited states of appropriate gelators can return to their ground states either radiatively or nonradiatively. In the former case, the frequencies of the emitted photons are characteristic of the energy differences between the excited and ground states in the range of nuclear coordinates where the "relaxed" excited state exists and the relative intensity of the emitted radiation at each frequency is a complicated convolution of several probability factors related to the overlap of the wave functions [22]. Figure 2 shows some possible photophysical processes of an isolated (i.e., unaggregated) molecule. They include absorption, fluorescence (i.e., emission involving transition between two states of the same spin multiplicity, usually the first excited singlet and the ground state; $S_1 \to S_0$), phosphorescence (i.e., emission between states of different multiplicity, usually the first excited triplet and the ground state; $T_1 \to S_0$), vibrational relaxation within one electronic state, internal conversion (i.e., nonradiative transition between two states of the same multiplicity), and intersystem crossing (i.e., nonradiative transition between two states of different multiplicity, usually the first excited triplet and the ground state). Changes in the efficiency and spectral distribution of emission from an appropriate lumophoric group (i.e., the part of a molecule in which the excitation energy is localized) provide a sensitive probe (or label) of local environments [23]. To be most informative, the emission should have a high quantum efficiency and be traceable from the original absorbing chromophore to the ultimate lumophore; radiationless decay processes should be low efficiency, and energy transfer, if present, should have a spectroscopic signature that is easily discerned.

The design of supramolecular structures or assemblies in which photoinduced or energy- or electron-transfer processes can proceed over significant distances is motivated by fundamental interests in areas such as energy storage and applications such as information processing. The efficiency of energy or electron transfer in strand-shaped assemblies of low-mass organogelators will be strongly dependent upon the rigidity of the connecting links. For instance,

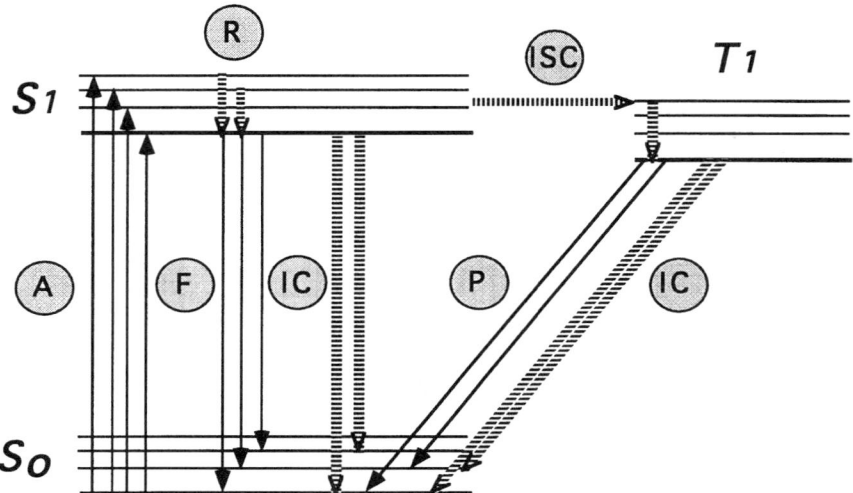

FIG. 2 Simplified energy level diagram with some photophysical processes. A: absorption; F: fluorescence; P: phosphorescence; R: vibrational relaxation; IC: internal conversion; ISC: intersystem crossing. Straight lines are radiative processes, while dashed lines are radiationless processes. (From Ref. 23.)

electron-transfer processes can occur via dipole–dipole resonance or electron exchange mechanisms. Examples of one-dimensional energy migration in rod-like structures are found in covalent complexes (i.e., dinuclear heterometallic bipyridine complex [24] of ca. 17 Å length) or columnar assemblies of liquid crystals of octasubstituted phthalocyanines [25,26]. In the latter example involving assemblies of low-mass molecules, a large path length for transfer, ca. 1 μm, has been found and is ascribed to a random walk of incoherent excitons. The efficiency of energy migration is higher in the liquid-crystalline than in the crystalline phase.

Similar studies with dilute and semidilute suspensions of luminescent fibers in organogels should be possible, but none to our knowledge has been conducted with the same level of detail. It is most probable that interchain hopping between gelator fibers will be much less efficient than transfer within and along the long axis of a fiber.

We now give basic expressions to describe the rate of radiationless energy transfer for isolated molecules where associations, if they occur, are transient and depend upon diffusion. Electron exchange (Dexter) is given by

$$k \approx \exp(-d) \tag{3a}$$

Dipole–dipole resonance (Förster) is given by

$$k \approx \frac{1}{d^5} \tag{3b}$$

d is the distance between the chromophores. Some other energy transfer processes with greater applicability to the aggregated systems in organogels will be presented in the following section.

IV. SOME USEFUL TECHNIQUES AND CONCEPTS FOR ELUCIDATING THE STRUCTURES OF ORGANOGELS

A. Rheology

The rheological (measure of the elastic modulus) and thermodynamic (determination of the enthalpy and entropy of the aggregation) measurements provide insights about the nature of the gelator aggregates and can be used to develop models for the interactions responsible for cohesion of molecules within the strands. The elastic and viscous components of the rheological measurements, mixed in the (non-Newtonian) gels, are determined from oscillatory experiments in which a small periodic strain/stress is applied within the appropriate amplitude range defining the linear regime. Information about the cohesion of gel fibers can also be obtained with steady shear, oscillation, stress relaxation, and creep mechanical procedures. For solidlike gels [27], the elasticity measured by the storage modulus $G'(\nu)$ is much larger than the dissipated energy (or viscous behavior as measured by $G''(\nu)$) and extends as a plateau from short to long times. These systems are characterized by a yield stress value σ^* at typical time scales of at least seconds that correspond to the minimum stress causing flow to occur. Rheological parameters are calculated from the phase lag between the applied shear stress and the related flow and from the ratio of the amplitudes of the imposed oscillation to the response of the gel. G', G'', and η^* are linked; ω is the angular frequency ($\omega = 2\pi\nu$), G' and G'' are the real and imaginary parts of the dynamic shear modulus:

$$G(\omega) = i\omega \int_0^\infty \exp(i\omega t)\, G(t)\, dt \tag{4a}$$

$$|\eta^*(\omega)| = \frac{[G'^2(\omega) + G''^2(\omega)]^{1/2}}{\omega} \tag{4b}$$

For a viscoelastic solid (like an organogel), any rheological description should give a constant finite elastic modulus and infinite viscosity at zero frequency or long times. The situation is somewhat comparable to that of a cross-linked network [2]. The equilibrium shear modulus for small deformations is proportional to the number n of network strands per unit volume that are terminated at both ends by the fixed cross-links (on which g depends):

$$G_e = gnRT \tag{5}$$

The dynamic moduli can be expressed as

$$G' = nRT\left(1 + \sum \frac{\omega^2\tau^2}{1+\omega^2\tau^2}\right)$$

$$G'' = nRT \sum \frac{\omega\tau}{1+\omega^2\tau^2} \tag{6}$$

in which one of the relaxation times τ must be infinite to account for the equilibrium modulus G_e contribution.

In actuality, organogels networks involve dangling ends, coupling entanglements (or their equivalent), and a sol fraction, which are all factors affecting Eq. (5). An important property of these networks is the stiffness of the elements involved in the cross-links. In other words, the fibers are "energetic" [28] rather than "entropic" in their elasticity. In such networks, the deformation is restricted to the bending of conformationally labile chains. It has been shown that the elastic modulus in physically cross-linked (usually polymeric) gels exhibits a ϕ^2 concentration dependence: the major contribution to elasticity comes from twists and bends of the stiff fibers (the energetic contribution); in entropic ideal networks, this is not the case.

B. Phase Diagrams

The phase diagram can be constructed and the pertinent sol-to-gel phase transition temperatures T_{SG} can be measured using different techniques. The length scale on which the transition is probed accounts for the variations of T_{SG} observed. A nonexhaustive list of methods includes differential scanning calorimetry (DSC), rheological experiments with temperature ramps, and various spectroscopies (such as fluorescence, IR, EPR, and NMR). Each of the spectroscopic methods uses the temperature dependence of an amplitude or frequency feature to follow the aggregation phenomenon during which gelation occurs. However, we note that the simple rheologically based "falling ball" method gives quite satisfactory T_{SG} values. In it, a tube with a metal ball on top of a gel is heated until the ball drops. The reproducibility of such measurements is usually ca. 0.5°C. The melting transition of physical gels at T_{GS} can be used to extract the thermodynamic parameters of the aggregate formation. Different models, as described by Eqs. (7) below, give similar trends. In the models, either the aggregates are assumed to result from a phase separation process (for T_{SG}), or the network melting (for T_{GS}) is assumed to occur when the crystalline fibers dissolve. These models have been applied to gelatin gels [29] and different low-mass organogels [1].

$$\Delta G_{ag}^0 = RT \ln(\phi_{ag})$$

$$\Delta H_{ag}^0 = -RT^2 \frac{\partial \ln(\phi_{ag})}{\partial T}$$

$$\Delta S_{ag}^0 = -R \ln(C_{ag}) - RT \frac{\partial \ln(\phi_{ag})}{\partial T} \qquad (7)$$

In Eqs. (7), T is the absolute temperature, ΔG_{ag}^0 is the standard free energy, ΔH_{ag}^0 is the standard free enthalpy, and ΔS_{ag}^0 is the entropy of formation of the aggregates; ϕ_{ag} is the mol fraction of gelator that participates in the aggregate. Different variations of this analysis have been proposed [29,30], and a similar equation has been derived [31] for a "fringed micelle" network of amorphous and crystalline regions (which involves an additional parameter for end-group energy).

In addition, the temperature T_{SG} from cooling and the temperature T_{GS} deduced from heating usually differ due to hysteresis effects, regardless of the technique of measurement. Temperatures based on rheology are more likely to be accurate representations of a gel-to-sol transition, since they are based upon measurements that reflect the macroscopic manifestations of the microscopic definition; they report either the point at which colloidal contacts and/or colloidal structures themselves are interrupted (T_{GS}) and thus when flow begins, or when colloids of strands enter into semipermanent contact with each other throughout the sample (T_{SG}). By contrast, any of the spectroscopic methods may be in error systematically, depending upon the difference between the temperature they report for a local (microscopic) change and the onset or cessation of macroscopic flow. Even rheologically based measurements may report incorrect transition temperatures if the shearing stress perturbs the fiber growth (T_{SG}) or melting (T_{GS}) processes. As mentioned, changes of temperatures based upon the onset or the maximum rate of a spectroscopic change need not correlate with the cessation or initiation of flow, since they measure alterations of aggregate units at the molecular level. These may (or may not) include the loss of junction zones and the separation of colloidal units into strands, complete melting of strands (and therefore, their colloids), and changes in local molecular packing.

Because of the complexity of the processes involved and the various methods of measuring them, phase diagrams of organogels can be envisioned by different theoretical approaches. In one option, network formation can be considered according to a spinodal decomposition mechanism. Briefly, the spinodal curve in a phase diagram represents the limit of metastability defined by the second derivative of the free energy with respect to concentration. Important features of such a mechanism are (1) the phase separation in solute-rich and solute-poor phases occurs without nucleation; (2) the (solute-rich) network that is produced is periodic; (3) since the connectivity above and below the phase transition temperatures is almost unchanged, the concentration dependence of the elastic modulus is weak. Experimentally, the random solidlike network of low-mass organogels (which may appear as statistically periodic) is sensitive to the usual nucleation parameters. In addition, secondary crystallization is suspected to be responsible for increases in the elastic modulus of certain fatty acid organogelators with time [32]. For these reasons, the spinodal decomposition mechanism provides an inappropriate description of organogel formation and can be eliminated from consideration. An approach more consistent with observation, the CSK model [33], combines thermodynamics and connectivity in a site-bonded correlation. It ''binds'' the liquid and the polymeric (solute) components so that the phase diagram is constructed in terms of temperature and ''connectivity'' (correlation and connectivity lengths describe the nonrandom distribution of liquid and the solute molecules) and applies mainly to systems for which bonds between monomeric species are continuously formed and broken on a short time scale. Figure 3 is an example of a simplified CSK diagram whose shape is dependent upon the ratio of the critical temperature T_c and the T_{GS} of the gel. At T_c, the equilibrium curve (corresponding to the minimization of the first derivative $\partial \Delta G/\partial \phi$ of the free energy with respect to the volume fraction in the two separate phases) and the stability curve $\partial^2 \Delta G/\partial \phi^2 = 0$ coincide (the criticality evaluated at constant temperature and pressure corresponds to $\partial^3 \Delta G/\partial \phi^3 = 0$). Depending on the type of liquid, T_c is in the sol phase, on the gelation curve, or in the gel phase. At T_c, the correlation length diverges, while at the gelation threshold defined by T_{SG}, the connectedness diverges; the two temperatures can be coincident for some gels. Experimentally, phase separation of low-mass organogelators has been observed to depend upon the liquid type. The low-temperature domain is experimentally difficult to investigate, since it involves very low gelator concentrations, and frequently gelation appear to occur at temperatures where the liquid alone can crystallize or form a glass.

C. Electron Microscopies

Determination of gelator microstructure is an important part of understanding how and why a gel forms. It provides indications of the relationship between the molecular structure of the gelator, its mechanism of autoassociation, and the topology and properties (from viscoelasticity to photoluminescence) of the aggregates. Electron microscopy combined with scattering techniques is a powerful analytical combination, since both real and reciprocal spaces are employed. Transmission electron microscopy (TEM) requires that native gels with their liquid be temperature quenched, fractured, etched, and metal shadowed. Sample preparation is protracted and potentially hazardous to the morphological integrity of the extremely fragile solidlike mesh that is to be imaged. In spite

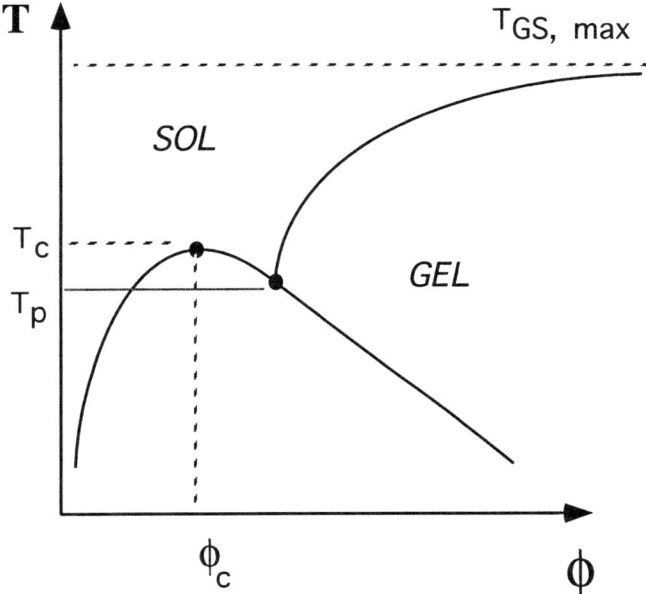

FIG. 3 Phase diagram according to the CSK statistical–mechanical model [33] showing the sol and gel domains and the coexistence curve. T_c is the critical temperature (corresponding to the volume fraction ϕ_c), T_p is the temperature at which the gelation curve meets the coexistence curve, and $T_{GS,max}$ is the limiting temperature above which there is no gelation. (From Ref. 33.)

of this, careful work has provided spectacular micrographs of the porous and fibrillar networks, including their junction zones, with ca. 20 Å resolution and fascinating views of the associated labyrinth of the colloids [34].

Scanning electron microscopy (SEM) provides similarly valuable information about the microstructures of solid samples, such as xerogels. Slow evaporation of the organic liquid from the gel gives a gelatinous mass that shrinks progressively to a solid residue, the xerogel. Capillary forces and liquid–vapor interfaces are responsible for a collapse of the fragile 3-D structure during migration of the meniscus in the sample. Owing to the collapse, SEM focuses on the general shapes and morphologies rather than on quantifiable parameters like diameters, lengths, or topologies. Direct imaging methods, originally used for biological systems [35,36], are now utilized for analogous aqueous systems and should become the method of choice for organic media also to minimize perturbations to the microstructures. Even with such a method, there is no guarantee that artifacts will always be absent, since they may arise from confinement of the gel in very thin layers. In such a case, the colloids may be destroyed or their sizes distorted, and the appearance of chain flexibilities and the topology of entanglements will be changed.

D. Small-Angle Scattering Techniques

Scattering techniques offer a distinct advantage in probing unadulterated gel samples. However, the structural information derived from scattering data is model-driven owing to loss of phase information and requires mathematical manipulation [37–41]. To apply small-angle neutron scattering (SANS) successfully, coherent scattering (generated by Fermi nuclear interactions—depending upon the isotopic composition—between the sample and the neutron particles) must be discernible above the flat, incoherent scattering background that originates mainly from hydrogenated substances. As a practical consequence, the use of deuterated liquids is necessity.

Low-angle scattering is based upon the laws of diffraction for large particles. The elastic deviation of the radiation trajectory is described by the momentum transfer Q (Å^{-1}), $\vec{Q} = \vec{k_i} - \vec{k_{sc}}$ (where $\vec{k_i}$ and $\vec{k_{sc}}$ are the wave numbers of the incident and scattered beams, respectively). The modulus of Q ($|Q| = (4\pi/\lambda)\sin\theta$, with θ being one-half the scattering angle and λ the radiation wavelength) relates inversely to real space ($\langle d \rangle \equiv 2\pi/Q$): low angles correspond to large correlation distances of the scatterers. With dilute systems, interferences between scattered neutrons or x-ray waves by single particles generate a form-factor intensity I_F that can provide information concerning the shapes, sizes, monodispersities, homogeneities, and masses per unit volume of the particles. With semidilute or concentrated systems, interferences between correlated and/or oriented particles generate a structure intensity I_S that reveals the ordering degree of the system. Despite the loss of phase information of the interfering waves in small-angle scattering, conditions that prevent the real-space modeling from being unequivocal, the technique has been successfully applied to numerous colloidal systems [42]. With gels from low-mass gelators, the concentrations employed are ca. 1 wt%, so that effects of interferences between the aggregates can easily be recognized on the scattering profiles. In such dilute systems, these interactions correspond to long-distance forces responsible for the typical viscoelastic behavior of the gels, and they can generate a very-low-angle contribution. Within appropriate correlation conditions of the positions of the interacting particles, the differential cross-section per particle $d\sigma/d\Omega$, to which the measured intensity is proportional, is given by Eq. (9).

$$I_F(Q) = \left\langle \left| F(\vec{Q}) \right|^2 \right\rangle_Q \tag{8}$$

$$\frac{d\sigma}{d\Omega} = I_S I_F \tag{9}$$

At reduced concentrations, so that $I_S \approx 1$, the intensity reduces to the form factor of cylindrical rods of length L (Eq. (10)).

$$F(\vec{Q}) = F(Q, \gamma) = 2\Delta\rho L \frac{\sin(QL/2 \cos \gamma)}{QL/2 \cos \gamma} \frac{J_1(Qr \sin \gamma)}{Qr \sin \gamma} \quad (10)$$

In this expression, γ is the angle between $\vec{Q}$ and the rod axis (Fig. 4) and J_1 is the Bessel function of the first kind.

The vertical plane of the detector $(\vec{x}, \vec{z})$ contains $\vec{Q}$ for large sample–detector distances (i.e., low-angle scattering). The scattered intensity is a time-average intensity emanating from the irradiated volume investigated within the length scale defined by Q. I_F is the average of Eq. (10) over all orientations of the long rod axis with respect to $\vec{Q}$. According to Eq. (10), rods with lengths much larger than their diameters [38] make a contribution to the scattering only when they lie nearly perpendicular to $\vec{Q}$. Then I_F is the product of two nearly independent terms, one axial (proportional to $1/Q$) and the other cross-sectional (proportional to J_1 for a circular shape).

In a random gel network, the long fibers are oriented completely randomly and there is no dependence of I_F with the angular position of the sample in the plane $(\vec{x}, \vec{z})$. The intensity reduces to Eq. (11), and the detector shows isotropic scattering with circular isointensity contours (Fig. 4).

$$I_F = \frac{\pi C}{Q} \overline{\Delta b}^2 M_L \left[2 \frac{J_1(Qr)}{Qr} \right]^2 \quad (11)$$

In this expression, $\overline{\Delta b}$ is the neutron specific contrast of the rod with respect to the solution, and M_L is the mass unit per length of the rod.

Conversely, with oriented systems, the rods are position and/or orientation correlated, and the scattering is dependent upon the angular position of the sample with respect to $\vec{Q}$ [43]. Two extreme situations for perfect orientation of the rods can be imagined:

(1) If the director $\vec{n}$ (i.e., direction of orientation) is orthogonal to the incident beam and if $L/2r >$ ca. 10, Eq. (10) gives a nonzero contribution only for $\gamma = \pi/2$. A horizontal line or band is observed on the detector (Fig. 4, image A). A real sample, in the absence of any external orientational forces (shearing stress, electric field, etc.) is usually not perfectly oriented and is composed of domains with a distribution of orientations around the main director $\vec{n}$. Each rod in a domain is referred in polar coordinates by angles θ and φ, and the direction $\vec{n}$ is assumed to coincide with $Q_{\parallel}$ (where $Q_{\perp}$ defines the orthogonal direction). With a sheared material, $Q_{\parallel}$ is conveniently taken to be parallel to the direction of shear. The orientational distribution of the rods is described by a probability function $p(\theta, \varphi)$. The signal evolves into an anisotropic scattering with elliptical isointensity contours (Fig. 4, image A') that reveal, for each ψ

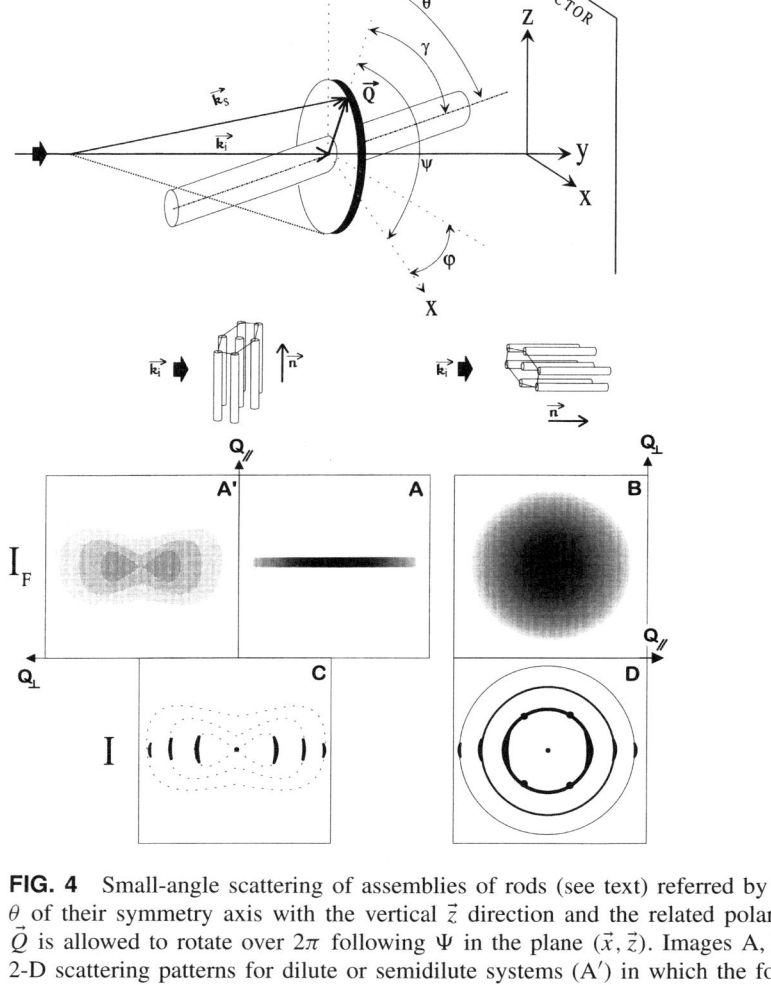

FIG. 4 Small-angle scattering of assemblies of rods (see text) referred by the angle θ of their symmetry axis with the vertical $\vec{z}$ direction and the related polar angle φ. $\vec{Q}$ is allowed to rotate over 2π following Ψ in the plane $(\vec{x}, \vec{z})$. Images A, B are the 2-D scattering patterns for dilute or semidilute systems (A′) in which the form-factor intensity I_F is a major contribution. Images C and D relate to the intensity for concentrated systems in which the interparticle structure factor intervenes. $Q_\parallel$ and $Q_\perp$ are indicated with reference to the director direction $\vec{n}$. Image A: rods oriented in a direction orthogonal to the direct beam; either a horizontal stripe or an elliptical scattering is observed (image A′ corresponds to an imperfect orientation of the rods). Image B: rods fully oriented parallel to the direct beam; a continuous isotropic scattering is observed. Image C: assembly of rods either fully of partially oriented orthogonal to the direct beam; "equatorial" spots or elliptical scattering with "equatorial" spots are spaced in a $1 : \sqrt{3} : \sqrt{4} : \sqrt{7}$ sequence. Image D: assembly of rods fully oriented parallel to the direct beam; rings are spaced in $1 : \sqrt{3} : \sqrt{4} : \sqrt{7}$ Q-sequence; six hexagonally located spots on the first ring are usually observed.

value in the $(\vec{x}, \vec{z})$ plane, the existence of a population of rods that satisfies the condition of orthogonality of $\vec{Q}$ to the related rod axes.

(2) The director can be parallel to the incident beam $\vec{k}_i$ (i.e., $\vec{n}$ is orthogonal to the $(\vec{x}, \vec{z})$ detector plane). In this case, Eq. (11) provides the isotropic scattering (circular isointensity contours) independent of the $\vec{Q}$ position in the $(\vec{x}, \vec{z})$ plane (Fig. 4, image B).

In the semidilute $(C > 1/L^3)$ and, especially, in the concentrated $(C > 1/2rL^2)$ regimes, the long $(L \gg 2r)$ rods interact. Depending upon the symmetry of the cross-sectional ordering of the rods, their related 2-D structure factor can add specific diffraction features. Again, two extreme orientational situations can be analyzed assuming that the rods are hexagonally packed. First, if $\vec{n}$ and the incident beam are orthogonal, spots, separated in a $1 : \sqrt{3} : \sqrt{4} : \sqrt{7}$ Q-sequence, can appear along the horizontal $\vec{x}$ direction (called the "equatorial line") in a "fiber diffraction pattern." Because of imperfect ordering, the elliptical intensity distribution can be superimposed to the spots (Fig. 4, image C). Second, if $\vec{n}$ is parallel to the incident beam, rings appear Q-spaced in a $1 : \sqrt{3} : \sqrt{4} : \sqrt{7}$ sequence and contain hexagonally located spots (usually on the first-order ring only). Additional spots along the horizontal line reveal the deviation from perfect ordering (Fig. 4, image D). The angular extension of these crescentlike spots shows the degree of ordering [44] described by the orientational distribution function $p(\theta, \varphi)$ (as for the measurement of the nematic order parameter of rodlike molecules) [45].

In the larger angle (wide-angle x-ray scattering, WAXS) domain, a method that relates the morphs of a neat solid gelator and the molecular packing in gel strands has been developed [14]. The method involves a comparison of the powder diffraction patterns of the fibers in a gel (by subtraction of the large, broad diffraction of the amorphous liquid component) and of the neat crystalline gelator (obtained from single-crystal diffraction data). A complementary analysis can be made with xerogels [46], assuming that the short-distance range ($\leq$ several tenths of an Å) of the degree of molecular ordering is not affected by the collapse of the network during the gel-to-xerogel formation. Such studies require powder diffractometers using high flux and resolution sources to provide diffractograms of sufficient quality for correlations with single-crystal data (i.e., application of the Rietveld method). It is interesting to observe that few single crystals of low-mass organogelators have been obtained: good solubility and gelation ability appear frequently to be conflicting properties in this class of systems.

E. Luminescence in Aggregated Systems

Any molecule, including an organogelator or its liquid component, will luminesce with a nonzero probability after absorbing a photon. In practice, the quan-

CAB

STRUCTURE 1

tum efficiency of luminescence Φ_L, defined as the number of photons emitted by a sample divided by the number of photons absorbed by it, and the sensitivity of modern fluorimeters and photodetectors, limit the types of molecules whose luminescence can be detected routinely. Other factors such as collection geometry of the fluorimeter, the intensity of the exciting source, the concentration of organogelator, the nature of the liquid and its optical/electronic properties also influence the practical detection of emitted electromagnetic radiation.

Generally, when Φ_L is $< 10^{-4}$, the intensity of emitted radiation will be insufficient to be useful on a routine basis. In the cases discussed here, "luminescent organogelator" refers to molecules with $\Phi_L > 10^{-4}$ when they are in dilute (unaggregated) solutions and the liquid is not a "quencher" (i.e., a species that shortens the excited-state lifetime). In the final analysis, the pragmatic determination of whether an organogelator is luminescent is provided by the ability to detect relatively easily emission of a sample, gelled or otherwise, using a good quality fluorimeter. However, small amounts of impurities with very high values of Φ_L mixed with an organogelator with a very low value of Φ_L can lead to erroneous conclusions and sometimes embarrassing consequences. High purity is a necessity if luminescence results are to be believable and reproducible!

Since organogelators are in crystalline or lyotropic aggregate states in a gel, the nature of their intermolecular interactions becomes a factor of paramount importance in determining the nature and the intensity of emitted radiation from a gelled sample. Molecular proximity opens possible reactive channels for the excited states that are not available in dispersed solutions. For example, it has been shown that **CAB** dimerizes in its neat solid, liquid-crystalline, and gelled (fiber) states when exposed to UV radiation [47,48]. (See Structure 1.) In dilute isotropic solutions, no photoreaction is observed because the time required for an

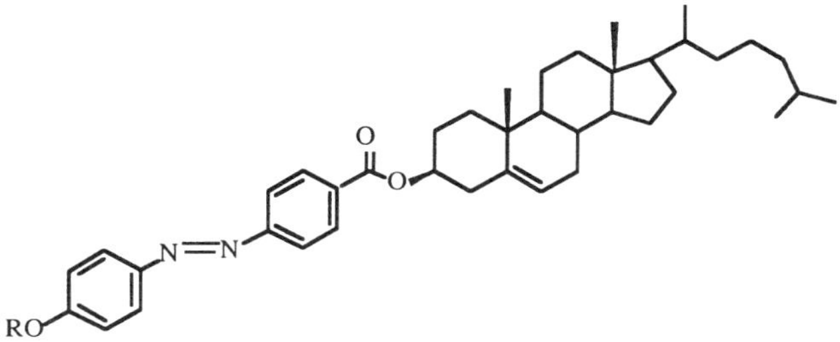

13

STRUCTURE 2

excited state molecule to collide with one in the ground state is much longer than the lifetime of the reactive excited singlet state ($< 10^{-8}$ s). As a consequence of photoreaction, the gel phase is destroyed—the photodimers, in addition to occupying less volume in the crystalline lattice of a fiber, may act as catalytic sites for transformation of the gel morph into another packing arrangement.

A fascinating practical consequence of a photoreactive pathway in an organogel containing an organogelator with an azobenzenyl group, trans **13** (R = Me), is its reversible transformation into a sol [49]. (See Structure 2.) Thus a trans azobenzene in normal solutions can be isomerized by selected UV/vis radiation to its cis isomer; the cis isomer can be reconverted either photochemically or thermally to the trans. Irradiation of gelled trans **13** (0.1 wt%) in 1-butanol and 15°C results in a gel-to-sol transition. The gel is re-established slowly in the dark (as the less thermodynamically stable cis **13** isomerizes thermally to the trans) or rapidly upon irradiation at wavelengths where only the cis absorbs; the gel-sol phase change can be repeated many times by employing the appropriate wavelengths of radiation.

Aggregation also makes more feasible the emission from various excited-state molecular complexes. When in dimeric units, these complexes may (or may not) [50,51] behave like dynamically formed excimers (i.e., excited complexes that are dissociative in the ground state and are formed upon collision of two molecules of the same kind after one has been electronically excited) [52]. Larger luminescent aggregate units are also known in nongelled samples [53]. Additionally, excitation of a molecule in a well-ordered aggregate can lead to emission from another molecule that is quite remote from the first, owing to numerous energy transfer steps (i.e., excitation-energy hopping) whose rate is faster than the rate of decay of the (localized) excited state [54,55]. The final locus of emission may be an energy ''hole'' (i.e., a molecular impurity whose excited state energy is lower than that of the bulk or a molecule of the bulk that resides at a defect site in the molecular array).

Aggregate units sometimes provide exciton bands in absorption and/or emission spectra that are characteristic of the mode of interaction of the constituent molecules [56–60]. That is, they depend upon the orientation of the molecules (i.e., their transition dipoles) with respect to each other, the number of molecules interacting, and the distance between the chromophoric/lumophoric centers. The excitonic interactions result from the excited state of one molecule being affected by the electronic distribution of neighboring molecules. The location and shape of exciton bands can provide useful information concerning the relative orientations of molecules within an aggregate.

We present here a short summary of the point–dipole model of Kasha that allows exciton bands to be associated with aggregate organization [56,57]. Related models, such as that employing extended dipoles, lead to the same qualitative conclusions [58,59]. Recent advances in semiempirical molecular mechanics and quantum techniques now allow very good calculations of the magnitude of exciton splitting in complicated aggregate systems [61]. Application of the methodology to correlate the electronic spectra and molecular packing of organogelators in their gel fibers is underway [62]. If successful, this approach will provide a powerful analytical tool to complement other structural and spectroscopic techniques now being employed. In fact, it bridges the gap between the molecular and nanoscopic probes presently available.

Both the model of Kuhn [58,59] and that of Kasha [56,57] predict that the magnitude of the exciton splitting will be proportional to the square of the transition moment M of the affected transition and will decrease as r_{ab}^{-3} (where r_{ab} is the center-to-center distance between transition dipoles on adjacent electronically interacting molecules a and b). In the case of dimeric interactions within an "infinite lattice" of single molecules in a unit cell [56,57], the degenerate excited states will be split into a lower E' and a higher E'' energy level whose splitting ΔE is given by Eq. (12) when both transition dipoles lie in the same plane. The cases in which there is more than one molecule per unit cell and in which the transition dipoles do not lie in the same plane have been solved also [56], but the simpler scenario illustrates the pertinent concepts. The angles α and θ are defined in Fig. 5 [56], where the transition dipoles are shown as double-headed arrows.

$$\Delta E = E'' - E' = \frac{2|M|^2}{r_{ab}^3}(\cos\alpha + 3\cos^2\theta) \tag{12}$$

Several interesting limiting cases arise for specific values of α and θ. For instance, if the transition dipoles are parallel (i.e., $\alpha = 0$ and $\theta = 90°$), transition to the lower energy state will be forbidden from the ground state and a hypsochromic (blueshifted) exciton band is predicted. When the transition

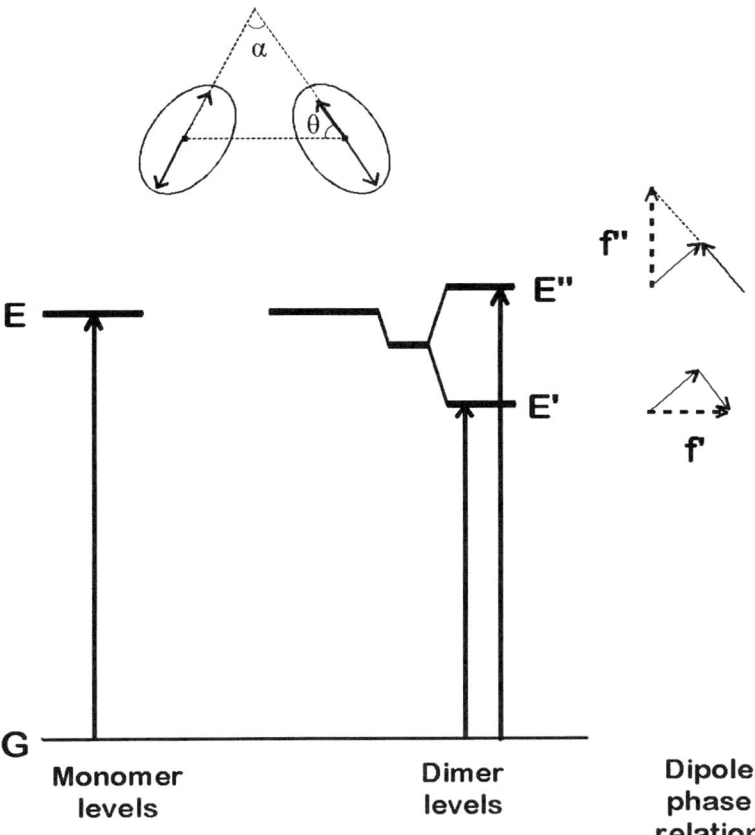

FIG. 5 Exciton band energy diagram for a molecular dimer, or a double molecule, with oblique transition dipoles. The polarization axis for the electronic transition is shown aligned with the long axis of the molecule represented in oval profile (top). The oscillator strengths (dotted arrows of the vector diagrams to the right) f' and f'' for transitions to the two exciton states are polarized mutually perpendicularly. α is the angle between polarization axes for the component absorbing units, and θ is the angle made by the polarization axes of the unit molecule with the line of molecular centers. An exciton splitting (singlet–singlet transition) is observed. (From Ref. 56.)

dipoles are colinear, only transition to the lower energy state is allowed, and a bathochromic (redshifted) exciton band is predicted. When the transition dipoles are orthogonal (i.e., $\alpha = \theta = 90°$) or parallel and offset at the magic angle (i.e., $\alpha = 0$ and $\theta = 54.7°$), no splitting will occur. Obviously, the magnitude of the splitting at constant r_{ab} will vary within the limiting cases as the values of α and θ change from 0 to 90°.

Furthermore, exciton splitting interactions can enhance intersystem crossing and, potentially, Φ_L of phosphorescence when the lower energy exciton split state is forbidden in absorption [56]. It follows that emission from it to the ground state (i.e., fluorescence) is forbidden also. If internal (nonradiative) conversion of the upper-to-lower exciton state is rapid (as it frequently is owing to the similarities of E'' and E'), intersystem crossing to produce a triplet state may be able to compete effectively with internal conversion to the ground state from the lower exciton state (since the rate of the latter is usually slow because of the large energy gap between the two states). The triplet states are not split because of the very low values of ΔE that follow from the very low values of M of singlet–triplet electronic absorptions.

In addition, the size of the lattice can be a factor in determining the magnitude of ΔE and the position of the new exciton band [63,64]. For a lamellar aggregation, a lattice cross section with at least 35 molecular units has been calculated as the minimum to lead to convergence in calculation using the model; smaller aggregates are not infinite in practical terms, and their degree of splitting will not follow the quantitative aspects of the model. Although the definition of convergence will vary depending on the system under study (N.B., cases in which as few as 6 molecules along any radian emanating from an excited state allows convergence are smaller than the lower limit cited; the excitonic band structures and positions may not correspond to those expected from the same molecular packing in a bulk crystal!

V. CLASSES OF LUMINESCENT ORGANOGELATORS

Discoveries of new luminescent gelators or thickeners remain mainly fortuitous for the same reason that the selection *a priori* of any low-mass molecule as a gelator is not possible. Only organogelators for which at least preliminary emission studies have been conducted are included, though many others contain structural units that should be luminescent. Correlations, when possible, are made between the molecular structures and electronic properties of the gelators, and some properties of their gels.

The number of known organogelators whose luminescent properties should be investigated is large and growing. For instance, the previously mentioned rods of metalloporphyrins in cyclohexane might have interesting electron-transfer (and luminescence) properties like those of micellar fibers of octopus porphyrin in aqueous media [66]. 1,3:2,4-Di-O-benzylidene-D-sorbitol is another organogelator that is probably highly fluorescent but whose optical spectroscopic properties appear not to have been investigated yet [67–69].

A. Anthryl Derivatives

2,3-*bis*-*n*-Decyloxyanthracene (**DDOA**) can gel various alkanes, alcohols, aliphatic amines, and nitriles [70,71]. The related anthraquinone **2** is also a gelator, but the 2,3-dialkoxynaphthalene is not. (See Structure 3.) Hydrogenation of one of the rings of **DDOA** does, however, yield molecules that gel a variety of liquids [72]. The sol–gel phase diagram of **DDOA** is typical of the class of low-mass organogelators (Fig. 6). Thermodynamic parameters for the transition $\Delta H° \approx 25.1$ kCal mol^{-1}, $\Delta S° \approx 89.7$ Cal mol^{-1} K^{-1}), deduced from variations of the electronic absorption spectrum, can be compared to values extracted from the phase diagram, Eqs. (7), as classically constructed [73,74].

The onset of formation of a 3-D elastically connected network of **DDOA** aggregates has been followed by rheological measurements [75]. There is an abrupt increase of the storage modulus G' at a characteristic time of the aggregation process that corresponds to the growth of the infinite network (Fig. 7).

In 1-octanol, the magnitude of the stabilized storage modulus value G' (ca. 10^4 Pa for $\phi \approx 0.01$) is much larger than the loss modulus G'' (ca. 1500 Pa), as expected for a gel (Fig. 8). Comparison of the magnitude of the elasticity of the gel network with that of other systems suggests that the interaction zones are permanent and highly ordered (crystalline) microdomains at a constant temperature

DDOA

2   **3**

STRUCTURE 3

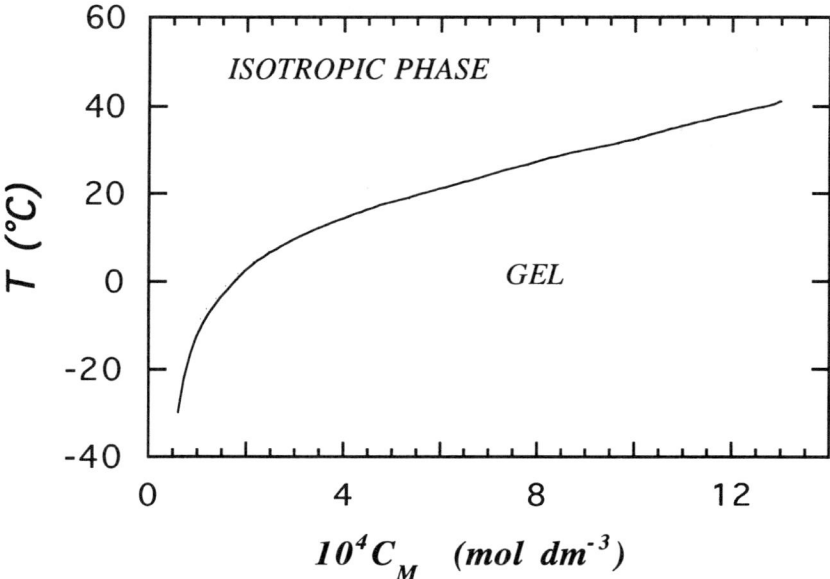

FIG. 6 Gelation temperature of methanolic solutions of **DDOA** vs. concentration. (From Ref. 70.)

in the gel. The rheological properties of **DDOA**/1-octanol gels exhibited marked hysteresis when cycled between 20 and 40°C (Fig. 9). The large yield-stress value ($\sigma^* \approx 550$ Pa for $\phi = 0.01$, not shown) [75] reveals a strongly structured network and is consistent with the self-supporting ability of the solidlike gel. Hysteresis is a commonly observed feature of low-mass organogels. Close analogies exist between gelation/melting transitions and the crystallization/dissolution behaviors of supersaturated solutions. Crystallization results from two simultaneous processes, the formation of nucleated aggregates of the new phase and the growth of crystals from them, while the melting transition is a simple first-order process. Germination processes can be the rate-limiting steps of the sol-to-gel phase transition. Especially, the rate of the imposed temperature ramp in a rheological experiment has to be slow enough to allow structural modifications to occur. If not, some hysteresis effects can also appear (Fig. 9).

DDOA gels ($\phi \approx 0.02$) in butanol-d_{10} and decane-d_{22} exhibit comparable isotropic SANS patterns with a monotonic decrease ($I = Q^{-3.7}$) from small-to-large angles, where diffraction peaks (for instance, at $Q \approx 0.1$ Å^{-1}) are observed. Since neutron-nuclear contrasts of **DDOA** aggregates in the two liquids are comparable, the experiments describe the part of the morphology that is

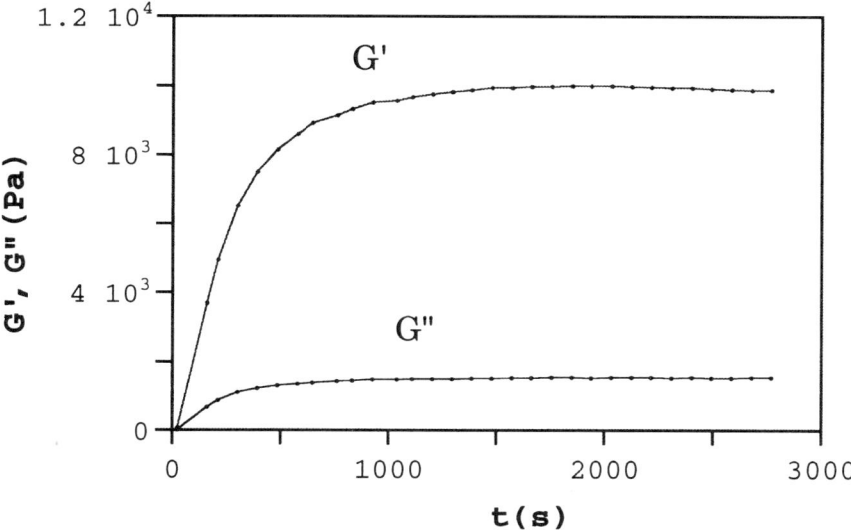

FIG. 7 Rheological time sweep experiment for a 0.98 wt% **DDOA**/1-octanol gel. The warmed solution was introduced between the cone and plate of the rheometer and cooled to 20°C. Frequency = 1 Hz, angular displacement $\alpha = 6.9 \times 10^{-4}$ rad, and strain amplitude $\gamma = 1\%$. (From Ref. 75.)

insensitive to the liquid type. A Porod analysis shows a broad bump ($0 < Q < 0.075$ Å^{-1}) attributed to the form-factor of the **DDOA** aggregates and whose width is related to the size polydispersity of the scatterers. The bump can be modeled as an interfacial scattering from a distribution of diameters ranging from ~ 100 to 500 Å. The peaks at larger angles ($Q \approx 0.1$ Å^{-1}) are due to the structure factor of ordered heterogeneities in the gel network.

Freeze-fracture electron micrographs of **DDOA**/propanol gels indicate the presence of a dense 3-D random network of bundles of rigid fibers with 600–700 Å diameters (Fig. 10). Syneresis can be a conformational response of thin fibrils to a mechanical stress, even low (i.e., a **DDOA** gel network, which shrinks in acetonitrile and produces thick fibers) [32]. The swelling or shrinking of the network results from a complex balance of forces, including the elasticity of the network, fiber–fiber affinity, and in some rare cases the polarity of liquids (which influences the osmotic pressure of the gel through the ionization degree of ionic gelators).

The crystal is always the reference state from which structural evolutions are discerned. The WAXS diffraction pattern of crystalline **DDOA** powder exhibits a sequence of Bragg peaks at 0.095, 0.166, 0.191, 0.254, 0.287, and 0.336 Å^{-1}

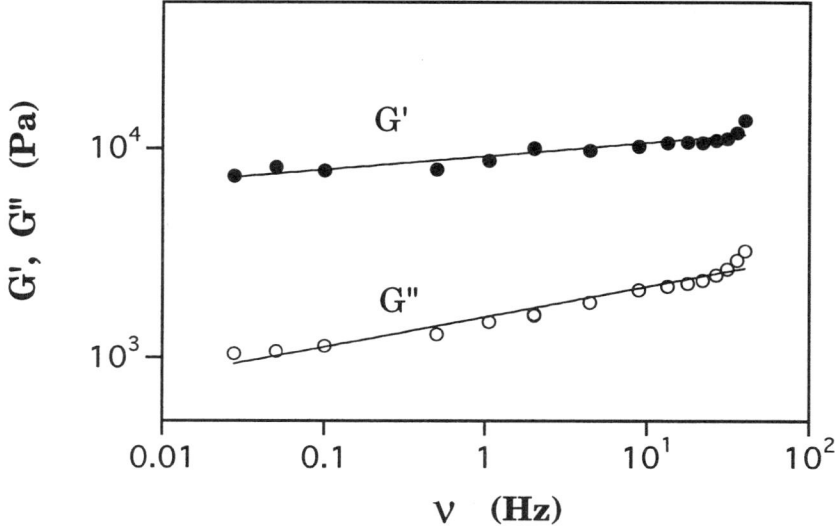

FIG. 8 Rheology ($\alpha = 6.9 \times 10^{-4}$ rad, $\gamma = 1\%$) of a 0.98 wt% **DDOA**/1-octanol gel sample. The variations of the storage modulus G' and the viscous modulus G'' as a function of frequency are shown. (From Ref. 75.)

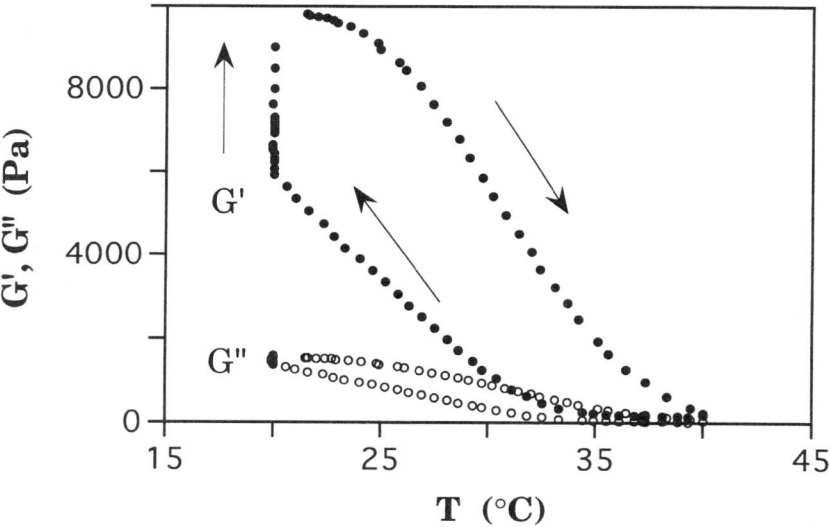

FIG. 9 Rheology of a 0.98 wt% **DDOA**/1-octanol gel: temperature sweep experiment at $\nu = 1$ Hz and $\gamma = 1\%$. The sample at equilibrium at 20°C is heated to 40°C in 1200 s (descending curve →) and cooled to 20°C (ascending curve ←). The vertical ascending part (↑) corresponds to the equilibration of the sample at 20°C after the temperature cycle. (From Ref. 75.)

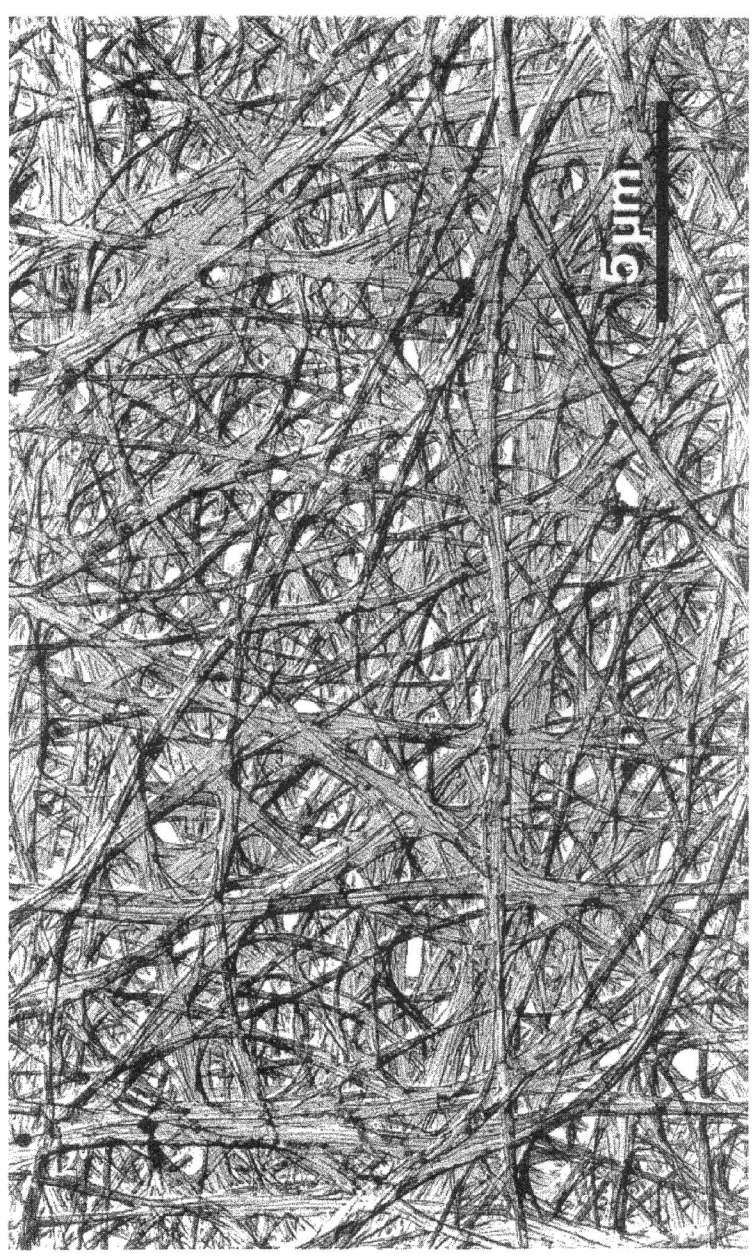

FIG. 10 Freeze-fracture electron micrograph of a **DDOA** gel in propanol. Rigid bundles of fibers are clearly seen. (From Ref. 70.)

whose spacings are in the $1 : \sqrt{3} : \sqrt{4} : \sqrt{7} : \sqrt{9} : \sqrt{12}$ ratios typical of hexagonal ordering. The ordering of **DDOA** in gel strands is also hexagonal, since Bragg peaks of the gel and neat solid are located at the same positions. The *[11]* reflection, associated with a reticular distance $\langle d \rangle \approx 66.1$ Å, corresponds to an intercolumnar distance of 76.3 Å in a hexagonal array. It arises from crystalline junction zones formed by bundles of fibers. The junction zones in **DDOA** organogels are "dry" crystalline microdomains, free of swelling by the liquid (in contrast to some other systems; vide infra). Derivatives like **3**, in which the polarity and length of the alkyl chains are varied, are being investigated in hopes of producing thinner fibrils [32].

The electronic properties of **DDOA** and its analogues can act as internal sensors of aggregation and deaggregation. The absorption spectrum in Fig. 11 shows two types of bands that change in opposite senses upon gelation. The lower energy band at 300–400 nm from ^1L_a and ^1L_b type transitions is redshifted, while the higher energy band at 200–300 nm from a $^1B_1$ type transition is blueshifted. In the gel phase, a 220 nm band appears at the expense of the band at 258 nm found in isotropic solution. These features have been attributed to antiparallel packing of gelator molecules in the fibers. The redshifted fluorescence from the gel also leads to the same conclusion: **DDOA** molecules are arranged head to tail with partially overlapping aromatic rings. This arrangement is also consistent with the scattering structural data, which did not reveal a simple relation between the size of the aggregates and the molecular length of **DDOA** (vide supra).

The temporal decay of the fluorescence from gelled **DDOA** in methanol at $-45°C$ remains biexponential regardless of the emission wavelength λ_{em} being monitored, but the magnitude of the decay constants τ_i becomes progressively longer as λ_{em} is monitored from the blue to the red: for instance, at $\lambda = 420$ nm, $\tau_1 = 5.0$ ns and $\tau_2 = 14$ ns; at $\lambda_{em} = 480$ nm, $\tau_1 = 11.3$ ns and $\tau_2 = 37$ ns. This behavior is rather curious and suggests further studies since it indicates the presence of a variety of aggregate structures, each of which must have a slightly different geometry. One would expect that their emission spectra would overlap in more than a pairwise fashion.

B. Anthryl and Anthraquinone Appended Steroid-Based Gelators

A class of molecules (**ALS**) functionalized at C_3 (which cannot be H-bond donors) are efficient gelators of many types of organic liquids. They consist of a group containing fused aromatic rings (**A**) connected to a steroidal moiety (**S**)

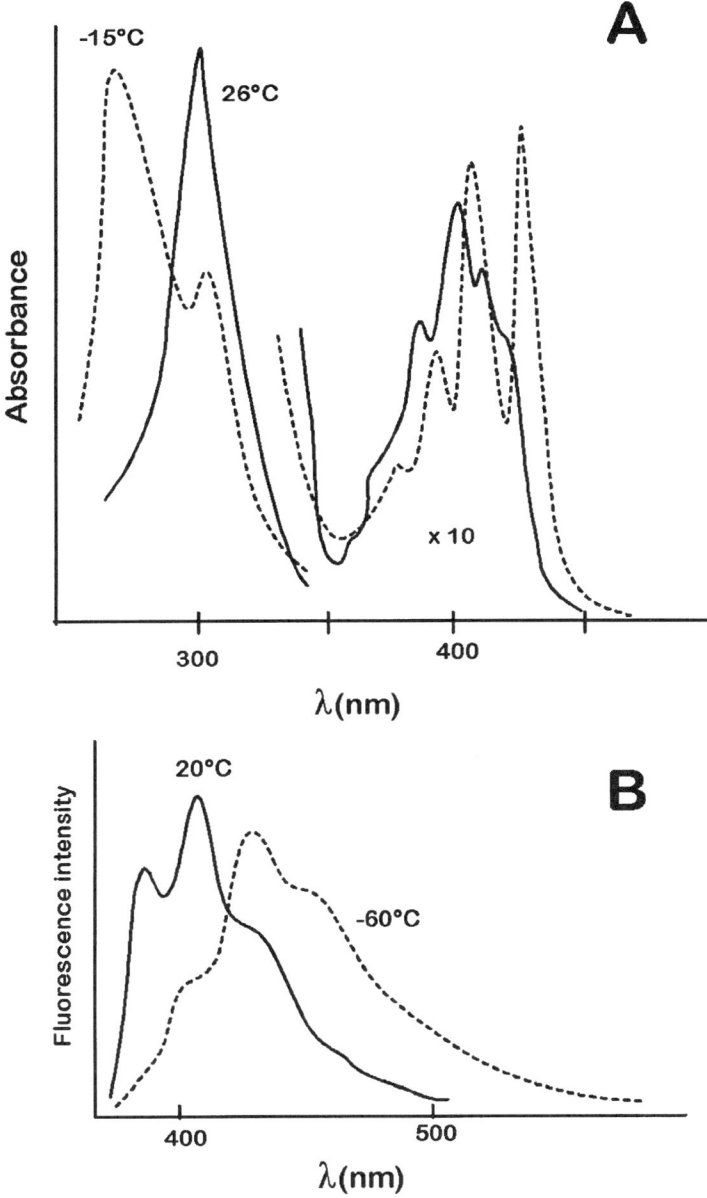

FIG. 11 Electronic absorption (A: $C = 10^{-3}$ mol dm^{-3}) and corrected emission spectra (B: $C = 10^{-4}$ mol dm^{-3}) of **DDOA** in methanol ($\lambda_{exc} = 370$ nm). Isotropic solution (——); gel: (– – –). (From Ref. 70.)

CAQ

4

7

CAB

5

6

Low-Mass Luminescent Organogels

CMAQ

STRUCTURE 4

via a linking group (**L**). The role played by the **A** and **S** autoassociative parts upon the mechanisms of aggregation and gelation can be studied using a family of **ALS** gelators. In particular, the competition between **A** and **S** to drive the aggregation process as well as specific properties that they are able to impart to gels on a mesoscopic level (e.g., changes in luminescence or sense of chirality) are of interest.

The linking groups of the **ALS** molecules, cholesteryl 4-(2-anthryloxy)butanoate (**CAB**) and cholesteryl anthraquinone-2-carboxylate (**CAQ**), allow them to adopt an overall rodlike shape. (See Structure 4.) **CAB** forms gels with hydrocarbons, alcohols, aldehydes, esters, amines, etc. (Table 1) [76], and both **CAB** and **CAQ** gel more complex fluids, like silicone oils [77]. Many **CAB** gels are luminescent and some are metastable, exhibiting a gel-to-liquid/solid phase separation after various periods of time.

A comparison of **DDOA**, a gelator with only an **A** part, with the comportment of another gelator, an androstanol derivative (**STNH**) that has only an **S** part (but can both donate and accept H-bonds due to a hydroxyl group at C3), illustrates the complexity of the interactions available in **ALS** molecules. (See Structure 5.) The range of liquids gelled by **STNH** is limited to saturated hydrocarbons, but its cyclohexane gel is more stable than those with n-alkanes and remains for more than 10 years! [78]. By contrast, cyclohexane gels of **CAB** persist at room temperature for no more than one hour and are much less stable than its gels with n-alkanes.

Gelation abilities have been correlated with each of the structural components of a wide variety of **ALS** molecules [48,79–81]. For instance, neither a 1-pyrenyl analogue of **CAB** (containing four nonlinearly fused rings in the **A** portion) nor a 2-naphthyl analogue of **CAQ** (containing only two fused rings) gelled any of a variety of liquids [76]; the shape and nature of the ring systems in the **A** part are important considerations in gelator design. By contrast, cholesteryl 4-(2-anthrylamino)butanoate, **CAB** in which the ether oxygen has been replaced by an isoelectronic N–H amino group, does not gel any of several liquids gelled efficiently by **CAB** [81]. However, several simpler **ALS** gelators with carbamate L-parts and naphthyl or even 4-n-alkylphenyl A-parts do gel some organic liquids! These observations illustrate the potential importance of H-bonding within strands and the delicate balance between the forces that control the nature of the solid aggregates at the time of nucleation.

From changes in luminescence, absorption, and dichroic intensities, the T_{SG} of **CAB**/n-alkane gels are independent of alkane chain length. However, the lifetime of a gel at room temperature in a sealed vessel is dependent upon gelator concentration and liquid type.

Absorption and fluorescence spectra from the anthryloxy group provide information on the local ordering of **CAB** molecules in aggregates of the organogels.

TABLE 1 Organic Liquids Tested for Gelation by **CAB**

Liquid	wt% of **CAB**	Phase formation[a]
1-bromopropane	1.4	N
1-bromo-3-phenylpropane	1.54	N
pentane	1.20	Y (5°C)
hexane	0.80	Y
heptane	0.80–2.7	Y
octane	0.81	Y
dodecane	0.10	N (sol)
	0.25–4.1	Y
tetradecane	2.23	Y
hexadecane	0.81–2.9	Y
heptadecane	3.00	Y
octadecane	2.70	Y
eicosane	1.57	Y
isooctane	0.81	Y
cyclohexane	1.41	Y
methylcyclohexane	0.80	Y
decalin	1.56	Y
1-tetradecene	1.50–4.8	Y
1-propanol	1.50	Y
1-dodecanol	1.52	Y
1-octanol	0.80	Y
1,3-propanediol	1.09, 1.33	N
benzyl alcohol	1.50	Y
4-heptanol	1.53	Y
2-butene-1,4-diol	1.33	N (ppt)
3,3-dimethyl-2-butanol	1.43	N (sol)
	0.72	Y (5°C)
heptanal	1.51	Y
nonanoic acid	1.61	Y
valeric acid	1.47	N (sol + ppt)
n-pentyl acetate	1.51	N (soln)
		Y (5°C)
n-butylamine	1.50	N (sol)
		Y (5°C)
benzylamine	1.46	N
		Y (5°C)
N-methylbenzylamine	1.25	N (soln)
		Y (5°C)
a-methylbenzylamine	1.49	N (soln)
		Y (5°C)

(*continued*)

TABLE 1 (*Continued*)

Liquid	wt% of **CAB**	Phase formation[a]
benzene	1.50	N (soln)
toluene	1.33–2.31	N (soln)
7-tridecanone	1.48	N (ppt)
2-octanone	0.99–1.55	N (ppt)
2-undecanone	1.54	N (ppt)
4-heptanone	1.53	N (ppt)
1,2-dimethoxyethane	1.46	N (ppt)
		Y ($< -10°C$)
1,3-pentadiene	1.46	N (soln)
		Y ($< -10°C$)

[a] Y = gel formed at room temperature unless noted otherwise; N = no evidence of gel formation; sol = sol formation; ppt = precipitate; soln = isotropic-like solution.

Fluorescence intensity can be used to detect the aggregation and deaggregation of **CAB** associated with the sol-to-gel transition (as a sample is cooled or heated; Fig. 12) as well as (isothermal) gel aging and hysteresis effects. New absorption, excitation, and, especially, CD bands, from excitation dipole coupling between anthracenyl groups stacked face-to-face within the aggregates, are present in the gelled state. They are ascribed to partially overlapping neighboring **CAB** molecules whose long molecular axes are nonparallel. Figure 13 shows the absorption, fluorescence, and excitation spectra of a **CAB**/n-dodecane sample in its isotropic fluid phase (67°C) and gel phase (room temperature). Gelation induces a marked increase in the amplitude of emission and dichroic absorption (not

STNH

STRUCTURE 5

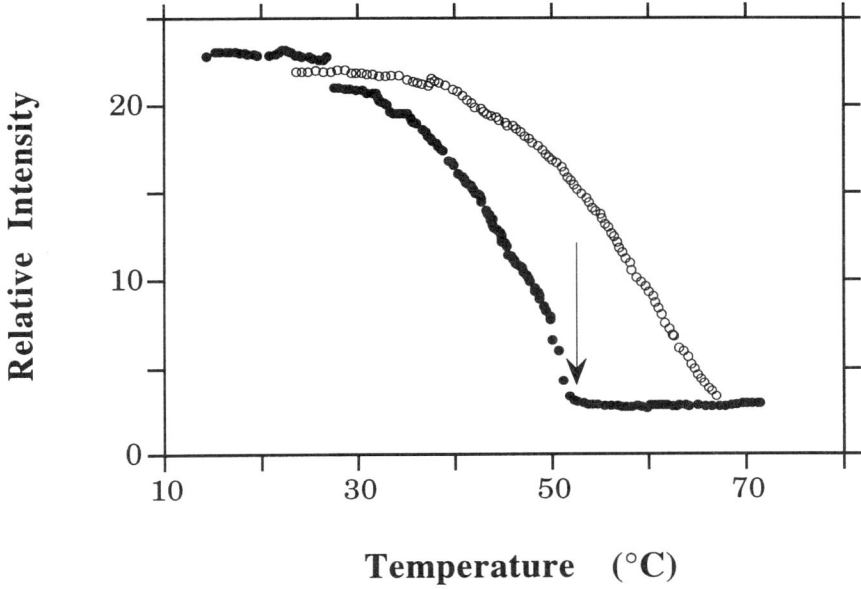

FIG. 12 Histogram of formation/disaggregation of **CAB** colloids in n-dodecane ($C = 1.43$ wt%). Fluorescence intensity at 422 nm ($\lambda_{exc} = 355$ nm). Cooling (●) and heating (○) cycles are shown. The arrow indicates the gelation temperature. (From Ref. 76.)

shown) [47,48,76]. At least in part, these increases are not due to photophysical changes of the molecules involved; Tyndall scattering and the Christiansen effect [82] may contribute strongly.

The position, shape, and intensity of absorption and emission features of the sol and gel phases of **CAB** in 1-octanol/n-hexadecane mixtures display remarkable features that depend upon the liquid composition, the width of the vessel in which the sample is held, and the rate at which it is cooled to below T_{SG} [83]. When the concentration of **CAB** is ca. 1.5 wt%, T_{SG} is the same in neat hexadecane (ca. 40°C) as in liquid mixtures containing 75 vol% 1-octanol, regardless of the cooling rate. At 90/10 1-octanol/n-hexadecane mixtures, T_{SG} is again constant and independent of cooling rate but is equal to that of samples containing neat 1-octanol as the liquid component (ca. 60°C). At liquid compositions containing 80–90 vol% 1-octanol, T_{SG} is that of hexadecane when samples are cooled rapidly (ca. 8 deg/min) or of 1-octanol when cooled slowly (ca. 0.5 deg/min)! The same sample can be cycled several times through the sol-to-gel transitions to give either T_{SG} value by controlling the rate of cooling. However, in no case was a value intermediate between the two T_{SG} limits observed: only

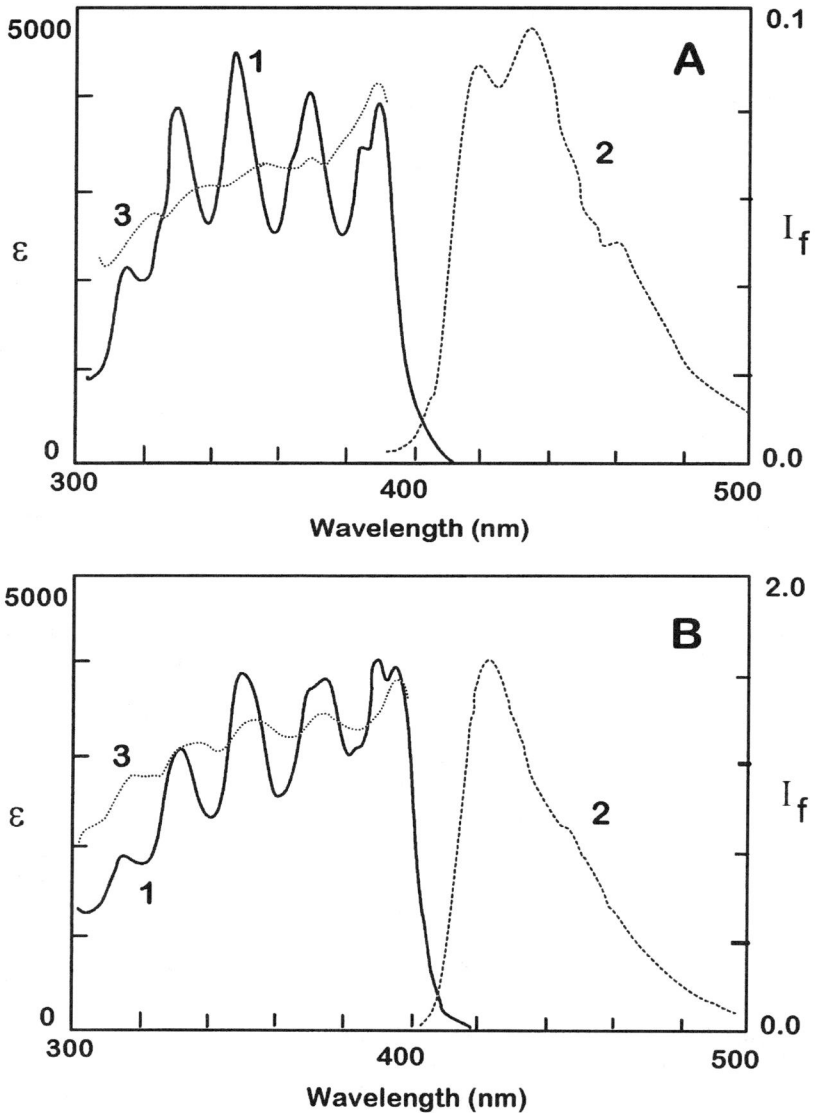

FIG. 13 Electronic absorption and emission spectra of **CAB** in n-dodecane ($C = 0.72\%$). Absorption (1: ——), fluorescence (2: ---) for $\lambda_{exc} = 346$ nm, and excitation (3: ···) ($\lambda_{emi} = 433$ nm in the solution and $\lambda_{emi} = 422$ nm in the gel phase). A: isotropic solution ($T = 67°C$); B: gel phase ($T = 25°C$). Relative fluorescence intensities are indicated between the upper and lower curves. (From Ref. 48.)

one of the two morphs of the **CAB** fibers (as seen by SEM) [76] are formed; a mixture of them is not obtained. These results indicate that bulk solvent properties, like polarity, rather than intimate liquid–gelator molecular interactions, control the nature of the morph being formed (i.e., the nucleation/assembly processes). However, interactions of the proton-donor liquid, 1-octanol, and specific parts of **CAB** may contribute to strand twists that are absent when n-alkanes are the liquid. Furthermore, when samples were placed in cells whose width is narrower than the diameters of the **CAB** colloids [76], no gel was formed [83]. Polytypism is the term applied to the related phenomenon in which the solvent type influences the mode of crystallization of molecules containing long alkyl chains that form bilayers [84].

These observations are corroborated by spectroscopic measurements on the same samples [83]. Although the electronic absorption spectra of **CAB** in hexadecane or 1-octanol above or below T_{SG} appear very similar in position and shape to that in benzene (a nongelled liquid in which **CAB** does not associate appreciably), the fluorescence and excitation spectra are very much dependent upon the phase and the liquid. Above and below T_{SG}, the excitation spectrum in either liquid exhibits a very strong narrow exciton coupling band (λ_{max} is 407 nm for hexadecane and 410 nm for 1-octanol) and a monomeric vibronic progression that is offset from the wavelengths of the corresponding absorption spectrum; some aggregation exists at temperatures above T_{SG}, and those species contribute disproportionately to their population to the fluorescence. Below T_{SG}, in the gel phases, the excitation spectra are similar to those above T_{SG}, but the exciton band in each case is broader, indicating a larger distribution of aggregate types. The shapes of the fluorescence spectra in hexadecane and 1-octanol are similar also, but are somewhat different from that in benzene. Several new peaks and shoulders appear on the red side of the fluorescence spectra in the gel phases. They are consistent with the presence of a variety of emitting aggregate units.

Transient birefringent textures (faint focal conic patterns, evolving to fanlike textures or Maltese crosses) have been observed by polarized optical microscopy for **CAB** and other **ALS** samples cooled from above T_{SG} [76]. They indicate that the steps leading to the dispersed gel state may include formation of a lyotropic cholesteric phase [85]. Neat **CAB** and some (but not all) of the other **ALS** gelators form thermotropic cholesteric liquid-crystalline phases [47]. This is consistent with the analysis of the distribution of the photodimers of **CAB**, which indicates that stacking of **CAB** in the gel and liquid crystal are not very different. Although many organogelators form thermotropic mesophases, others do not. Mesomorphism is not a condition for a molecule to be a gelator. In fact, several mesomorphic **ALS** molecules are unsuccessful gelators of a variety of common liquids, while some nonmesomorphic **ALS** molecules are good gelators.

CD measurements suggest the presence of helical packing of molecules within fibers of **CAB** gels in n-alkanes and 1-octanol. Above T_{SG}, the CD intensity is negligible; below it, very strong dichroic absorption is observed, and the exciton coupling band seen in the excitation spectra (but not in the absorption spectra) is quite prominent [76]. In fact, changes in the intensity of the exciton coupling band in the CD spectra have been employed in an analogous fashion to fluorescence to measure T_{SG}. At the same time, ^{1}H NMR experiments provide no evidence for impaired liquid mobility on the molecular scale [76]; the same conclusion was reached from shorter-time-scale observations using EPR and a paramagnetic analogue of **STNH** as gelator [86].

An interesting and apparently unique (to date) example of the use of phosphorescence to study an organogel has employed **CAQ** in poly(dimethylsiloxane)s (PDMS) at 77 K [77]. Like other anthraquinone derivatives, including those that are gelators, **CAQ** exhibits no discernible fluorescence [79], nor does it phosphoresce at room temperature in normal solutions or in the gelled state [87]. Gels cooled quickly to 77 K may or may not have the same strand morphology as at room temperature, but the fractions of the gelator incorporated within the strands and remaining dissolved in the liquid component must be similar. Rations of phosphorescence intensity from "free" and fiber-included **CAQ** can be measured, since the 0–0 bands of emissions (i.e., between the lowest vibronic levels of the triplet and ground states) from the two environments are offset by 16 nm, but the overall shapes of the phosphorescence curves are nearly the same. The vibronic spacings, 1670 cm^{-1}, clearly identify the sources of both emissions (Fig. 14). Additionally, the excitation spectra are nearly superimposable on the absorption spectrum of **CAQ**. The data indicate that the equilibrium fraction of free **CAQ** decreases in gels as the viscosity of the PDMS liquid is increased from 2 to 100 mPa s. They also show, somewhat surprisingly, that the free/fiber-included ratios of phosphorescence reach a plateau near 2–3 wt% **CAQ** regardless of the PDMS viscosity. Although phosphorescence requires some adulteration of gel samples, it can supply valuable spectroscopic information concerning gel microstructures that may not be available using other analytical techniques. Clearly, it has been underutilized.

Some thermally reversible low-mass organogels exhibit such a low yield stress that they are thixotropic [88,89]: beyond a sufficient stress, isothermally applied, a sol phase is formed that reverts to the gelled state at rest. For example, thixotropic gellike suspensions are obtained in cyclohexane in the presence of a zinc(II) trisubstituted porphyrin that can form short length rods [90]. These observations are important, since the yield stress can be considered a measure of the internal strength of structures developed in a suspension. Studies of thixotropy involve the underlying mechanisms of structural recovery and shear thinning phenomena. Thixotropy can also be induced: although gels of **CAQ** with 1-octanol of n-alkanes are not thixotropic, they can become so if a small amount

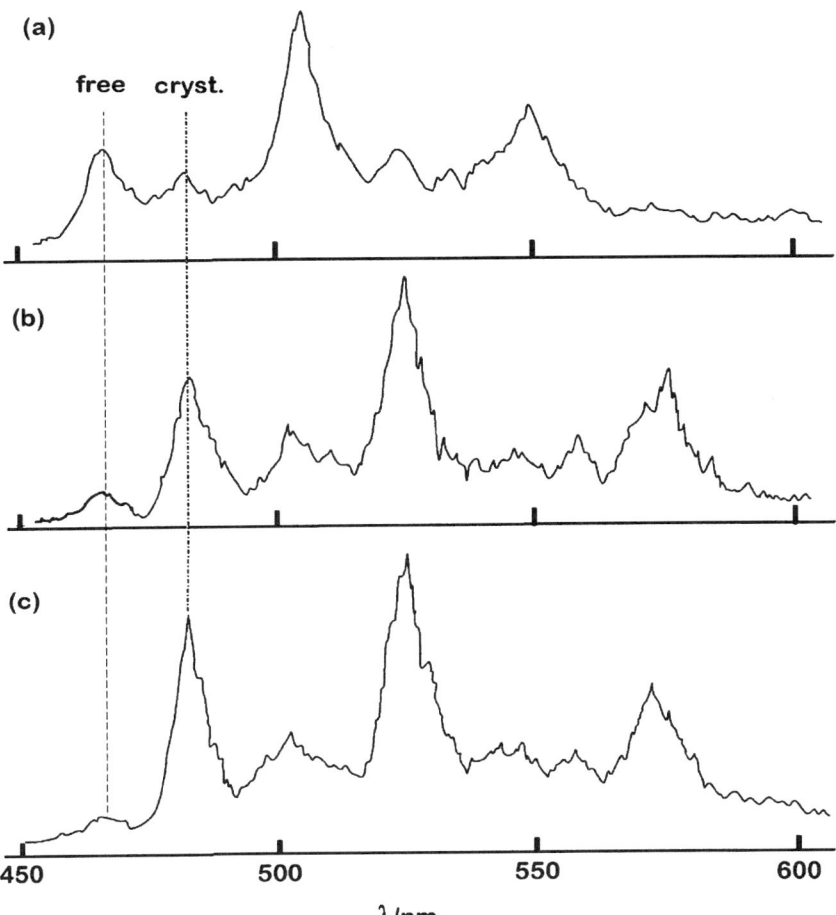

FIG. 14 Phosphorescence emission spectra at 77 K of **CAQ** in the gel consisting of PDMS with a viscosity of 20 mPa s, containing different amounts of **CAQ**: (a) 0.2, (b) 0.33, and (c) 2 wt% of **CAQ** in PDMS. The dotted vertical lines indicate phosphorescence bands characteristic of the amount of free **CAQ** species (463 nm) and crystalline **CAQ** (479 nm). (From Ref. 77.)

($\ll$ 1 wt%) of cholesteryl 9,10-dimethoxyanthracene-2-carboxylate (**CMAQ**), a nongelator, is added [91]. As observed with their neat solid mixtures, **CAQ** and **CMAQ** form a charge-transfer complex whose characteristic absorption can be detected visually. The incorporation of **CMAQ** into the **CAQ** fibers appears to be a kinetic phenomenon, since quick cooling of sols is required to induce

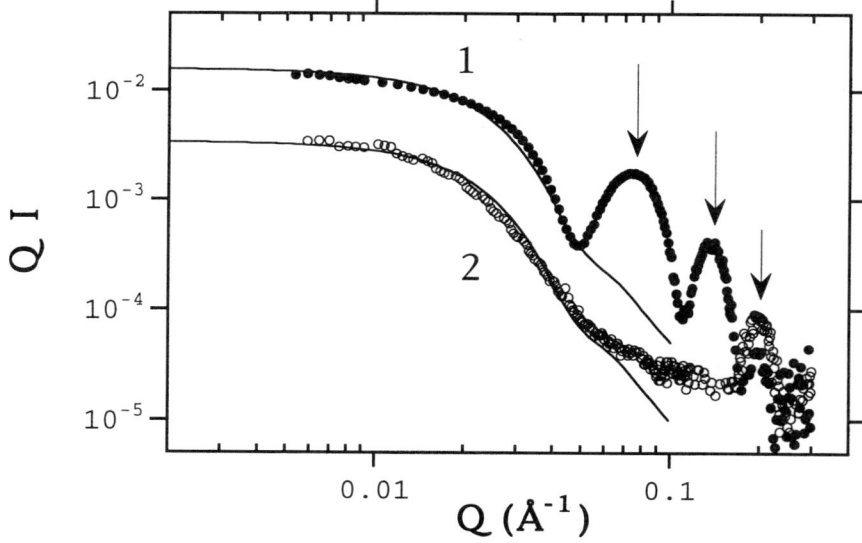

FIG. 15 SAXS cross-sectional intensity QI versus Q plots of **CAB** organogels. (1) (●) n-decane, $C = 0.0192$ g cm^{-3}; (2) (○) 1-octanol, $C = 0.01058$ g cm^{-3}. Intensity for n-decane is multiplied by 13 for the sake of clarity. Lines are best fits according to Eq. (11): (1) $r_o = 82$ Å, $\varepsilon = 0.24$; (2) $r_o = 85$ Å, $\varepsilon = \Delta r/r = 0.24$. ε is the radial Gaussian dispersity of the cross sections. The diffraction peaks of the lyotropic microdomains in the n-decane network are marked by arrows. (From Ref. 93.)

thixotropy. SANS patterns from quickly cooled thixotropic and slowly cooled nonthixotropic **CAQ/CMAQ** gels differ; the latter provide patterns like those from **CAQ** gels without **CMAQ** [92].

Thixotropy in "crystalline gels" (Fig. 1, image 1, i.e., those with high-energy junction zones) is thought to arise via preferential formation of the charge-transfer complexes at and/or near the junction zones [91]. When these mixed crystalline segments are disturbed mechanically, they cleave, yielding a sol, but they reattach themselves with time.

Data from SANS and SAXS scattering techniques are consistent with nanoscopic TEM, CD, and other spectroscopic measurements [60,61]. The SANS scattering profiles (Fig. 15), coupled with the similarity of the neutron contrast values of **CAB** aggregates in deuterated decane and 1-butanol, support the previous conclusions that the liquid type can cause significant structural changes in the gels. QI vs. Q scattering curves of **CAB**/1-butanol and **CAB**/decane gels exhibit a low-angle plateau typical of the form factor of long and rigid fibers (Eq. (11)). Values of the diameter ($2r$) of the **CAB** fibers can be estimated from $\ln(QI)$ ver-

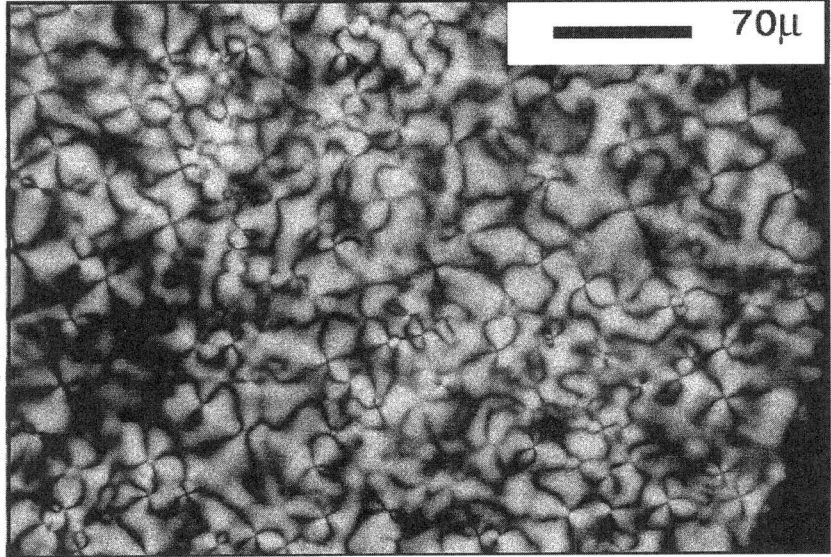

FIG. 16 Schlieren optical texture observed for a thin gel of ca. 2 wt% **CAQ** in 1-octanol between crossed polarizers at ambient temperature. (From Ref. 94.)

sus Q^2 "Guinier plots" (at low Q, $QI = \phi(\pi r \Delta\rho)^2 \exp(-Q^2 r^2/4)$), assuming a homogeneous circular symmetry (192 Å in alcohols and 160 Å in alkanes); of course, this is a rough approximation to the rectangular cross sections found by TEM. r values deduced from SANS and SAXS data are slightly different (167 vs. 153 Å, respectively, for **CAB**/decane gels), which may be indicative of heterogeneous cross-sectional contrast profiles. At larger angles of the scattering curves, broad and intense peaks characterize the structure factor of ordered heterogeneities in the network. The full scattering curve mixes the fibrillar form factor (low-angle part) and the structure factor of ordered domains (wide-angle part). The good fit to Eq. (11) combined with the Guinier and Porod analysis [38] gives a consistent picture of the **CAB** aggregates in alcohols. The number of **CAB** molecules per unit length ($n_L = M_L/M$) of fiber in alcohols is estimated from absolute neutron intensities [93]. The value $n_L \approx 18$ molecules Å^{-1} is slightly lower (by ca. 10%) than that for the packing in the crystalline solid (ca. 19.1 mol Å^{-1}). This is an indication that the neutron scattering length density of the fibers in the gels is close to that in the bulk solid. The small fraction of junction zones, organized as swollen lyotropic microdomains (see Fig. 16), does not affect significantly either the n_L determination or the isotropic character of the scattering of such dilute gels (Fig. 4).

Crystalline **CAB** powder exhibits Bragg reflections (at 0.096, 0.163, 0.188, 0.248, and 0.281 Å^{-1}) in the ratio 1 : $\sqrt{3}$: $\sqrt{4}$: $\sqrt{7}$: $\sqrt{9}$ typical of hexagonal ordering. The intercolumnar distance in the neat solid is $D_0 = (2/\sqrt{3})2\pi Q_{(11)} = 75.7$ Å, where $Q_{(11)}$ is the first reflection generated by a 2-D hexagonal array of fibers. The **CAB** gel network is heterogeneous and composed of fibers interconnected by swollen lyotropic microdomains (whose intercolumnar distance is 102.2 Å; the diameter of the connecting fibers is ca. 160–170 Å). Peaks with a spacing ratio 1 : 1.92 : 2.75 characterize a swollen lyotropic organization of the junction zones. The SAXS diffraction patterns of **CAB**/1-butanol gels clearly show that the molecular packing in the aggregates is different from that in both the neat solid and the **CAB**/decane aggregates.

The junction zones of **CAQ** organogel networks are lyotropic microdomains that vary from hexagonal in decane to more compact, lamellar-like ordering in 1-octanol [94]. Schlieren optical textures confirm the inhomogeneity of the orientation of the threadlike structures (Fig. 16), and CD spectra resemble those of neat **CAQ** solid, cooled from its mesophase. However, the solid morph obtained from solvent crystallization is packed differently from the gel strands [14].

Comparison of the molecular length of **CAB** (ca. 38 Å) and the characteristic lengths calculated from SAS data (vide ante) indicate that molecular pairs are involved in the columns of the solid state while, in gels, association of swollen columns might be involved. Micrographs of freeze-fractured and etched **CAB** gels (Fig. 17) show a 3-D network of fibrous bundles. The dimensions of the rectangular cross sections of the nontwisted fibers in dodecane, 209 × 104 Å, and the twisted ones in 1-octanol, 263 × 82 Å [48], correspond approximately to the cross-sectional areas determined by SANS in which a circular cross section model was employed.

Alkane liquid molecules are included in some strands whose swelling may be facilitated by the molecular shapes of the gelators and electrostatic compatibility with the lipophilic external shell of **ALS** and other steroid-based aggregates [95]. When swelling is not promoted by the gelators (for instance, by **DDOA**, which lacks a steroidal part), gels with ''dry crystalline nodes'' (involving stacks of oriented **DDOA** anthryl groups) are formed, and diffraction patterns of the bulk solid and gel phases are almost superimposable. Moreover, since the diffraction patterns of **DDOA** and **CAB** solids are similar, it is reasonable to expect that factors associated with interactions of the aromatic parts play a crucial role in determining the structure of the ordered condensed states. For networks with nodes involving lyotropic microdomains, the diffraction peaks are broadened and shifted to an extent that depends on the liquid types and its concentration within the strands [93,94].

A ''molecular engineering'' approach to gelation has been attempted in order to identify the influence of the different segments of **ALS** molecules [79,96]. Parameters of the **S** part, such as the α/β stereochemistry at C_3 and the nature

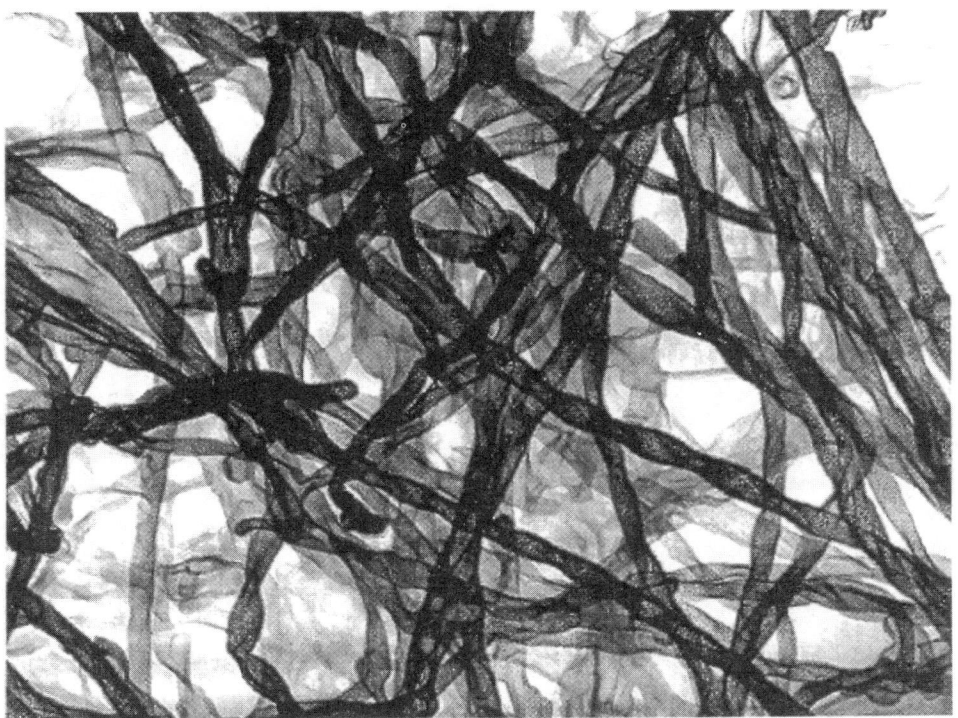

FIG. 17 Electron micrograph of **CAB** in 2-octanol [21]. The cross-sectional dimensions of the twisted fibers are virtually the same as those for 1-octanol (263 × 82 Å).

of the alkyl chains on C-17, and the length and functionality of the **L** part, have been examined; see, for example **4–9**. Gelation ability seems favored by β stereochemistry at C_3, while changes at C_{17}, as long as they are minor (e.g., introduction of unsaturation or an ethyl substituent), do not have a large impact. Again, this contrasts with the behavior of **STNH**-type gelators, where gem-*n*-propyl chains on the steroid D-ring provide a good gelator, but gem-2-propenyl groups do not.

C. Azobenzene Steroid-Based Gelators

A variety of alkoxy-azobenzenes has been appended to C3 of cholesterol to yield a new variety of **ALS** gelators (e.g., **12–14** named **AZOC** gelators) [49]. (See Structure 6.) 4-((Cholesteryloxy)carbonyl)azobenzene (**AZOC**) derivatives can gel liquids such as ethanol, acetic acid, amines, dichloromethane, ethers,

AZOC

STRUCTURE 6

and esters (Table 2). An interesting selectivity is observed with the α- and β-anomers, which preferentially gel apolar and polar liquid, respectively. The observation contrasts with most other low-mass organogelators, for which no single physical property (polarity, permittivity, etc.) can determine so sharply the gelation ability. It was also observed that the number of carbon atoms of an alkyl group attached to the azobenzene moiety, constituting its lipophilic counterpart, influences gelation behavior: chains with odd numbers of carbon atoms are more efficient than those with an even number.

Test-tube tilting, the ball-drop methods, and turbidity have been used to determine the gel-to-sol phase transition temperatures [49]. The transition can be envisioned as a dissolution process of crystalline-like aggregates in which the melting energy ($\Delta H° \approx 9.8$ kCal mol^{-1}) can be deduced from measurements of T_{GS} as a function of concentration (Eqs. (7)). A comparison of the $\Delta H°$ values for **DDOA** and **AZOC** shows that a strong interaction between the **A** segments results in more crystalline gels with highly associated fibers (bundles).

As mentioned above, the sol-to-gel phase transitions can be induced reversibly by trans–cis photoisomerization of the azobenzene groups. UV irradiation ($330 < \lambda < 380$ nm) transforms a part of the trans isomers to the cis. As a consequence of the structural change, the gel state is ''switched'' to the sol. Visible irradiation (at $\lambda > 460$ nm) isomerizes the cis isomers to their trans form and allows the gel to be re-established. The reversible photocontrol of the sol-to-gel phase transition can be monitored by CD spectroscopy.

A related gelator system is based upon cyclic monosaccharide esters of the **ALS**-type molecule **15** (with a carbamate **L** part and phenylboronic acid **A** part) [97]. (See Structure 7.) The 1 : 2 condensate products with both enantiomers of lyxose, xylose, mannose, and galactose gelled at least some of the liquids tested. Only the condensate with D- and L-glucopyranose was unsuccessful with all. In some cases, only one member of a pair of condensates with different enantiomers of the same saccharide (i.e., diastereomers, owing to the inclusion of the chiral **S** part) gelled a particular achiral liquid. An investigation of the derivatives with structural variations of the functional groups has shown that some of them were capable of chiral recognition. The T_{GS} of chloroform gels with lyxose as the saccharide of the condensate gelator are much higher when the D-form is employed, but the L-form of xylose has the higher T_{GS} (Fig. 18) [97].

The growth of fiberlike aggregates of optically active molecules [48], like those encountered with some optically active organogelators, is known to be accompanied by important CD modifications. CD data, free of linear dichroic effects, so as to avoid the contribution of macroscopic anisotropy, reveal chirality effects typical of helical structures. The electronic absorption spectra of the gelators indicate the wavelengths at which the electronic transitions occur and where dichroic effects have to be sought (unless selective reflection components

TABLE 2 Organic Liquids Tested for Gelation by **AZOC** Homologues[a]

Liquid	(R = MeO)	(R + EtO)	(R = n-PrO)	(R = n-BuO)	(R = n-PeO)	(R = n-DecO)
n-hexane	Gf	G	G	G	G	G
n-heptane	Gf	G	G	G	G	G
n-octane	Gf	G	G	G	G	G
paraffin	G	G	G	G	G	G
cyclohexane	S	S	S	S	S	G
methyl-cyclohexane	S	Gf	Gf	Gf	G	G
decalin	S	S	S	Gf'	Gf'	G
carbon tetrachloride	S	S	S	S	S	S
carbon disulfide	S	S	S	S	S	Gf'
benzene	S	S	S	S	S	S
toluene	S	S	S	S	S	Gf
p-xylene	S	S	S	S	S	S
nitrobenzene	S	S	S	S	S	G
m-cresol	S	S	S	S	S	Gs
1,2-dichloro-ethane	S	S	S	S	S	Gs
dichloro-methane	S	S	S	S	S	S
chloroform	S	S	S	S	S	S
diethyl ether	S	Gs	S	Gs	G	G
dipropyl ether	S	Gf'	Gf'	Gf'	Gf'	G
diphenyl ether	S	S	S	S	S	G
tetrahydrofuran	S	S	S	Gf	S	S
1,4-dioxane	S	S	S	S	S	G
ethyl formate	Gf	G	G	G	G	G
methyl acetate	S	G	G	G	G	G
ethyl acetate	S	Gf	Gf	Gf	G	G
ethyl malonate	S	G	G	G	G	G
acetone	G	SG	G	G	G	SG
methyl ethyl ketone	S	G	G	G	G	G

Low-Mass Luminescent Organogels

Solvent					
N,N-dimethyl-formamide	G	SG	G	G	SG
N,N-dimethyl-acetamide	G	G	G	G	G
dimethylsulfoxide	SG	G	G	G	SG
N-methyl-2-pyrrolidone	S	Gf	Gf	Gf	G
acetonitrile	SG	SGc	SG	SG	SG
methanol	SG	SGc	SG	SG	SG
ethanol	SG	SGc	SG	SG	SG
1-propanol	G	SG	SG	SG	SG
1-butanol	G	SG	SG	SG	SG
1-octanol	G	G	G	G	SG
benzyl alcohol	G	G	G	G	G
acetic acid	G	SGc	SG	SG	SG
hexanoic acid	S	S	S	S	G
acetic anhydride	G	SG	SG	SG	SG
propylamine	S	Gf	S	S	G
diethylamine	S	S	Gf'	Gf'	G
triethylamine	S	Gf'	Gf'	Gf'	G
aniline	S	Gs	s	s	G
pyridine	s	s	s	Gs	G
triethylsilane	G	G	G	G	G
trimethyl-chlorosilane	S	Gf'	Gf'	Gf'	G
dimethylpoly-siloxane(1cs)	G	SG	SG	SG	SG
dimethylpoly-siloxane(6cs)	G	SG	SG	SG	SG
cyclomethicone	G	SG	SG	SG	SG
trifluoro-ethanol	G	SG	SGs	G	R
glycerol	I	I	I	I	I
water	I	I	I	I	I

[a] [Gelator] = 1–7 wt%; G = gel formed when cooled at 2–20°C and stable at room temperature; SG = gel was formed at [gelator] < 1 wt%; Gc = gel formed but became crystals at room temperature; Gf = gel formed when cooled in a refrigerator (at −6°C) and stable at room temperature; Gf' = gel formed when cooled in a refrigerator (at −6°C) but not stable at room temperature; gel not formed because of crystallization (R); gelator dissolved at room temperature (S); or gelator insoluble upon heating (I).

STRUCTURE 7

are present, too). The absorption at 360 nm in the isotropic phase spectrum of **AZOC** is blueshifted upon gelation. The dichroic signature above and below the gelation temperature (Fig. 19) is remarkable: an asymmetric exciton coupling band appears in the gel with an intersection at zero ellipticity at λ_{max} of the absorption spectrum. The positive exciton coupling band observed for the

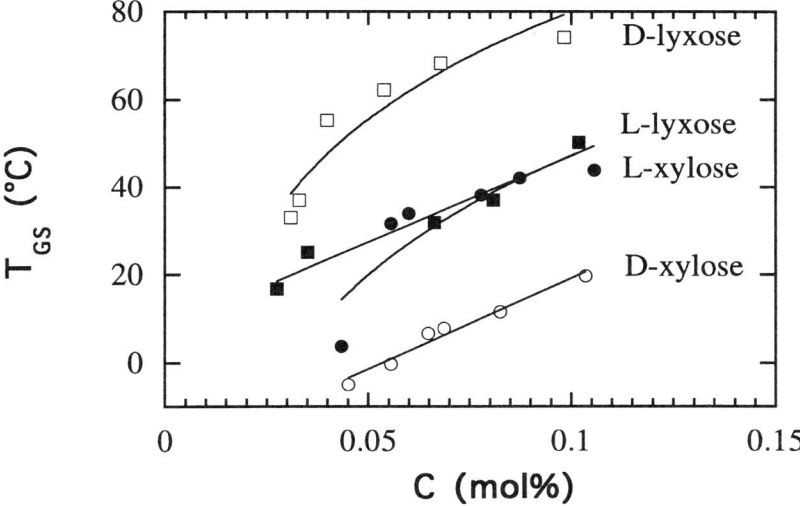

FIG. 18 Discrimination of monosaccharide enantiomers by gel formation of their esters with **15** in chloroform. Gel-solution phase transition temperatures versus the mol% of ester are shown. (From Ref. 97.)

(α)-anomer indicates that the azobenzenyl groups are associated in the aggregate in a clockwise fashion ((R)-chirality). Owing in part to the (α)-anomeric form of the gelator, metastable structures can be formed whose nature depends upon the specific cooling rate of initial solutions. The different phases contain strands of opposite helicities (SEM observations) and signs of the CD bands. Similar observations have been made with **CAQ** gels in hexadecane/1-octanol mixtures [83].

The role of the absolute configuration at C_3 is determinant not only for aggregation/gelation but also for the aggregation modes of monolayers developed at the air–water interface [98].

Association of the cholesterol moieties is thought to constitute the primary driving force for aggregation. The (β)-anomer of the 4′-methoxy derivative of **AZOC**, with R = Me, has an extended rodlike conformation with orthogonal cholesterol and azobenzene planes. The stacking of cholesterol groups of the (β)-anomer, with (S)-chirality at C-3, places the azobenzenyl parts in face-to-face orientations that favor additional interactions and strengthen the aggregates. Consequently, linear aggregation is favored, and the stability of gels form the (β)-anomers is greater than that from the (α)-anomers. Conversely, the solubility of the (β)-anomers is the lower. With (R)-chirality at C-3, **12** adopts an L-shaped bent conformation that inhibits aggregation.

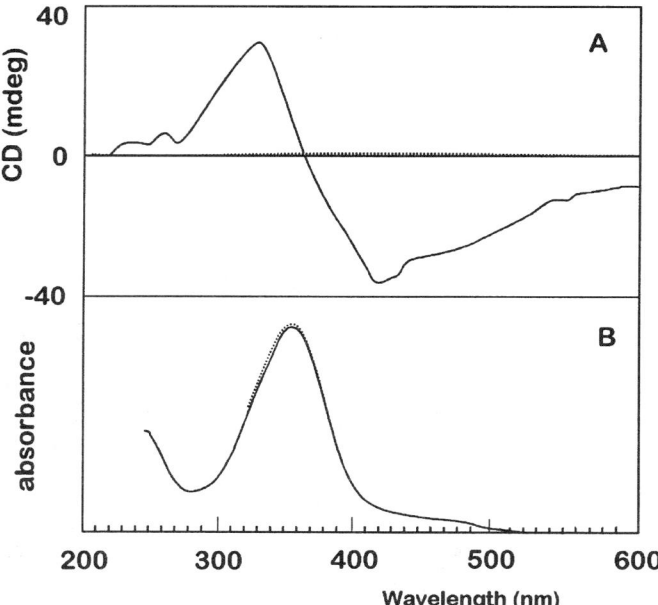

FIG. 19 CD (A) and absorption (B) spectra for **13** ($R = Et$) (0.4 wt%) in methanol. Solid line (——) represents the gel at 25°C and dotted line (···) is for the solution state at 65°C. The dotted line is almost horizontal at 0-degree ellipticity in (A) and almost follows the full-line absorbance curve in (B). (From Ref. 49.)

Interestingly, reflectance spectroscopy studies indicate that the (β)-anomers form monolayers with J-aggregation but are transformed into an H-aggregation when the layer is compressed (Fig. 20) [98]. Since they are better defined structurally than many 3-D gelator fibers, 2-D monolayers, when preparable from gelator molecules, can provide insights into the organization of aggregates in gels. Also, studies with monolayers of nonluminescent fatty acid organogelators have illustrated the important role that interfacial interactions can have in controlling the bulk properties of gel fibers [99,100].

It is noteworthy that although WAXS data indicate that **DDOA** and **CAB** exhibit comparable types of molecular organization in their bulk solid states [93], their gel fibers are very different. Junction zones of **DDOA** gels are dry and crystalline, while those of **CAB** and **CAQ** are swollen lyotropic microdomains whose organization is clearly different from that of the solid [14]. **ALS** and **AZOC** gelators confirm that the cholesteryl group is an important element in

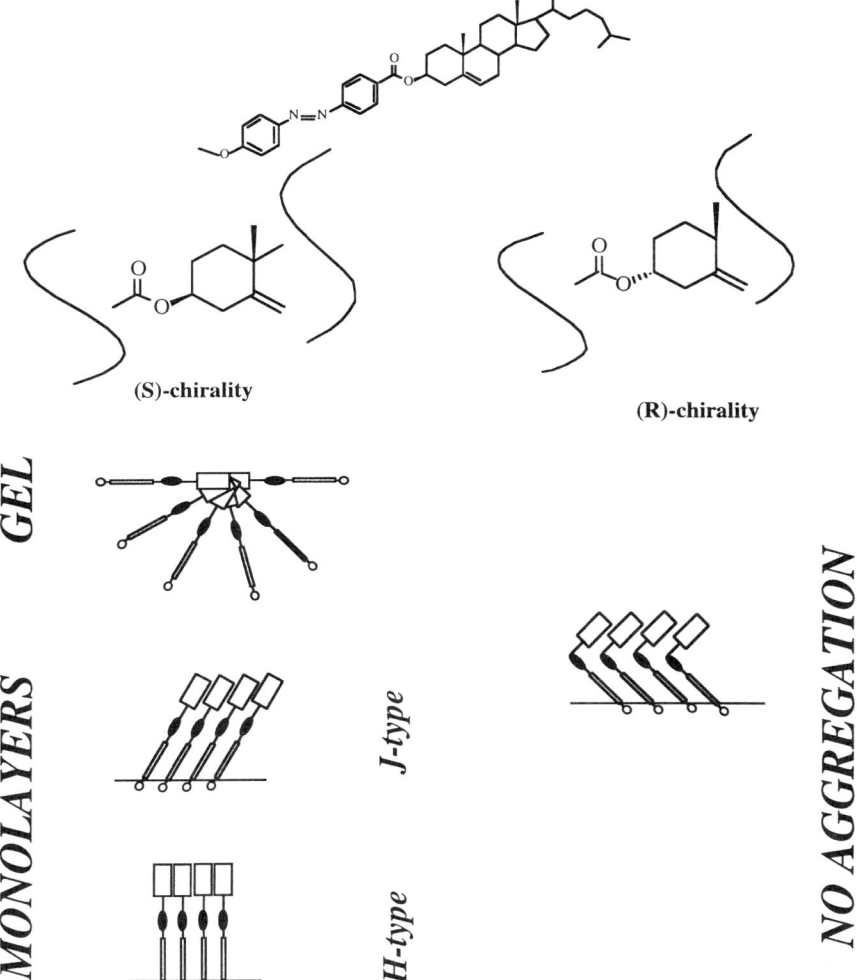

FIG. 20 Schematic representation of the **AZOC** aggregation according to the chirality at C-3. Left: The stacking of the cholesterol moieties in the (S)-derivative put the azobenzene moieties into a favorable orientation to persist in the stacking of the fibers of the gel and the monolayers (J and H aggregates) at air–water interface. Right: The inverted configuration ((R)- chirality at C-3) cannot lead to the face-to face orientation, and no aggregates are obtained. (From Refs. 49 and 98.)

DBS

STRUCTURE 8

inducing liquid-crystalline-like properties in the aggregates of the gels (as well as in monolayers and films) [97].

D. Sorbitol and Polyol Derivatives

1,3:2,4-di-O-Benzylidene-D-sorbitol (**DBS**) is a chiral polyol whose enantiomers can gel numerous liquids (varying from hexadecane and dimethyl phthalate to ethanol and ethylene glycol) [67–69,101] but whose racemate does not. (See Structure 8.) Although no luminescence studies appear to have been performed with **DBS**, it is an excellent candidate for future work. The gelation time, determined by rheological measurements of the storage modulus (G') versus time during the sol-to-gel transition [101], varies with **DBS** concentration in dimethyl phythalate according to a power law with a 9/2 exponent! The kinetics and visual detection of gelation are dependent upon the liquid type. The rigidity of the gels is extremely sensitive to **DBS** concentration; the elastic modulus exhibits a surprising fourth-power dependence on gelator concentration ($G' \approx \phi^4$). Initially clear gels in ethylene glycol [68] with $\phi < 0.15$ are metastable and phase separate. The DSC-determined phase diagram with ethylene glycol shows three distinct gelator concentration regimes (below $\phi \approx 0.05$, 0.15, and 0.65). Solid-state NMR and freezing-point depression studies confirm the interaction of liquid and **DBS** molecules in the gel.

The relative orientation of the aromatic rings in the aggregates of the gel has been estimated from electronic absorption spectra. A transition from linear (corresponding to a hypsochromic shift) to parallel stacking (corresponding to a bathochromic shift) occurs at $\phi \approx 0.15$ and is accompanied by an increase of gel crystallinity. By optical microscopy between crossed polarizers, spherulitic textures are observed as extinction crosses (negative optical sign) and rings, suggesting the existence of supramolecular helical arrangements. IR spectra show the presence of hydrogen bonds in the aggregates, and UV spectra support side-by side arrangement of the phenyl rings [69].

VI. APPLICATIONS

The great diversity of microscopic and mesoscopic structures that gels can display [102] accounts for their numerous industrial applications. Thermoreversibility and chemical sensitivity of organogels are intrinsic characteristics of organogels that make them excellent candidates for future technological uses. As yet, there are no specific applications that exploit specifically the luminescence of low-mass organogels. However, as indicated by this chapter, luminescence is a useful tool in investigating the structures of the gels and may provide the key to monitoring mechanical (as well as morphological) perturbations of materials caused by external forces. Additionally, the fibrous nature of the gel networks makes them quite attractive for studying the mechanisms of energy migration in the context of quasi-one-dimensional incoherent excitons. Although crystals and columnar liquid crystals have been employed to study such energy transfer processes, fibrillar networks of thermoreversible organogels await study. There is a very high probability that applications involving energy transport and optical control (e.g., lasing) will be found that exploit, as a combination, the luminescence and/or specific properties of chromophoric groups and the percolation and reversibility of low-mass organogels. A brief and nonexhaustive summary of some applications of low-mass organogels follow, and we mention also some prospects for the future.

Hydrometallurgy, cosmetics, food processing, and lubrication are among the areas in which gels are used. Methods for the recovery or disposal of spilled oils also use gels or complex microemulsions. Active enzymes and bacteria can be entrapped in apolar gels of gelatin [103] and may be a new drug delivery domain. Highly concentrated oil-in-water (O/W) emulsions and various gels have been used as aviation fuels and in cosmetics, while highly concentrated water-in-oil (W/O) emulsions are used in some explosives. Macromolecular separations, crystallizations [104,105], etc. specific to organic fluids may be aided by the hindered transport in pores of gel networks [106]. The key property of gels in this case is the drag effect they induce, due to steric and hydrodynamic interactions, between permeating macromolecules and the disordered fibrous media. The advantages of using gels in place of liquid systems lie in that hydrodynamic convection is reduced and various species can be immobilized in a medium that mimics the situation in space with a reduced gravity.

If a supercritical drying process is used at high temperatures and pressures (i.e., above T_c and P_c), an extremely porous structure called an aerogel can be obtained. The remarkable damping of sound and suppression of heat transport by some aerogels portends insulation and acoustic applications. Presently, only organometallic oxides (silica) have been utilized to form aerogels [107], but luminescent organogelators are promising candidates for lasing materials. Aerogel networks that can provide small- and large-diameter holes may also become useful as purification/separation tools. Recently, "reverse aerogels" have been constructed by polymerizing styrene or alkyl methacrylates while gelled by N,N,N,N-tetraoctadecylammonium bromide [108], or a gluconamide [109], and then removing the gelator. Aerogels (like organogels), with their networks of microchannels, may also be useful as membranes, drug delivery agents, sensors, etc.

Gels composed of anisotropic networks and mixed with low-molecular-mass liquid crystals can be used as active or passive optical components [110]. Anisotropic gels obtained with organogelators forming finite rods [90] might be of interest as molecular electronic devices or semiconductors if appropriate donor–acceptor conjugated systems are used. Long-range electronic or magnetic interactions (N.B., molecular-based ferromagnetism) are expected with low-dimensional linear systems of stacked molecules with appropriate properties (e.g., phthalocyanine derivatives, cation radicals of tetrathiafulvalene and its selenium analogues, or more complex structures with two-center bridging elements of a metal) [108,111–113].

The detection of specific functionalities by donor–acceptor interactions, or detection of morphologies that exploit spatial recognition by cavities, grooves, and channels of the sensor host material, is the basis for many sensor technologies [114]. Nanochemistry developments utilize mostly host systems in the solid state [115] to yield materials with novel electronic, mechanical, and other properties [116]. In these contexts, gels offer an attractive medium for new applications. For instance, the sensitivity of some gelling systems to water or other polar "poisons" can make them candidates for specialized sensors [117]. While luminescent gels may be useful in linear, anisotropic energy transfer, the chirality of optically active organogels can also be used for molecular and chiral recognition. Molecular transducers formed by cholesterol-based organogelators [98] that can translate chiral recognition interactions into readable outputs are examples of host–guest signal-responsive chemistry. In such gels, the gel–solution phase transition temperature T_{GS} differentiates the two optical isomers of a monosaccharide (for instance, xylose) added to the gelling system. The locally ordered structures of gels and xerogels are able to act as nucleating agents or polyolefins [118] or are interesting in templating for the syntheses of materials [109,119].

ACKNOWLEDGMENTS

RGW gratefully acknowledges the U.S. National Science Foundation for its support of the gel research performed at Georgetown. Coauthors cited in the references are thanked for their contributions. The Large Instruments facilities (neutron and synchrotron sources: ILL, LLB, ESRF, LURE-France, NIST-USA) are acknowledged for providing technical support during the scattering experiments.

REFERENCES

1. P. Terech, in *Specialist Surfactants* (I. D. Robb, ed.), Chapman and Hall, London, 1997, pp. 208–268.
2. J. D. Ferry, in *Viscoelastic Properties of Polymers*, John Wiley, New York, 1980.
3. J. W. Cahn. J. Phys. Chem. *42*:93 (1965).
4. M. Avrami. J. Chem. Phys. *7*:1103 (1939).
5. M. Avrami. J. Chem. Phys. *8*:212 (1940).
6. P. G. De Gennes, in *Scaling Concepts in Polymer Physics*, Cornell University Press, Ithaca, N.Y., 1979.
7. P. J. De Gennes. Reviews Modern Physics *57*:827 (1985).
8. E. Ruckenstein and R. Nagarajan. J. Phys. Chem. *84*:1349 (1980).
9. P. H. Hermans, in *Colloid Science, Reversible Systems*, II, Elsevier, Amsterdam, 1969.
10. D. M. Keller, R. E. Massey, and O. E. Hileman. Can. J. Chem. *56*:3096 (1978).
11. J. C. Heughebaert and G. H. Nancollas. J. Phys. Chem. *88*:2478 (1984).
12. P. Terech. J. Colloid Interface Sci. *107*:244 (1985).
13. J. N. Israelachvili, in *Intermolecular and Surface Forces*, 3d ed., Academic Press, London, 1992, pp. 341–435.
14. E. Ostuni, P. Kamaras, and R. G. Weiss. Angew. Chem. Int. Ed. Engl. *35*:1324 (1996).
15. P. S. Russo, in *Symposium Series* 350 (P. S. Russo, ed.), ACS, Washington D.C., 1987, Chapter 9.
16. D. Stauffer, A. Coniglio, and M. Adam. Adv. Polym. Sci. *44*:103 (1982).
17. M. E. Cates. J. Phys. France *49*:1593 (1988).
18. M. E. Cates and S. J. Candau. J. Phys.: Condens. Matter *2*:6869 (1990).
19. A. A. Lamola, in *Creation and Detection of the Excited State*, Marcel Dekker, New York, 1971.
20. A. I. Levin and I. Tinoco. J. Chem. Phys. *66*:3491 (1977).
21. H. De Voe. J. Chem. Phys. *41*:393 (1964).
22. A. S. Marfunin, in *Spectroscopy, Luminescence and Radiation Centers in Minerals*, Springer-Verlag, Berlin, 1979.
23. A. Mayer and S. Neuenhofer. Angew. Chem. Int. Ed. Engl. *33*:1044 (1994).
24. F. Vögtle, M. Frank, M. Nieger, P. Belser, A. von Zelewsky, V. Balzani, F. Barigelleti, L. De Cola, and L. Flamigni. Angew. Chem. Int. Ed. Engl. *32*:1643 (1993).
25. B. Blanzat, C. Barthou, N. Tercier, J.-J. Andre, and J. Simon. J. Am. Chem. Soc. *109*:6193 (1987).
26. D. Markovitsi, I. Lecuyer, and J. Simon. J. Phys. Chem. *95*:3620 (1991).
27. K. Almdal, J. Dyre, S. Hvidt, and O. Kramer. Polymer Gels and Networks *1*:5 (1993).
28. J. L. Jones and C. M. Marques. J. Phys. France *51*:1113 (1990).
29. J. E. Eldridge and J. D. Ferry. J. Phys. Chem. *58*:992 (1954).
30. P. J. Flory. J. Chem. Phys. *17*:223 (1949).
31. A. Takahishi, T. Nakamura, and I. Kagaura. Polym. J. *3*:207 (1972).
32. P. Terech, unpublished data.

33. A. Coniglio, H. E. Stanley, and W. Klein. Phys. Rev. B 25:6805 (1982).
34. R. H. Wade, P. Terech, E. A. Hewat, R. Ramasseul, and F. Volino. J. Colloid Interface Sci. 114:442 (1986).
35. J. Dubochet, M. Adrian, J. Lepault, and A. McDowall. Trends Biochem. Sci. 10:143 (1985).
36. R. H. Wade and D. Chretien. J. Structural Biol. 110:1 (1993).
37. A. Guinier and G. Fournet, in *Small Angle Scattering of X-Rays*, John Wiley, New York, 1955.
38. O. Glatter and O. Kratky, in *Small Angle X-Ray Scattering*, Academic Press, London, 1982.
39. C. Cabane, in *Surfactant Science Series*, 22d ed., Marcel Dekker, New York, 1987, pp. 57–145.
40. J. S. Higgins and H. C. Benoit, in *Polymers and Neutron Scattering*, Clarendon Press, New York, 1994.
41. P. Lindner and T. Zemb, in *Neutron, X.-Ray and Light Scattering: Introduction to an Investigative Tool for Colloidal and Polymeric Systems*, North-Holland, Bombannes, France, 1990.
42. O. Kratky. Prog. Colloid Polymer Sci. 77:1 (1988).
43. J. B. Hayter and J. Penfold. Colloid Polym. Sci. 261:1022 (1983).
44. A. J. Leadbetter and E. K. Norris. Mol. Phys. 38:669 (1979).
45. P. Davidson, D. Petermann, and A. M. Levelut. J. Phys. II France 5:113 (1995).
46. P. Terech and V. Rodriguez. Prog. Colloid Polym. Sci. 97:151 (1994).
47. Y.-C. Lin and R. G. Weiss. Liquid Crystals 4:367 (1989).
48. Y.-C. Lin, B. Kachar, and R. G. Weiss. J. Am. Chem. Soc. 111:5542 (1989).
49. K. Murata, M. Aoki, T. Susuki, T. Harada, H. Kawabata, T. Komori, F. Ohseto, K. Ueda, and S. Shinkai. J. Am. Chem. Soc. 116:6664 (1994).
50. E. A. Chandross. J. Chem. Phys. 43:4175 (1965).
51. E. A. Chandross and J. Ferguson. J. Chem. Phys. 45:397 (1966).
52. J. B. Birks, in *Photophysics of Aromatic Molecules*, John Wiley, New York, 1970.
53. H. Chen, M. S. Farahat, K.-Y. Law, and D. G. Whitten. J. Am. Chem. Soc. 118:2584 (1996).
54. A. A. Lamol and N. J. Turro, in *Energy Transfer and Organic Photochemistry* (A. Weisberger, ed.), Techniques of Chemistry Series, Vol. 14, Interscience, New York, 1969.
55. R. Englman, in *Nonradiative Decay of Ions and Molecules in Solids*, North-Holland, Amsterdam, 1979.
56. M. Kasha, H. R. Rawls, and M. A. El-Bayoumi. Pure Applied Chemistry 11:371 (1965).
57. R. M. Hochstrasser and M. Kasha. Photochem. Photobiol. 3:317 (1964).
58. V. Czikkely, H. D. Forsterling, and H. Kuhn. Chem. Phys. Lett. 6:207 (1970).
59. V. Czikkely, H. D. Forsterling, and H. Kuhn. Chem. Phys. Lett. 6:11 (1970).
60. A. S. Davydov, in *Theory of Molecular Excitons*, Plenum, New York, 1971.
61. H. Chen, K.-Y. Law, J. Perlstein, and D. G. Whitten. J. Am. Chem. Soc. 117:7257 (1995).
62. J. Perlstein, L. Lu, and R. G. Weiss, work in progress.

63. R. M. Hexter. J. Chem. Soc. *37*:1347 (1962).
64. D. P. Craig and J. R. Walsh. J. Chem. Soc. 1613 (1958).
65. B. M. Sheikh-Ali, M. Rapta, G. B. Jameson, C. Ciu, and R. G. Weiss. J. Phys. Chem. *98*:10412 (1994).
66. T. Komatsu, K. Yamada, E. Tsuchida, U. Siggel, C. Böttcher, and J.-H. Fuhrhop. Langmuir *12*:6242 (1996).
67. S. Yamasaki and H. Tsutsumi. Bull. Chem. Soc. Jpn. *67*:906 (1994).
68. S. Yamasaki and H. Tsutsumi. Bull. Chem. Soc. Jpn. *67*:2053 (1994).
69. S. Yamasaki, Y. Ohashi, H. Tsutsumi, and K. Tsujii. Bull. Chem. Soc. Jpn. *68*:146 (1995).
70. T. Brotin, R. Utermöhlen, F. Fages, H. Bouas-Laurent, and J. P. Desvergne. Chem. Soc., Chem. Comm. 416 (1991).
71. J.-L. Pozzo, G. M. Clavier, M. Colomes, and H. Bouas-Laurent. Tetrahedron *53*:6377 (1997).
72. F. Placin, M. Colomes, and J.-P. Desvergne. Tetrahedron Letters *38*:2665 (1997).
73. K. Hanabusa, K. Okui, K. Karaki, T. Koyama, and H. Shirai. J. Chem. Soc., Chem. Commun. 1371 (1992).
74. C. M. Garner, B. Mistrot, J. J. Allegraud, and P. Terech. J. Am. Chem. Soc., submitted.
75. P. Terech. J. P. Desvergne, and H. Bouas-Laurent. J. Colloid Interface Sci. *174*:258 (1995).
76. Y.-C. Lin and R. G. Weiss. Macromolecules *20*:414 (1987).
77. T. Itoh, D. E. Katsoulis, and I. Mita. J. Mater. Chem. *3*:1303 (1993).
78. P. Terech. J. Phys. France *50*:1967 (1989).
79. R. Mukkamala and R. G. Weiss. J. Chem. Soc., Chem. Commun. 375 (1995).
80. L. Lu, T. M. Cocker, R. E. Bachman, and R. G. Weiss, Langmuir (in press).
81. L. Lu, thesis, Georgetown University, 1997.
82. C. Christiansen and J. Liebigs. Ann. Chem. *23*:289 (1884).
83. I. Furman and R. G. Weiss. Langmuir *9*:2084 (1993).
84. F. Kaneko, O. Shirai, H. Miyamoto, M. Kobayashi, and M. Suzuki. J. Phys. Chem. *98*:2185 (1994).
85. D. Demus and L. Richter, in *Textures of Liquid Crystals*, Verlag Chemie, Veinheim, 1978.
86. P. Terech. R. Ramasseul, and F. Volino. J. Colloid Interface Sci. *91*:280 (1983).
87. T. Itoh. Spectrochim. Acta Part A 1083 (1986).
88. J. Mewis. J. Non-Newtonian Fluid Mechanics *6*:1 (1979).
89. E. A. Toorman. Rheol. Acta *36*:56 (1997).
90. P. Terech. G. Gebel, and R. Ramasseul. Langmuir *12*:4321 (1996).
91. E. M. S. Ostuni, thesis, Georgetown University, 1995.
92. O. Schurr, C. Glinka, E. Ostuni, and R. G. Weiss, manuscript in preparation.
93. P. Terech. I. Furman, and R. G. Weiss. J. Phys. Chem. *99*:9558 (1995).
94. P. Terech, E. Ostuni, and R. G. Weiss. J. Phys. Chem. *100*:3759 (1996).
95. P. Terech, I. Furman, H. Bouas-Laurent, J. P. Desvergne, R. Ramasseul, and R. G. Weiss. J. Chem. Soc., Faraday Discussions *101*:345 (1995).
96. R. Mukkamala and R. G. Weiss. Langmuir *12*:1474 (1996).

97. T. D. James, H. Kawabata, R. Ludwig, K. Murata, and S. Shinkai. Tetrahedron *512*:555 (1995).
98. H. Kawabata, K. Murata, T. Harada, and S. Shinkai. Langmuir *11*:623 (1995).
99. N. Shinde and K. S. Narayan. J. Phys. Chem. *96*:5160 (1992).
100. T. Tachibana and K. Hori. J. Colloid Interface Sci. *61*:398 (1977).
101. G. B. McKenna, F. Kern, and S. J. Candau. Polymer Preprints *32*:455 (1991).
102. J. Prost and F. Rondelez. Nature *350*:11 (1991).
103. G. Haering and P. L. Luisi. J. Phys. Chem. *90*:5892 (1986).
104. P. Andreazza, F. Lefaucheux, and B. Mutaftschiev. J. Crystal Growth *92*:415 (1988).
105. M. C. Robert and F. Lefaucheux. J. Crystal Growth *90*:358 (1988).
106. R. J. Phillips, W. M. Deen, and J. F. Brady. J. Colloid Interface Sci. *139*:363 (1990).
107. J. Fricke and A. Emmerling. Adv. Mater. *3*:504 (1991).
108. P. A. Albouy. J. Phys. Chem. *98*:8543 (1994).
109. R. J. H. Hafkamp, B. P. A. Kokke, I. M Danke, H. P. M. Geurts, A. E. Rowan, M. C. Feiters and R. J. M. Nolte. J. Chem. Soc., Chem. Commun. 545 (1997).
110. A. Rifat and A. M. Hikmet. Adv. Mater. *4*:679 (1992).
111. M. Rosenblum. Adv. Mater. *6*:159 (1994).
112. M. Ricco and E. Dalcanale. J. Phys. Chem. *98*:9002 (1994).
113. P. Weber, D. Guillon, and A. Skoulios. J. Phys. Chem. *87*:2242 (1987).
114. F. L. Dickert and A. Haunschild. Adv. Mater. *5*:887 (1993).
115. G. Ozin. Adv. Mater. *410*:612 (1992).
116. B. M. Novak, D. Auerback, and C. Verrier. Chem. Mater. *6*:282 (1994).
117. M. Tata, V. T. John, Y. Y. Waguespack, and C. L. McPherson. J. Am. Chem. Soc. *116*:9464 (1994).
118. A. Thierry, C. Straupe, B. Lotz, and J. C. Wittmann. Polymer Comm. *31*:299 (1990).
119. W. Gu, L. Lu, G. B. Chapman, and R. G. Weiss. J. Chem. Soc., Chem. Commun. 543 (1997).
120. P. Terech. Prog. Colloid Polym. Sci. *102*:64 (1996).
121. C. Fee, Y.-C. Lin, B. Kashar, and R. G. Weiss, unpublished data.

11
Chromatographic Methods for Measurement of Antibody–Antigen Association Rates

CLAIRE VIDAL-MADJAR and ALAIN JAULMES Laboratoire de Recherche sur les Polymères, CNRS, Thiais, France

I.	Introduction	346
II.	Modeling of the Chromatographic Process	347
	A. Description of the model	347
	B. Numerical solutions for frontal analysis	349
	C. Simplified analytical models	350
III.	Chromatographic Method	356
	A. Experimental procedure	356
	B. Instrumentation	356
	C. Immunoadsorbent	358
IV.	Kinetic Studies	359
	A. Adsorption of β-galactosidase on immobilized anti-β-galactosidase	359
	B. Adsorption of lysozyme on immobilized antilysozyme	360
	C. Adsorption of immunoglobulin G on immobilized protein A	361
	D. Adsorption of human serum albumin (HSA) on immobilized anti-HSA	362
V.	Conclusion	369
	References	371

I. INTRODUCTION

The determination of the kinetic parameters of the antigen–antibody interaction is important for characterizing the antigen–antibody immunoreaction [1]. There are numerous precise experimental techniques to determine the equilibrium binding constants in solution [1–3], but most of the methods for association rate measurement study the binding to molecules immobilized on a surface [4,5]. The potentialities of affinity chromatography for characterizing the biological interactions were recognized long ago [6], and several methods have been developed to determine the association–dissociation rate constants of the immunoreactions occurring at the liquid–solid interface [7,8].

The instrument based upon surface plasmon resonance detection [5,9–11] is now routinely used to measure the binding of an antigen (or antibody) to an immobilized antibody (or antigen) in a flow cell. The technology relies on the covalent immobilization of one of the interacting species and the detection of the adsorbed analyte. The sophistication of this expensive instrumentation makes its use difficult for routine investigations in many laboratories. Furthermore, the models used to extract the rate constants are not always appropriate to the kinetic data analysis [12–14].

Immunochromatography is a highly selective method [15] that combines the principles of immunological interactions with the advantages of high-performance liquid chromatography (HPLC). In principle, the immobilized affinity ligand (antigen or antibody) should bind specifically to a single solute. The adsorbed molecules are then eluted by means of a change in pH or composition of the eluent. The technique [16] has been mainly used for the recovery and purification of such biological macromolecular substances as proteins, antigens, and antibodies present in complex mixtures.

Immunochromatography is now increasingly used in assays involving the injection into an immunoreactor [17,18]. These immunoassays are based on direct or competitive recognition. The analytical procedure is rapid, is compatible with on-line chromatographic separations, and can be fully automated by using modern HPLC instrumentation. A fundamental knowledge of the kinetics of the adsorption process is important to optimize this methodology. The chromatographic approach is well suited to study the kinetics of the immunoreaction occurring at the liquid–solid interface.

Several theoretical models were constructed to describe the chromatographic process in the frontal [16,19] and the zonal elution mode [20]. The conventional method of obtaining the kinetic parameters consists in fitting the model to the experimental breakthrough curves. Another method based on the split-peak effect is a direct measurement of the apparent association rate constant [7,21]. Because of the slow adsorption process, a fraction of the solute injected as a pulse into the immunochromatographic column is eluted as a nonretained peak. This behavior is observed at high flow rates, with very short or low-capacity columns [21–25].

II. MODELING OF THE CHROMATOGRAPHIC PROCESS

In affinity chromatography [8], the adsorption kinetics are studied by two techniques differing by the shape of the injection signal. In the frontal elution mode, the injection signal is a sharp concentration step, with the solute continuously introduced into the column. The resulting breakthrough curve, which is a plot of solute concentration against time, is analyzed by fitting a theoretical model to the experimental elution front. In the zonal elution mode, the shape of the injection signal is a pulse, with a small amount of solute introduced into the column. In the case of a linear adsorption isotherm, the plate height theories can be applied to study the mass transfer kinetics by following the elution band broadening as a function of flow rate.

A. Description of the Model

The theories of chromatography are based on material balance equations that describe the migration of the solute through a packed bed [26,27]:

$$\varepsilon \frac{\partial C}{\partial t} + u_s \frac{\partial C}{\partial z} + (1-\varepsilon) \frac{\partial C_s}{\partial t} = D \frac{\partial^2 C}{\partial z} \tag{1}$$

where z is the distance along the column, t is the time, C is the solute concentration, and u_s is the superficial velocity or flow rate divided by the cross section of an empty column. The volume void fraction is ε, and the axial dispersion coefficient is D. The average concentration of the adsorbate in the particle is C_s and includes the amount of solute in the pores.

The solution of such an equation is complicated because the equilibrium is generally described by a nonlinear adsorption isotherm. Moreover, the mass transfer kinetics are extremely complex and include the film mass transfer resistance, the diffusion of the protein into the pores of the support, and the interaction with the immobilized species.

The formulation of the mass transfer kinetic processes was first presented by Kucera [28] to describe gas adsorption chromatography. This approach was then adapted for modeling in liquid chromatography [29,30] and affinity chromatography [31] experiments. The model used to describe the adsorption of the protein on the affinity support is given in Fig. 1 and includes the following steps:

1. Transfer across the stagnant fluid layer around the particle
2. Diffusion into the pores of the support
3. Surface reaction and adsorption at the liquid–solid interface

The adsorption process is a combination of these steps, and the slowest one will determine the overall rate of adsorption.

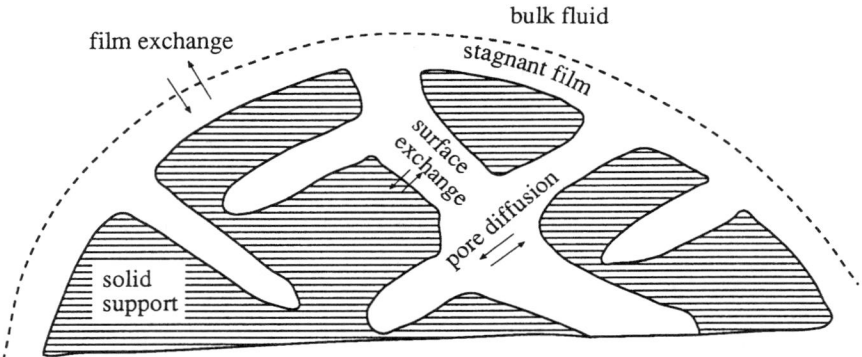

FIG. 1 Mass transfer kinetics into porous particles.

For spherical particles, the solute mass transfer rate from the bulk to the solute inside the pores is described by a linear kinetic law:

$$(1-\varepsilon)\frac{\partial C_s}{\partial t} = \frac{6(1-\varepsilon)}{d_p} k_f [C - C_p(d_p)] \tag{2}$$

where $C_p(d_p)$ is the solute concentration inside the pores at the outer surface of a spherical particle of diameter d_p. The liquid film mass transfer coefficient is k_f. Several correlations exist to estimate this parameter, valid either for stirred tank experiments or for packed bed operations.

The diffusion of the solute into the porous spherical particle is described by the equation

$$\frac{\partial q}{\partial t} = \frac{\varepsilon_p D_p}{r^2}\frac{\partial}{\partial r}\left(r^2 \frac{\partial C_p}{\partial r}\right) - \frac{\varepsilon_p \partial C_p}{\partial t} \tag{3}$$

where q is the amount adsorbed per unit volume of adsorbent, D_p is the effective particle diffusion coefficient, and ε_p is the particle porosity.

The rate of formation of the affinity complex is often described by the second-order Langmuir adsorption rate equation

$$\frac{\partial q}{\partial t} = k_a C_p (q_x - q) - k_d q \tag{4}$$

where k_a is the second-order adsorption rate constant and k_d the first-order rate constant for the desorption process. The maximum loading capacity per unit volume of adsorbent is q_x. At equilibrium, a Langmuir-type adsorption isotherm is obtained.

The system of differential equations is too complex to be solved analytically. Assumptions of a linear adsorption isotherm can be used to obtain analytical solutions, but this approach is generally not applicable to describe affinity chromatography experiments. Several numerical techniques are used to solve the system of partial differential equations. The other method is to use an analytical solution with simplifying approaches [32] that describe the adsorption process with a single step and a lumped mass transfer coefficient [27].

B. Numerical Solutions for Frontal Analysis

Arve and Liapis [33] presented a numerical solution of the breakthrough curve that takes into account the different rate-controlling mechanisms, i.e., the resistances due to pore diffusion and film diffusion and the kinetics of the biospecific interaction step. Their model expressions are applicable to systems involving single and multicomponent adsorptions and also monovalent or multivalent adsorption. The numerical method of orthogonal collocation was used to solve the partial differential equations that describe the mass transfer terms, and the resulting nonlinear differential system was integrated by using the Runge–Kutta method. The value of the axial dispersion coefficient is so low that D is considered negligible in packed bed simulations.

Arve and Liapis [34] suggest estimating the parameters characterizing the intraparticle diffusion and the adsorption–desorption step mechanisms of affinity chromatography from the experimental data obtained in a batch system. The numerical simulations of the chromatographic process will use the values of the parameters of the adsorption isotherm and those of the effective pore diffusion as determined from stirred tank experiments together with the film mass transfer coefficients calculated from chemical engineering expressions found in the literature.

Horstmann and Chase [35] have used the mass transfer parameters determined in stirred tank experiments to simulate the breakthrough curves of affinity chromatography experiments. Numerical methods using different computer packages were carried out to solve the differential equations of the stirred tank adsorption and to predict the performances of a packed bed chromatographic column.

McCoy and Liapis [36] used two different kinetic models to represent the column affinity process. In both models the transport of the adsorbate in the adsorbent particle is considered to be governed by the diffusion into the pores. In model 1 the adsorption is assumed to be completely reversible with no interaction between the adsorbed molecules. In model 2, it is assumed that the biomolecule may change conformation after adsorption. Although these two models represent different overall adsorption mechanisms, the differences between the simulated breakthrough curves is very small.

The parameter estimation and the kinetic discrimination studies obtained from the numerical simulations are tedious, complex, and time-consuming. Complex softwares are required, and lengthy strategies are used to fit to the data. Moreover, the number of parameters describing the model complicates the interpretation of the experimental results. Simplifying approaches are recommended to make easier the analysis of the chromatographic profiles [37].

C. Simplified Analytical Models

1. Frontal Breakthrough Curve

Another approach consists in simplifying the system of differential equations by assuming that one of the rate-limiting steps dominates the others. As generally found in the above numerical simulation models, the effect of axial diffusion can be considered as negligible for proteins [$D = 0$ in Eq. (1)]. The migration of the solute through the chromatographic column is given by the material balance equation

$$\varepsilon \frac{\partial C}{\partial t} + u_s \frac{\partial C}{\partial z} + (1 - \varepsilon) \frac{\partial q}{\partial t} = 0 \tag{5}$$

with a kinetic mass transfer term described by the binding rate equation alone. The second-order rate equation is given by

$$\frac{\partial q}{\partial t} = k'_a C(q_x - q) - k'_d q \tag{6}$$

where k'_a and k'_d are the apparent adsorption and desorption rate constants.

The analytical solution of this set of differential equations [Eqs. (5) and (6)] was first given by Thomas [38] for the frontal breakthrough curve. The concentration profile is

$$C(t) = \frac{C_i J(nr, nT)}{J(nr, nT) + [(1 - J(nr, nT))\exp((1 - r)(n - nT))]} \tag{7}$$

where C_i is the feed concentration and the dimensionless parameter r is given by

$$r = (1 + KC_i)^{-1} \tag{8}$$

The adsorption equilibrium constant is K, with $K = k'_a/k'_d$.

The dimensionless throughput parameter is defined as

$$T = \left(t\frac{F}{V} - \varepsilon\right) \frac{1}{Krq_x(1 - \varepsilon)} \tag{9}$$

where F is the flow rate and V the volume of an empty column. The function $J(x, y)$ of the two variables x and y is given by

$$J(x, y) = 1 - e^{-y} \int_0^x I_0(2\sqrt{\tau y}) e^{-\tau} \, d\tau \qquad (10)$$

The number of transfer units is related to the apparent adsorption rate constant by

$$n = \frac{k'_a q_x (1 - \varepsilon) V}{F} = \frac{k'_a q_x (1 - \varepsilon) L}{u_s} \qquad (11)$$

Hiester and Vermeulin [39] extended the model of Thomas by deriving a breakthrough curve for a slow diffusion rate-limiting step with a nonlinear Langmuir equilibrium isotherm. The model was further adapted by Arnold and Blanch [40] to describe affinity chromatographic experiments:

$$(1 - \varepsilon) \frac{\partial q}{\partial t} = K_{OL} \frac{A_s}{V} (C - C^*) \qquad (12)$$

where C^* is the solute concentration at equilibrium with the solute bound to the affinity sites. The overall forward rate constant of the diffusion process is K_{OL}, and the total surface area of the adsorbent is A_s. With some approximations ($q \approx 0.5 q_x$), the equation of the breakthrough curve is similar to that of Eqs. (7)–(10) except for the number of transfer units given by

$$n = n_{mt} = \frac{2 K_{OL} A_s}{(r + 1) F} \qquad (13)$$

If the mass transfer limited sorption is of comparable magnitude to the binding rate process, the number of transfer units is given by [40]

$$\frac{1}{n} = \frac{F}{k_a (1 - \varepsilon) q_x V} + \frac{1}{n_{mt}} \qquad (14)$$

where k_a is the effective adsorption rate constant for the sorption exchange step. This equation is useful in determining the governing kinetic process from the parameters defining the experimental system. The value of n_{mt} can be calculated or directly determined from the plate height contribution of an unadsorbed tracer injected as a pulse into the column [41]:

$$\frac{1}{n_{mt}} = \frac{(r + 1) F}{2 K_{OL} A_s} = \frac{(r + 1) F}{2 V (1 - \varepsilon)} \left(\frac{d_p^2}{60 D_i} + \frac{d_p}{6 k_f} \right) \qquad (15)$$

Chase [32] used the adsorption rate-limited model [Eqs. (7)–(11)] to analyze the experimental breakthrough curves in affinity chromatography. This empirical approach assumes that all the rate-limiting processes can be represented by an apparent single second-order Langmuir adsorption rate equation in which k'_a is considered a ''lumped'' parameter.

2. Zonal Elution Profile

The analytical solution of Thomas was further developed by Goldstein [42] for a rectangular pulse injection signal of duration t_i. The corresponding boundary conditions are

$$C(0, t) = C_i \quad \text{for} \quad 0 < t \leq t_i$$
$$C(0, t) = 0 \quad \text{for} \quad t_i < t \tag{16}$$
$$C(z, 0) = 0 \quad \text{and} \quad q(z, 0) = 0$$

For a rectangular pulse injection, the concentration of solute at the column outlet, for a time t larger than t_i larger than $t_i + V\varepsilon/F$, is given by

$$C(t) = [J(nr, nT) - J(nr, nT_i)]/E(t)$$
$$E(t) = J(nr, nT) - J(nr, nT_i) + (1 - J(n, nrT))e^{(1-r)(n-nT)} \tag{17}$$
$$+ J(n, nrT_i)e^{(1-r)(n-nT_i)}$$

where T_i is a dimensionless parameter given by

$$T_i = \left(t\frac{F}{V} - \varepsilon - t_i\frac{F}{V}\right)\frac{1}{Krq_x(1-\varepsilon)} \tag{18}$$

When $t \leq t_i$, the solution corresponds to the frontal elution curves [Eq. (7)], because the solute is still present in the mobile phase.

Figure 2 illustrates the elution peaks obtained with slow adsorption kinetics in nonlinear chromatography [20]. Important band broadenings are generated because of the slow adsorption–desorption process. Large variations in peak shapes are observed that depend upon the number of transfer units, n. This dimensionless number is related to the kinetic effects leading to the band broadening of an elution peak. The kinetic contribution (proportional to $1/n$) is generally expressed in terms of the plate height equation [30] valid in the domain of a linear adsorption isotherm (small amounts injected).

With very low adsorption rates (small n values), a fraction of solute is eluted at the retention time of the nonretained compound. This effect is increased by the column overloadings; this phenomenon, called the split-peak effect [7,21], can be used to determine the apparent adsorption rate constant [21–25].

Immunochromatography is a process involving high affinity interactions with slow adsorption–desorption rate constants leading to important band broadenings. In principle the rate constants of formation or dissociation of the complex could be measured from a study of these effects in isocratic zonal elution [7]. Hethcote and DeLisi [43] have derived equations valid in the linear domain of the adsorption isotherm that permit one to predict the statistical moments of retained solutes eluted isocratically using competitive inhibitors. The plate height theory and the studies of the band broadening effects are not applicable to immunochromatographic experiments. With the high-affinity systems used,

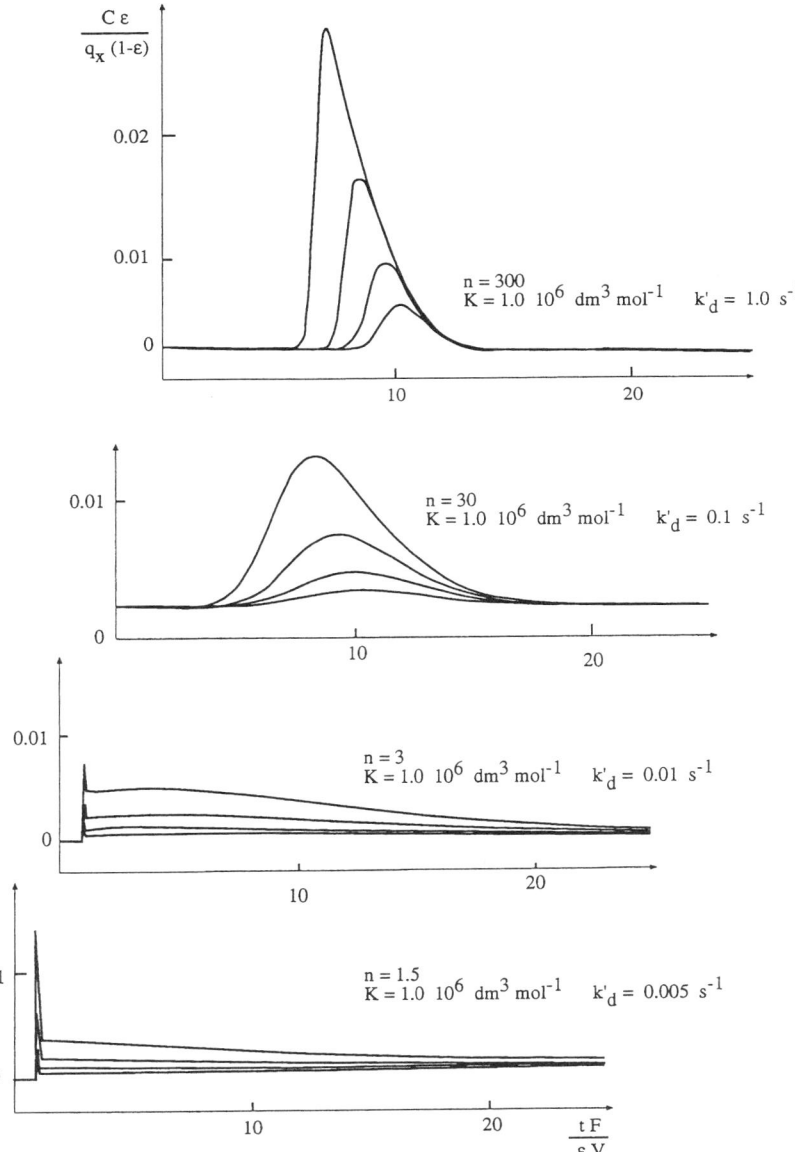

FIG. 2 Influence of the mass transfer kinetics on the shape of the elution peak in nonlinear chromatography [Eqs. (7) and (17)]. Simulations: column length 5 cm, $F = 16.67$ mm$^3 \cdot$s^{-1}, $\varepsilon V = 0.5$ cm^3, $Q_x = 5$ nmol. $k' = [K q_x (1 - \varepsilon)/\varepsilon] = 10$. Amounts injected $Q_i/Q_x = 0.08, 0.04, 0.02, 0.01$.

3. The Split-Peak Effect

Hage et al. [21] demonstrated the applicability of the split-peak effect to evaluate the binding rate constant between the soluble protein and the immobilized species (antigen or antibody). This phenomenon occurs when a fraction of the pure solute is eluted at the retention time of a nonretained solute, while the other fraction is strongly retained in the chromatographic column.

This effect was first predicted by Giddings and Eyring [44] by considering the random migration of a solute molecule through the chromatographic column. The stochastic theory gives an expression of the elution peak, valid for very small amounts injected, in the linear domain of the adsorption isotherm. Because the adsorption process is slow, a fraction of the solute molecules will pass through the column without being adsorbed. This effect will increase if the time spent in the column is shorter (high flow rates and short columns). The model predicts the probability for a solute molecule to elute at the column dead volume. This approach was further developed by Denizot and Delaage [45] for affinity chromatography.

The occurrence of the split-peak effect is also predicted in the model for zonal elution assuming a rectangular injection creneau, a negligible axial dispersion term, and a second-order Langmuir adsorption rate equation. Jaulmes and Vidal-Madjar [20] have shown that the integration of Eq. (7) for the duration time of the creneau gives the expression of the split-peak effect in nonlinear chromatography. If the desorption is very slow compared with the time scale of the experiment ($k'_d = 0$), the expression of the unretained fraction is

$$f = \frac{Q_x}{nQ_i} \ln[1 + e^{-n}(e^{nQ_i/Q_x} - 1)] \tag{19}$$

where Q_x is the total column capacity and Q_i is the amount injected. The number of transfer units is given by Eq. (14) for experiments in which the kinetics are controlled simultaneously by the mass transfer between the mobile phase and the stagnant fluid and by a second-order Langmuir kinetic law for the sorption exchange process. The model combining Eqs. (19) and (14) is useful in analyzing the variations of the nonretained fraction with the various experimental conditions (density of immobilized binding sites, flow rate, particle diameter) and in evaluating the governing step of the adsorption process.

The adsorption yield, $\rho = (1 - f)$, can be predicted from Eq. (19). Its variation with sample size is shown in Fig. 3. The values of the adsorption yield are close to 1 for large values of n, and the dotted line illustrates the asymptotic value of ρ. As predicted from Eq. (14), the yield will increase for larger column capacities and lower flow rates. These are the optimal conditions for preparative

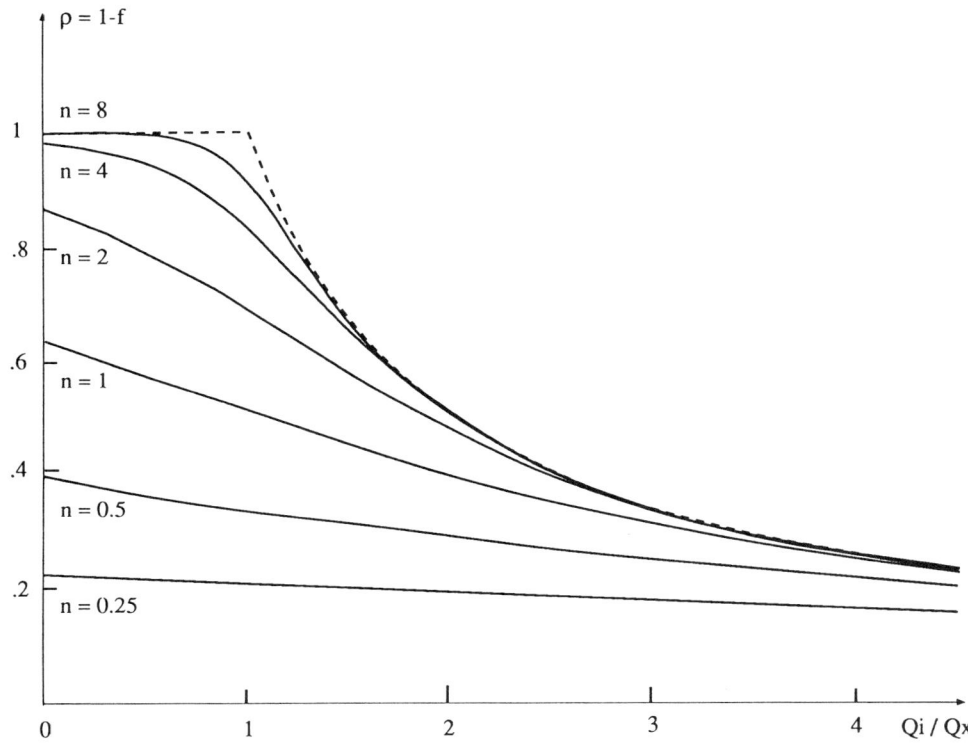

FIG. 3 Adsorption yield as a function of the amount injected.

chromatography or for most assays in immunoreactors. The opposite is true for the kinetic measurements of the adsorption rate, since the nonretained fraction will increase with increasing flow rate and decreasing density of immobilized ligand.

The limit of the split-peak fraction for zero amount injected is $f_0 = e^{-n}$. Combining with the value of the mass transfer unit term for a small amount injected [$r = 1$ with $C_i = 0$ in Eq. (15)] gives the expression of the split-peak effect in linear chromatography [21]:

$$-\frac{1}{\ln(f_0)} = F\left[\frac{1}{k_a Q_x} + \frac{1}{K_{OL} A_s}\right] \tag{20}$$

Using the expression of the overall adsorption rate coefficient [see Eq. (15)], one can use this equation to analyze split-peak effect experiments carried out at very low amounts injected.

III. CHROMATOGRAPHIC METHOD

There are several advantages in using the chromatographic method for adsorption rate measurements. The problems of mixing found in batch studies are eliminated in the packed bed format. With the development of HPLC instrumentation, the method is fast, highly sensitive, and easy to automate.

The use of nonporous supports for protein immobilization will eliminate the problems associated with the intraparticle diffusion limitations. With columns regularly packed with small spherical particles of limited size distribution, the mass transfer effects associated with a nonuniform flow distribution will be minimized. Extra column effects due to diffusion in the stagnant liquid pockets will be reduced by using short and properly designed connecting lines to the injector and the detector.

A. Experimental Procedure

The chromatographic procedure involves
1. Equilibration of the column with the buffer to be used in adsorption experiments.
2. Injection of the protein sample (antigen or antibody). In frontal analysis the sample is applied continuously to the column by switching to the eluent containing the protein at the desired concentration. With the split-peak method, small pulses are repeatedly injected into the column.
3. On-line detection (UV, fluorescence, etc.) by continuously recording the unbound protein concentration at the column outlet.
4. Desorption by selecting conditions promoting the dissociation of the antigen–antibody complex (nature of the eluent, pH, temperature). The bed should be regenerated by selecting conditions that do not alter irreversibly the biospecific properties of the immunoadsorbent.

The operating conditions needed for adsorption rate measurements differ considerably from the requirements of the preparative packed bed operation mode. Kinetic mass transfer effects will increase as the flow rate is increased and the column capacity decreased. Furthermore, the major problem is to differentiate between a diffusion rate-limited process and the kinetic-limited one. The contribution of the diffusional mass transfer to the overall adsorption process will be reduced by using an immunoadsorbent with a low density of binding sites immobilized on a nonporous support.

B. Instrumentation

Figure 4 illustrates an example of the HPLC equipment that can be used either for frontal elution mode or for pulse injection analysis. For the pulse injection mode, an injection valve with a small volume sample loop (20 mm^3) is placed at

Chromatographic Methods for Measurement

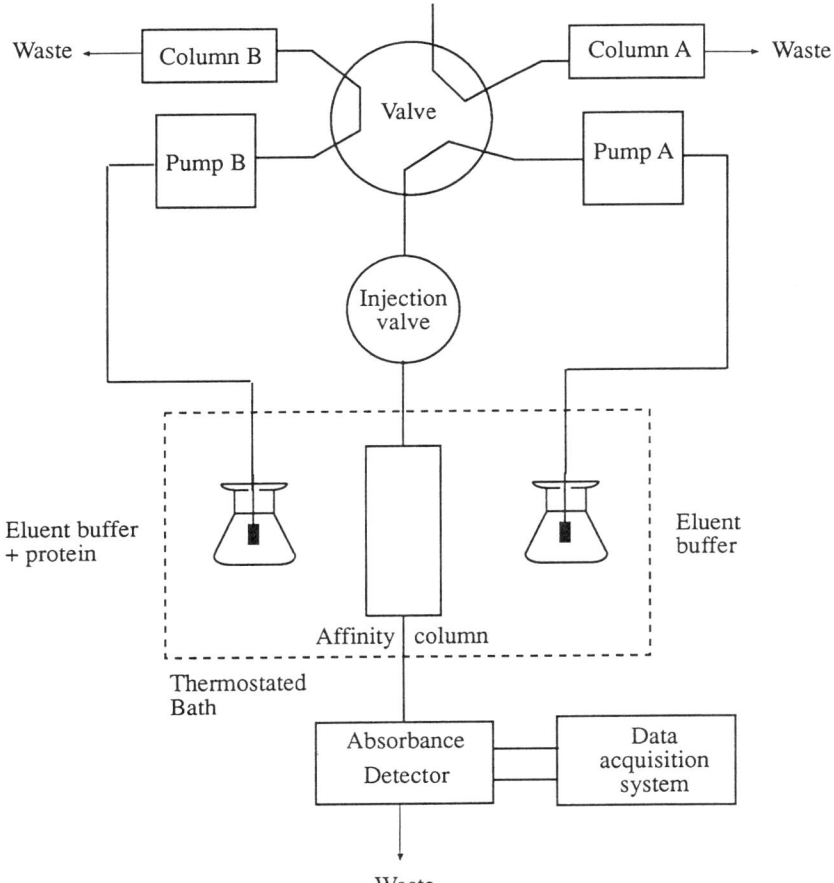

FIG. 4 Schematic representation of the HPLC system for frontal experiments.

the column inlet. The detector (e.g., a U.V. detector set at 280 nm) connected to a data acquisition system continuously records the signal at the column outlet. The columns packed with the immunoadsorbent HPLC supports are typically 2–5 cm in length and 2–4 mm in diameter. Minicolumns of 6.35 mm length [46] were used in several immunoassays [18,47] and in some kinetic studies based on the split-peak effect [21].

The two HPLC pumps and the six-port switching valve permit one to switch from one eluent to the other, and the immunoadsorbent column is automatically regenerated. In addition, for the frontal elution mode, this system permits one to commute from the eluent used for adsorption experiments to the other containing the protein. The two columns A and B placed in the circuit have the same pressure drop as that of the immunoadsorbent column in order to avoid pressure disturbances. This system with two different pumps requires large quantities of protein during the equilibration and feeding stages. To decrease this amount, a single pump and two six-port switching valves equipped with large sample loop volumes (0.5–1.5 cm^3) can be used as described in the miniaturized liquid chromatograph for frontal elution experiments [48].

C. Immunoadsorbent

The role of the matrix for protein immobilization is also very important in the design of the chromatographic experiment for reducing mass transport limitations. Several examples of kinetic measurements using soft gels as chromatographic supports have been published in the past [32,35]. Their relatively low rigidity limits their application at relatively high flow rates. Moreover, with these highly porous supports, it has been shown that a diffusion-limited mechanism in the pores of the support governs the adsorption process. Because of the large capacities achieved, these porous matrices are well suited for preparative purposes. They cannot be used for adsorption rate measurements because the contribution of adsorption kinetics is decreased while the diffusion resistance increases.

The mass transfer effects due to the diffusion into the intraparticle medium will be eliminated with nonporous supports [49,50]. One can also use pellicular supports by immobilizing the protein on the external surface. This approach is possible [21–25] with silica adsorbents of small pore size (< 6 nm) because the large biomolecules of antibodies cannot penetrate into the pores of the support, and the immobilization takes place on the external surface of the particles.

The contributions due to the mass transfer at the particle boundary will be decreased [49] with monodisperse spherical particles of small diameter (1–2 μm). However, with very small particles a regular packing of the columns is difficult to achieve. Moreover, the corresponding increase of the column capacity will lead to a decrease of the contribution for the binding rate process [Eq. (14)]. A compromise is thus necessary between these two contradictory requirements. Experiments by varying particle size are needed to ascertain the validity of adsorption rate measurements at the gas–liquid interface.

Another problem in model applications is the adsorption on a nonuniform adsorbent. The immobilization of polyclonal antibodies will lead to different populations of binding sites, and the measurements will only give an apparent adsorption rate constant [22]. The properties of the adsorbent surface are also greatly affected by the procedure used for protein immobilization. It may be important to select coupling methods that orient the covalently attached protein and reduce the diffusion-limited resistances. The nonspecific interactions will also have to be eliminated by using passivated matrices free of active sites.

Different methods are used to immobilize covalently the proteins to HPLC supports. The coupling procedures imply various activation chemistries via epoxide-, diol-, or aldehyde-silica [51,52]. Generally, the immobilized antibodies are randomly oriented. Dihydrazide-silica supports are used for an oriented immobilization of the antibody through its carbohydrate residues [53]. Another approach is to bind the antibody to protein A or protein G surfaces [17]. The antibody will bind through the Fc portion, leaving the antigen combining sites oriented away from the support.

IV. KINETIC STUDIES

A. Adsorption of β-Galactosidase on Immobilized Anti-β-Galactosidase

The "lumped" parameter approach [32] was applied to study the adsorption of *E. coli* β-galactosidase to an immobilized monoclonal antibody against β-galactosidase using batch experiments and frontal elution chromatography. Two different supports were used for antibody immobilization: a soft gel matrix, Sepharose 4B, an agarose of highly porous structure, and a silica-based adsorbent (100–200 μm particle sizes and 300 nm porosity). Important differences are observed in the adsorption parameters of the two immobilized antibody systems. It is shown that the value of the binding constant greatly depends on the support material for antibody immobilization. The equilibrium binding constants are respectively $K = 7.1 \times 10^7$ mol·dm^{-3} and $K = 2.1 \times 10^9$ mol·dm^{-3} for the association with the antibody immobilized to Sepharose 4B and to the silica support. The apparent adsorption rate constant for the interaction with the antibody immobilized to Sepharose ($k'_a = 1.4 \times 10^4$ dm^3·mol^{-1}·s^{-1}) is five times as large as that with the antibody immobilized to silica.

The effect of the amount of monoclonal antibody immobilized to the Sepharose matrix was further investigated by Fowell and Chase [54]. Batch experiments were used to determine the apparent adsorption rate constant k'_a. A marked increase in the k'_a value is observed as the amount of antibody immobilized is decreased. The values of k'_a vary from 1.3×10^3 dm^3·mol^{-1}·s^{-1} for the support of high capacity to 7.0×10^5 dm^3·mol^{-1}·s^{-1} for the support of low capacity. A marked increase of the efficiency of the adsorption process was also observed when using the frontal analysis techniques. This effect has been explained by the improved adsorption properties of the low-capacity immunoadsorbent.

The batch data of Chase [32] for the adsorption of β-galactosidase on anti-β-galactosidase immobilized to porous silica were discussed by Arve and Liapis [34] in terms of the different numerical simulation models of batch experiments:

"lumped" kinetic model, pore-diffusion model and interaction rate given by a second-order reversible equations, pore diffusion and local equilibrium at each point in the pores (Langmuir isotherm). It is difficult to discriminate between the various models, and additional experiments are needed to estimate the parameters of the interaction. Nevertheless, the numerical simulation method of Arve and Liapis [33,34] permits one to conclude that the adsorption of β-galactosidase onto the monoclonal antibody immobilized on the porous particles is controlled by the binding rate process and by the intraparticle diffusion mechanism. The effect of the film mass transfer resistance is small.

The mechanisms of biospecific adsorption will dominate for the adsorption on nonporous supports, because the effects of diffusion in the stagnant fluid are minimized with no diffusion into the pores [49]. The theoretical models based on the numerical solution of the adsorption process were used by McCoy and Liapis [36] to analyze the batch data [55] for the adsorption of β-galactosidase onto anti-β-galactosidase immobilized to nonporous glass beads ($d_p = 172$ μm). It is found that the adsorption process is controlled by both the film mass transfer and the rate of the adsorption step. The contribution to diffusion in the stagnant fluid mobile phase is assumed to be controlled by only the film mass transfer term. The value of this contribution was calculated from a theoretical chemical engineering expression. This approach permits one to estimate the value of the adsorption rate constant ($k_a = 2.9 \times 10^4$ dm$^3 \cdot$mol$^{-1} \cdot$s^{-1}) with an equilibrium constant equal to 8.8×10^6 dm$^3 \cdot$mol^{-1}. These parameters determined in stirred tank experiments were used to simulate the theoretical breakthrough curves of the chromatographic process.

Important differences of several orders of magnitude are thus reported for the value of the equilibrium binding constants, that depend upon the matrix to which the antibody was attached. These results are explained [36] by the alteration of the structure of the antibody when immobilized to the different supports.

B. Adsorption of Lysozyme on Immobilized Antilysozyme

Liapis et al. [50] studied the adsorption of lysozyme onto a monoclonal antilysozyme antibody immobilized to spherical nonporous particles 1.5 μm in diameter. Several breakthrough curves were measured on both high- and low-density ligand systems. The equilibrium isotherm is of the Langmuir type and was determined from a whole set of breakthrough curves at various concentrations. The amount absorbed is calculated from the area behind the breakthrough curve. The numerical simulation model for nonporous supports was used to analyze the experimental elution profiles. It assumes that the transport to the binding sites is controlled by the transfer through the film layer around the particle, estimated from chemical engineering expressions. The value of the adsorption rate constant k_a is obtained by fitting the model to the elution profile, and good agreement between theory and experiment is obtained for the initial part of the breakthrough curve. At a 11.7 mm^3·s^{-1} flow rate, the k_a values are respectively 4.5×10^4 dm^3·mol^{-1}·s^{-1} and 4.3×10^4 dm^3·mol^{-1}·s^{-1} for the low- and the high-density antilysozyme immunoadsorbent. At a larger flow rate (23.3 mm^3·s^{-1}), a value of 5.9×10^4 dm^3·mol^{-1}·s^{-1} is found for the adsorption rate on the high-density immobilized ligand. The difference in the k_a values determined at two different flow rates with the same affinity column demonstrates the difficulty of evaluating the various contributions implied in the overall adsorption process.

When the chromatographic data are analyzed with numerical methods, theoretical expressions are generally applied to describe the mass transfer resistances due to the transport to the binding site. Several nonideal effects that are not accounted for by the model, such as irregularities in the column packing and extra column contributions, may also increase the contribution for mass transfer in the stagnant fluid. Even if an appropriate design of the experiment can minimize these nonideal effects, it is highly desirable to have an estimation of this contribution from direct experimental measurements.

C. Adsorption of Immunoglobulin G on Immobilized Protein A

1. Frontal Analysis

For their study, Hortsmann and Chase [35] selected the interaction of human polyclonal immunoglobulin G(IgG) with Protein A immobilized to different agarose-based matrices. As the interaction occurs with a constant Fc region of the antibody molecules, the assumption of an interaction of the Langmuir type still holds. The study enables one to identify the rate-governing mass resistances. The film mass transfer coefficients in a stirred tank and in a packed bed were calculated from the corresponding expressions found in the literature. The effective diffusion coefficient in the adsorbent particles and the parameters of the equilibrium isotherms were determined from batch stirred tank experiments. This analysis indicates that the model in which pore diffusion and film mass transfer govern the adsorption is sufficient to model the kinetics of the adsorption process. The surface reaction step is fast compared to the other limiting steps. The model predicts reasonably well the frontal breakthrough curves when using the parameters determined from stirred tank experiments.

2. Analysis from the Split-Peak Effect

Another way of performing kinetic measurements is to inject the solute into short immunochromatographic columns and measure the amount that binds to the immobilized antibody (split-peak effect). Compared to the approaches previously described, the advantage is simplicity with a technique easy to automate that permits the rapid evaluation of the various adsorption rate-limiting steps [21]. Sportsman et al. [56,57] have indicated the possibility of deriving rate constants from such experiments performed on very short immunoaffinity columns. By this method, they investigated the interaction of two antigen proteins (human immunoglobulin and bovine insulin) with their respective antibodies immobilized on diol-bonded glass spheres.

Hage et al. [21] used the split-peak method to analyze the adsorption kinetics of IgG using mini columns packed with protein A immunoadsorbents. The protein was immobilized to silica-based supports of different porosities. The measurement of the fraction of nonadsorbed IgG as a function of flow rate enables one to evaluate the various adsorption processes involved. Important differences in the kinetic behavior are observed that vary with the protein immobilization method and with the porosity of the silica support selected. Slow adsorption kinetics are observed when using for protein immobilization a silica matrix of 50 nm mean porosity. In this case, the adsorption is mainly governed by a diffusion-limited process. With a support of 5 nm mean porosity, the process is mainly adsorption-limited by the binding rate, contributing about 97% to the overall adsorption process. Indeed, the large IgG molecules cannot penetrate into the pores of this immunoadsorbent, and the interaction takes place at the external surface of particles. As the contribution for transport to the binding site is small, the effective adsorption rate constant ($k_a = 1.2 \times 10^5$ dm^3·mol^{-1}·s^{-1}) can be determined from split-peak experiments. For studies using as a matrix the 50 nm silica supports, important differences in the values of the adsorption rate constants are observed, depending on the chemical procedure employed for the antibody immobilization.

D. Adsorption of Human Serum Albumin (HSA) on Immobilized Anti-HSA

For this study, Vidal-Madjar et al. [22–25] used different immobilized anti-HSA polyclonal and monoclonal antibodies. The matrix for antibody immobilization is an HPLC silica (10 μm particle diameter and 6 nm mean porosity). The large antibody molecules are attached to the external surface of the particles, and this support behaves as a pellicular immunoadsorbent. The monoclonal antibodies used differ by their specificities toward several HSA fragments and by their affinities for HSA. This high affinity support strongly retains HSA when a phosphate-buffered saline eluent (pH = 7.4) is used. The regeneration of the immunoadsorbent was generally possible by desorbing the protein under acidic conditions. The column can then be reused several times.

1. Frontal Analysis

Frontal analysis was used to study the adsorption behavior of HSA on a monoclonal anti-HSA immunoadsorbent [23]. Figure 5 illustrates the typical break-

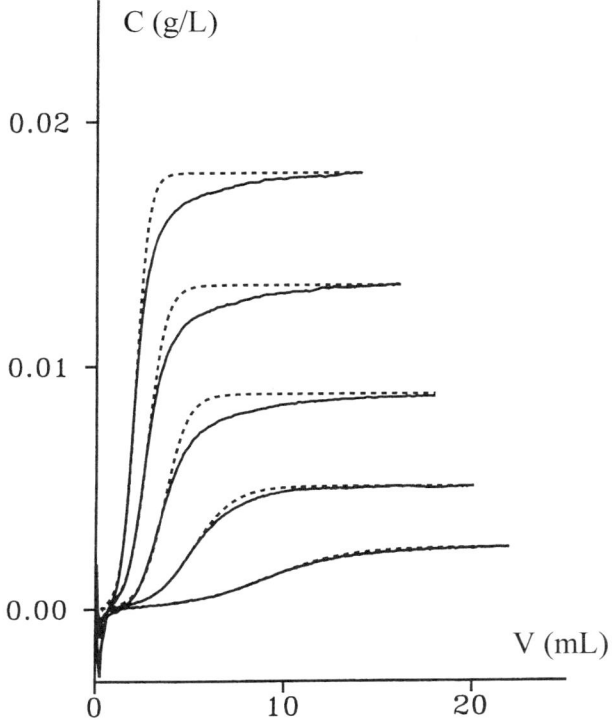

FIG. 5 Breakthrough curves as a function of HSA concentration in the eluent on a monoclonal anti-HSA immunoadsorbent. Eluent: 10 mM potassium phosphate buffer, pH 7.4, 150 mM NaCl, 20°C; flow rate: 8.33 mm$^3\cdot$s^{-1}; 30 × 4.6 mm column; dead volume: 0.3 cm^3. Solid line: experimental breakthrough curve. Dotted line: fit of the theoretical model [Eqs. (7)–(11) with $k_d = 0$]. (From Ref. 23, with permission from Academic Press, Inc.)

through curves observed at various HSA feeding concentrations. The simplified theoretical model [Eqs. (7)–(11)] with $k'_d = 0$ was fitted to the experimental front. With this high-affinity system, the $k'_d = 0$ hypothesis holds, as there was no HSA desorption after several hours of column rinsing with the buffer used for adsorption experiments.

At low HSA concentrations the model fits well the whole breakthrough curve (Fig. 5). Two parameters are necessary to describe the model, the column capacity Q_x and the number of transfer units n. The overall adsorption process is described by an apparent association rate constant that includes the transport to the active sites and the biospecific interaction [Eq. (6)].

TABLE 1 Analysis of the Breakthrough Curves at Different Flow Rates for the Adsorption of HSA on a Monoclonal Anti-HSA Immunoadsorbent Under the Same Experimental Conditions as in Fig. 5

Flow rate (mm^{3}s^{-1})	Q_x (μg)	$1/n$	k'_a (dm^{3}mol^{-1}s^{-1})	k_a (dm^{3}mol^{-1}s^{-1})
4.2	38	0.067	113,000	252,000
8.3	36	0.104	154,000	238,000
12.5	35	0.151	163,000	216,000
16.7	34	0.178	190,000	240,000
20.8	34	0.200	211,000	259,000
25.0	32	0.247	218,000	257,000

At larger protein concentrations, the model fits well only the initial part of the elution front (Fig. 5). This deviation between the fitted model and the experimental front is explained by a slower second adsorption step due to the restricted access to the binding sites in the presence of large HSA concentrations in the eluent.

The contribution to the transport to the binding sites was determined from studies as a function of flow rate, and Table 1 lists the values of the two parameters n and Q_x used to fit the theoretical model to the experimental profile. The apparent adsorption rate constant was calculated from n values. An important increase of nF and thus of k'_a is observed at larger flow rates. The contribution for the mass transfer in the stagnant fluid can be calculated by plotting $1/n$ as a function of flow rate. In agreement with Eq. (14), a straight line is obtained that intercepts the ordinate axis at $1/n_{mt} = 0.036$. The method assumes that n_{mt} does not vary with flow rate. This hypothesis was checked by measuring the theoretical plate height of the HSA nonretained peak. In the range of flow rate studied, there was no significant variation of the peak band broadening when HSA was injected into a column already saturated with the protein.

The n_{mt} contribution can also be determined from experiments with columns of the same size packed with immunoadsorbents of varying capacities [22]. Because of the low amount of monoclonal antibody available, these experiments were performed by immobilizing various quantities of polyclonal anti-HSA antibody on the silica matrix. The results for an 83.3 mm^3·s^{-1} flow rate are summarized in Table 2. An important decrease of the apparent adsorption rate constant is observed when the column capacity is increased.

As predicted from Eq. (14), the reciprocal of the number of transfer units n varies linearly with the reciprocal of q_x, the column capacity per unit volume of adsorbent. The value of $1/n_{mt}$ obtained by extrapolating the straight line to $1/q_x = 0$ is 0.038 and agrees with the results of Table 1 performed

TABLE 2 Comparison of Methods for the Adsorption of HSA on Anti-HSA Immunoadsorbents with Various Densities of Immobilized Polyclonal Antibodies; Flow Rate: 8.3 mm^{3}s^{-1}

Column	Q_x (μg)	$1/n$	k'_a (dm^{3}mol^{-1}s^{-1})	k_a (dm^{3}mol^{-1}s^{-1})	Method
A	16	0.100	359,000	570,000	Frontal analysis
B	43	0.067	200,000	446,000	($n_{mt} = 27$)
C	68	0.050	169,000	650,000	
D	70	0.051	161,000	587,000	
A	12	0.13	368,000	532,000	Repeated pulse
B	29	0.08	248,000	495,000	injections
C	44	0.08	163,000	327,000	($n_{mt} = 25$)
D	47	0.07	175,000	408,000	

Source: Refs. 22 and 23.

at different flow rates, with a monoclonal antibody column. The mean value ($1/n_{mt} = 0.037$) was used to calculate the effective association rate constant k_a for the biospecific interaction at the gas–liquid interface. When accounting for this corrective term, k_a values for HSA adsorption on the immobilized monoclonal antibody (Table 1) were calculated from k'_a measurements at different flow rates (mean value: $k_a = 2.4 \times 10^5$ dm^3·mol^{-1}·s^{-1}). The effective association rate constant is about twice as large for HSA adsorption on the polyclonal anti-HSA immunoadsorbent (Table 2).

In contrast with the results of Fig. 5, important deviations between the theoretical profile and the experimental one are observed for adsorption studies on the polyclonal antibody [23], even at low HSA feeding concentrations. The frontal elution model given by Eq. (7) with $k'_d = 0$ correlates well with the first part of the breakthrough curve, but later a deviation is observed even at very low feeding HSA concentrations. In this case, the simplified model assuming a uniform adsorbent surface is not appropriate. The polyclonal antibody is made of different populations of antibodies of various affinities. With polyclonal immunoadsorbents, the values of k_a in Table 2 are to be considered as apparent adsorption rates.

2. Repeated Pulse Injection Mode

In this approach, repeated injections of small HSA amounts are made into the immunoadsorbent column at intervals of a few minutes [22]. After column saturation, the regeneration is achieved with a washing step that uses an acidic buffer eluent. A second adsorption experiment is performed after equilibrating the column with the initial buffer. The model describing the split-peak effect in nonlinear chromatography [20] was used to analyze such experiments [Eq. (19)].

In this equation, the total injected amount is Q_i, and the nonretained fraction f is calculated from the total amount irreversibly adsorbed. By this method, the variation of f vs. Q_i is obtained, without any need of column regeneration after each injection.

Figure 6 illustrates the progressive overloading of an anti-HSA polyclonal antibody column after repeated injections of 2 μg of HSA. At first injections, impurities elute from the column at the dead volume, while HSA is totally adsorbed. The gradual emergence of the nonretained HSA elution peak is due to two different effects, the saturation of the support and the slow adsorption kinetic process. The unretained fraction is calculated from peak area measurements, subtracting the area of the impurity response peak.

Table 2 lists the two parameters n and Q_x necessary to describe the model as determined with columns differing by the density of immobilized polyclonal antibody. As previously described, from the variation of the column capacity one can evaluate the contribution to the transport to the binding sites ($1/n_{mt} = 0.040$) and calculate the effective adsorption rate constant k_a. The results agree with those obtained from frontal analysis. The value of the apparent adsorption rate constant k'_a is close to the value of k_a for experiments carried out both at high flow rates and with an immunoadsorbent column of low capacity [22]. In this case, the rate-controlling step is the biospecific interaction.

The results of Renard et al. [23] show that the effective adsorption rate constant can be determined either from the analysis of the breakthrough curves or from the repeated pulse injection mode. The advantage with the latter method is simplicity, because it is based on peak area measurements, with minute amounts of protein consumed. Another advantage is the standard HPLC instrumentation used for such experiments.

The repeated pulse injection method was applied to determine the association rate constant of the antigen-immobilized monoclonal antibody reaction [24,25]. These monoclonal antibodies differ by their specificity and affinity for HSA. The epitope recognized by the HA6 antibody is located between residues 1 and 128. The epitope of the commercial antibody (m-anti-HSA) is located between residues 124 and 298 of the HSA molecule. They differ by their affinity for HSA, with equilibrium binding constants of respectively 6.7×10^8 mol·dm^{-3} and 1.6×10^8 dm^3·mol^{-1} for the HSA/HA6 and the HSA/m-anti-HSA association in solution [25]. Close binding rate constants are obtained on both immobilized monoclonal antibodies with $k_a = 2.5 \times 10^5$ dm^3·mol^{-1}·s^{-1}.

3. Analysis from the Split-Peak Effect

On the basis of the split-peak equations, Hage et al. [18] developed a model to interpret competitive binding immunoassays and applied the method to the determination of low levels of albumin in urine [47]. An equation was derived to predict the assay response as a function of flow rate, amounts of analyte and

Chromatographic Methods for Measurement

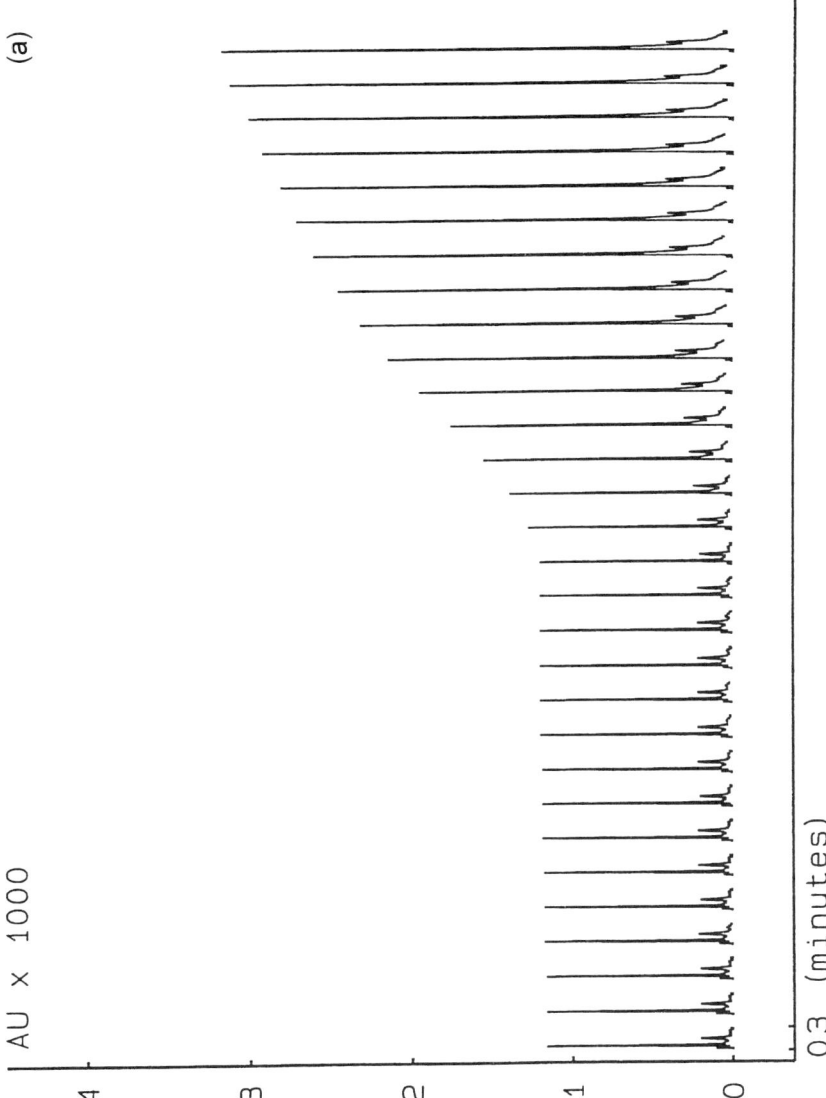

FIG. 6 Repeated injection of HSA on a polyclonal anti-HSA immunoadsorbent. Influence of flow rate. Same experimental conditions as in Fig. 5. Injected volume: 20 mm^3; HSA concentration: 0.1 g·dm^{-3}; flow rate: (a) 8.33 mm^3·s^{-1}; (b) 25 mm^3·s^{-1}. (From Ref. 22, with permission from Elsevier Science B.V.)

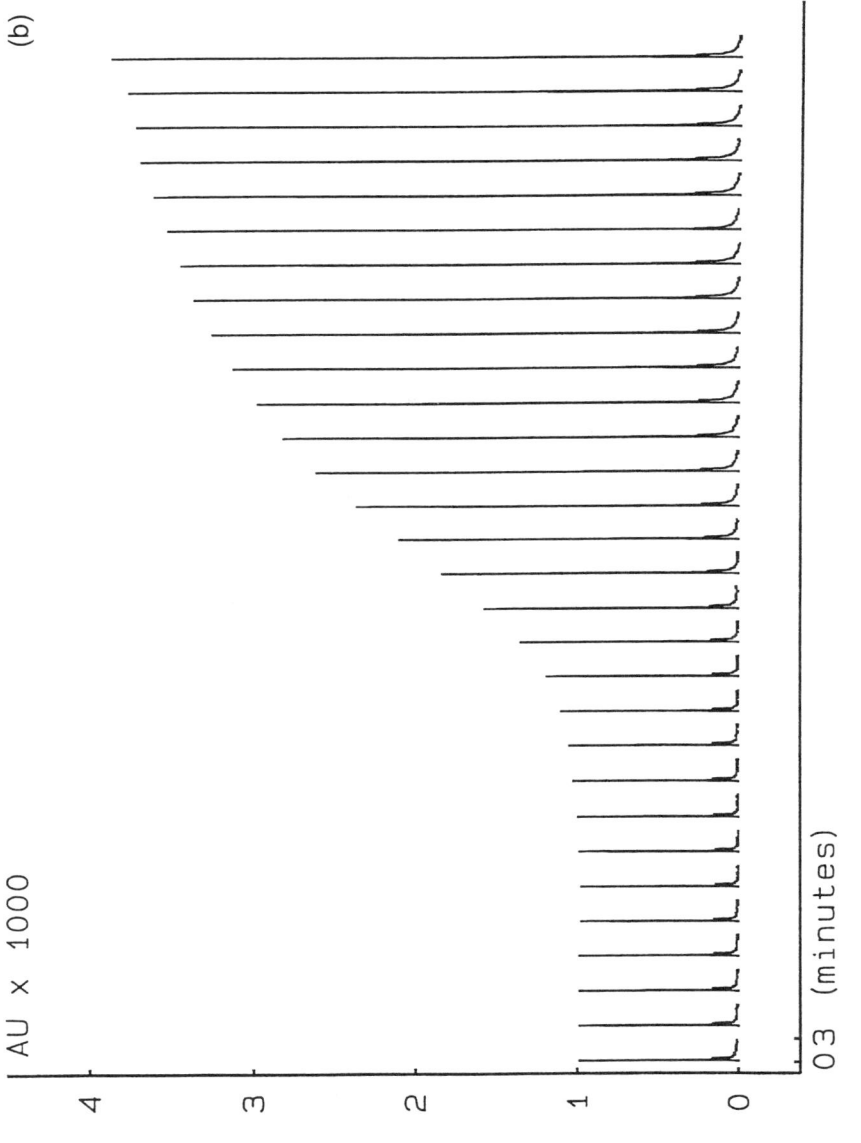

FIG. 6 *(Continued)*

label injected, and adsorption kinetics of the system. A porous silica matrix of 100 nm mean pore diameter was used for the polyclonal anti-HSA antibody immobilization.

It was shown [18] that the binding process is the rate-limiting step, with an adsorption rate constant of $k_a = 4.8 \times 10^4$ dm^3·mol^{-1}·s^{-1}. The calculated mass transfer coefficient for the diffusion into the pores of the support contributes 11% to the overall adsorption process. The value of k_a is an order of magnitude lower than that reported in Table 2. The binding properties of the polyclonal immunoadsorbents used in these two studies may differ because of the different methods employed for protein immobilization. Another possible explanation may be an underestimation of the contribution for the diffusion rate-limiting step as the polyclonal anti-HSA antibody was attached to a silica matrix of large pores [18].

V. CONCLUSION

Sophisticated mathematical models based on the numerical simulation of the chromatographic process consider different kinetic and thermodynamic mechanisms [19]. The theoretical approaches describe the biospecific adsorption of monovalent and multivalent adsorbates. They also account for the film mass transfer and pore diffusion contributions to the adsorption process and can be applied to analyze various complex experimental situations. Thus, ideally, the appropriate model will have to be selected to describe the actual chromatographic system.

There are several important difficulties in the application of such numerical models to analyze the chromatographic data for extracting the kinetic parameters. First, the computer package needed to describe adequately the chromatographic process has to be available. Second, long computer times are often required to fit the theoretical model to the experimental profiles. Third, with many variables used in a model, any combination may fit well the experimental measurements but may not be adequate to represent the physical reality of the process. Therefore it is important to design a chromatographic system that permits simplifications for the applicability of the model.

The main problem in the determination of association rates at the gas–liquid interface is the interplay of the mass transport effects and the biospecific sorption process. The experimental studies show that both effects are involved in the binding of antigen to the antibody attached to a surface. The variations of the value of the apparent adsorption rate constant with various experimental conditions reveal the importance of the nonideal effects in such experiments. To determine the effective rate of interaction, it is important both to minimize the diffusion resistances and to estimate this contribution by increasing the amount of information. Studies with varying flow rates, particle sizes, ligand densities, and immobilization methods will permit us to identify the various contributions to the mass transfer process.

The applications of the split-peak effect to adsorption rate studies enables one to perform such experiments rapidly. This approach can greatly help to determine the governing kinetic mechanisms of the antigen–antibody reaction at the gas–liquid interface. The expression of the number of mass transfer units based on simplified assumptions reveals important guidelines for designing experimental systems that permit one to minimize the transport effects in the overall adsorption process. As shown by Eq. 14, a decrease in the column capacity will increase the contribution to binding kinetics. This trend was observed in the above studies in which the lower was the density of immobilized ligand the larger was the apparent adsorption rate constant. The effect of flow rate is more difficult to predict as the film mass transfer term is also dependent on the flow rate.

HPLC is ideally suited to examine adsorption kinetics in the working conditions of column immunoassays. In this technique, high flow rates and minimized column volumes are required to perform rapid on-line immunodetections. The column capacities and residence times in the column are parameters that influence the efficiency of the immunoreactor. Kinetic studies using the chromatographic format will be useful to understand the process better and optimize the methodology.

In the kinetic studies of the adsorption process, the mass transport of the analyte to the binding sites is an important parameter to account for. Several theoretical descriptions of the chromatographic process are proposed to overcome this difficulty. Many complementary experiments are now needed to ascertain the kinetic measurements. Similar problems are found in the applications of the surface plasmon resonance technology (SPR) for association rate constant measurements. In both techniques the adsorption studies are carried out in a flow system, on surfaces with immobilized ligands. The role of the external diffusion limitations in the analysis of SPR assays has often been mentioned, and the technique is yet considered as giving an estimate of the adsorption rate constant. It is thus important to correlate the SPR data with results obtained from independent experiments, such as those from chromatographic measurements.

Immunochromatography is a convenient method for analyzing the kinetics of the antigen–antibody reaction at the gas–liquid interface. The technique, which uses a standard HPLC instrumentation, is relatively inexpensive and can easily be automated. Sophisticated theoretical models exist to analyze various experimental systems. They also could be applied to validate the simplified approaches. Some results show that it is possible to select the experimental conditions for which the kinetic limitations due to the transport of the protein to the binding sites can be considered negligible. More work is required to optimize the method and validate this approach.

REFERENCES

1. C. J. van Oss and D. R. Absalom, in *Molecular Immunology* (M. Z. Atassi, C. J. van Oss, and D. R. Absolom, eds.), Marcel Dekker, New York, 1984, p. 337.
2. B. Friguet, A. F. Chaffotte, L. Djavadi-Ohaniance, and M. E. Goldberg. J. Immunol. Methods 77 (1985): 305.
3. M. E. Goldberg and L. Djavadi-Ohaniance. Current Opinion in Immunology 5 (1993): 278.
4. M. Stenberg and H. Nygren. J. Immunol. Methods 113 (1988): 3.
5. R. Karlsson, A. Michaelsson, and L. Mattsson. J. Immunol Methods 145 (1991): 229.
6. I. M. Chaiken. Anal. Biochem. 97 (1979): 1.
7. R. R. Walters, in *Analytical Affinity Chromatography* (I. M. Chaiken, ed.), CRC Press, Boca Raton, FL, 1987, p. 117.
8. A. Jaulmes and C. Vidal-Madjar, *Adv. Chromatogr.* 28 (J. C. Giddings, E. Grushka, and P. R. Brown, eds.), Marcel Dekker, New York, 1989, p. 1.
9. I. M. Chaiken, S. Rose, and R. Karlsson. Anal. Biochem. 201 (1991): 197.
10. G. Zeder-Lutz, D. Altschuh, H. M. Geysen, E. Trifilieff, G. Sommermeyer, and M. H. V. Van Regenmortel. Mol. Immun. 30 (1993): 145.
11. P. R. Edwards, A. Gill, D. V. Pollard-Knight, M. Hoare, P. E. Buckle, P. A. Lowe, and R. J. Leatherbarrow. Anal. Biochem. 231 (1995): 210.
12. D. J. O'Shannessy. Current Opinion in Biotechnology 5 (1994): 65.
13. G. W. Oddie, L. C. Gruen, G. A. Odgers, L. G. King, and A. A. Kortt. Anal. Biochem. 244 (1997): 301.
14. D. J. O'Shannessy and D. J. Winzor. Anal. Biochem. 236 (1996): 275.
15. M. de Frutos and F. E. Regnier. Anal. Chem. 65 (1993): 17A.
16. H. A. Chase. Chem. Eng. Sci. 39 (1984): 1099.
17. S. A. Cassidy, L. J. Janis, and F. E. Regnier. Anal. Chem. 64 (1992): 1973.
18. D. S. Hage, D. T. Thomas, and M. S. Beck. Anal. Chem. 65 (1993): 1622.
19. A. I. Liapis. Sep. Purif. Methods 19 (1990): 133.
20. A. Jaulmes and C. Vidal-Madjar. Anal. Chem. 63 (1991): 1165.
21. D. S. Hage, R. R. Walters, and H. W. Hethcote. Anal. Chem. 58 (1986): 274.
22. J. Renard and C. Vidal-Madjar. J. Chromatogr. A661 (1994): 35.
23. J. Renard, C. Vidal-Madjar, and C. Lapresle. J. Colloid. Interface Sci. 174 (1995): 61.
24. J. Renard, C. Vidal-Madjar, B. Sebille, and C. Lapresle. J. Molecular Recognition 8 (1995): 85.
25. C. Vidal-Madjar, A. Jaulmes, J. Renard, D. Peter, and P. Lafaye. Chromatographia 45 (1997): 18.
26. C. M. Yang and G. T. Tsao. Adv. Biochem. Eng. 25 (1982): 1.
27. B. Lin, S. Golshan-Shirazi, and G. Guiochon. J. Phys. Chem. 93 (1989): 3363.
28. E. Kucera. J. Chromatogr. 19 (1965): 237.
29. C. Horváth and H. J. Lin. J. Chromatogr. 126 (1976): 401.
30. C. Horváth and H. J. Lin. J. Chromatogr. 149 (1978): 43.
31. F. H. Arnold, H. W. Blanch, and C. R. Wilke. Chem. Eng. J. 30 (1985): B9.
32. H. A. Chase. J. Chromatogr. 297 (1984): 179.

33. B. H. Arve and A. I. Liapis. Biotechnol. Bioeng. *32* (1988): 616.
34. B. H. Arve and A. I. Liapis. AIChE J. *33* (1987): 179.
35. B. J. Horstmann and H. A. Chase. Chem. Eng. Res. Des. *67* (1989): 243.
36. M. A. McCoy and A. I. Liapis. J. Chromatogr. *548* (1991): 25.
37. Q. M. Mao, A. Johnston, I. G. Prince, and M. T. W. Hearn. J. Chromatogr. *548* (1991): 147.
38. H. C. Thomas. J. Am. Chem. Soc. *66* (1944): 1664.
39. N. K. Hiester and T. Vermeulen. Chem. Eng. Prog. *48* (1952): 505.
40. F. H. Arnold and H. W. Blanch. J. Chromatogr. *355* (1986): 13.
41. F. H. Arnold, H. W. Blanch, and C. R. Wilke. J. Chromatogr. *330* (1985): 159.
42. S. Goldstein. Proc. R. Soc., ser. A., *219* (1953): 151.
43. H. W. Hethcote and C. DeLisi. J. Chromatogr. *248* (1982): 183.
44. J. C. Giddings and H. Eyring. J. Phys. Chem. *59* (1955): 416.
45. F. C. Denizot and M. A. Delaage. Proc. Natl. Acad. Sci. USA *72* (1975): 4840.
46. R. R. Walters. Anal. Chem. *55* (1983): 1395.
47. P. F. Ruhn, J. D. Taylor and D. S. Hage. Anal. Chem. *66* (1994): 4265.
48. J. X. Huang and C. Horváth. J. Chromatogr. *406* (1987): 275.
49. M. Hanson and K. K. Unger. LC-GC Int. *9* (1996): 741.
50. A. I. Liapis, B. Anspach, M. E. Findley, J. Davies, M. T. W. Hearn, and K. K. Unger. Biotechnol. Bioeng. *34* (1989): 467.
51. P. O. Larsson, M. Glad, L. Hansson, M. O. Manson, S. Ohlson, and K. Mosbach, *Adv. Chromatogr.* 21 (J. C. Gidding, E. Grushka, and P. R. Brown, eds.), Marcel Dekker, New York, 1983, p. 41.
52. K. Ernst-Cabrera and M. Wilchek. J. Chromatogr. *397* (1987): 187.
53. P. F. Ruhn, S. Garver, and D. S Hage. J. Chromatogr. *669* (1984): 9.
54. S. L. Fowell and H. A. Chase. J. Biotechnol. *4* (1986): 1.
55. M. A. McCoy, B. J. Hearn, and A. I. Liapis. Chem. Eng. Comm. *108* (1991): 225.
56. J. R. Sportsman and G. S. Wilson. Anal. Chem. *52* (1980): 2013.
57. J. R. Sportsman, J. D. Liddll, and G. S. Wilson. Anal. Chem. *55* (1983): 771.

12
The Acid–Base Behavior of Proteins Determined by ISFETs

WOUTER OLTHUIS and PIET BERGVELD   MESA Research Institute, University of Twente, Enschede, The Netherlands

I.	Introduction	373
	A. Acid–base properties of proteins	374
	B. The ISFET	376
II.	Methods of Protein Determination Using ISFETs	378
	A. External pH step	379
	B. Constant current coulometry	385
	C. Sinusoidal current coulometry	387
	D. External ion step	394
III.	Conclusions	401
	References	402

I. INTRODUCTION

Proteins form the specific selector in many biochemical sensors. The applications of proteins fall into two groups according to the principle employed. One group of sensors employs the catalytic properties of proteins. The most successful and extensively studied proteins in this group are enzymes. The other group utilizes affinity reactions, of which the immunosensor based on the formation of an antibody/antigen complex is an example [1]. A unique feature of using proteins as a part of the sensor is the high selectivity, either in a catalytic reaction or in a molecular binding (recognition).

The sensors that incorporate catalytic proteins measure the concentration change in substrate or products rather than the proteins themselves. The sensors that use proteins as binding ligands mostly detect a change in some intrinsic property of the protein itself such as charge density, molecular weight, or refractive index [2]. The transducer used in these cases can be an electrochemical or an optical one. As has been pointed out before [3], a practical problem that needed to be solved in the construction of a biosensor was the effective transfer of the highly specific signal, either from an enzymatic reaction or from immunological binding, into a sensor-detectable one. The pH ISFET-based enzyme sensors (ENFETs) have actually employed the proton-associated detection [4]. However, this application is limited, since only a few enzymes, such as urease and glucose oxidase, can catalyze a reaction that gives rise to an associated pH change.

An alternative and possibly more effective way is the determination of the change in the acid–base property of the protein itself. If the binding of a substance to a protein can significantly change this behavior, the detection of this change enables the construction of a new type of biosensor. This chapter focuses on the use of ISFETs for the determination of the protein acid–base behavior.

A. Acid–Base Properties of Proteins

1. Amino Acids

Treating protein solely as large molecules with certain acid–base properties necessitates the localization of the groups responsible for these properties. Amino acids can be designated as the basic structural elements of any protein [5]. All the existing twenty common natural amino acids (except for proline) have similar structures, consisting of an amino group ($-NH_2$), a hydrogen atom, a carboxyl group ($-COOH$), and a side chain R, all bonded to a central carbon atom:

$$\begin{array}{c} NH_2 \\ | \\ H-C-COOH \\ | \\ R \end{array}$$

Apart from possible reactions of the R group, treated shortly, amino acids are neutral at intermediate pH (~ 7) but appear as dipoles (or zwitterions). The positive center of charge is formed by the protonated amino group ($-NH_3^+$), while the dissociated carboxyl group ($-COO^-$) forms the negative charge center of the dipolar form. At low pH, the carboxyl group becomes neutral, while the amino group remains ionized, whereas at high pH values the amino group of neutral and the carboxyl group stays negative:

The Acid–Base Behavior of Proteins

$$\underset{\text{low pH}}{\text{H}-\underset{\underset{R}{|}}{\overset{\overset{NH_3^+}{|}}{C}}-\text{COOH}} \leftrightarrow \underset{\text{intermediate pH}}{\text{H}-\underset{\underset{R}{|}}{\overset{\overset{NH_3^+}{|}}{C}}-\text{COO}^-} \leftrightarrow \underset{\text{high pH}}{\text{H}-\underset{\underset{R}{|}}{\overset{\overset{NH_2}{|}}{C}}-\text{COO}^-}$$

Of the twenty distinct side chains, seven have ionizable groups that contribute to the acid–base behavior. Apart from amino and carboxyl groups, these are sulfhydryl ($-SH$) and hydroxyl ($-OH$) groups. Due to the terminal carboxyl and amino groups as well as the ionizable groups in the side chains, pK values for the acid–base groups in amino acids range from ~ 2 to ~ 12.

2. Peptides

If the carboxyl group of one amino acid binds with the amino group of another, a water molecule is released and an amide linkage (also called a peptide bond) is formed between the two amino acids. The compound formed is known as a dipeptide, and the incomplete amino acid molecules in the peptide chain are called amino acid residues [5]. The general representation for this reaction is

$$\text{HNH}-\underset{\underset{R}{|}}{\text{CH}}-\text{COOH} + \text{HNH}-\underset{\underset{R'}{|}}{\text{CH}}-\text{COOH} \rightarrow$$

$$\text{NH}_2-\underset{\underset{R}{|}}{\text{CH}}-\text{CO}-\text{NH}-\underset{\underset{R'}{|}}{\text{CH}}-\text{COOH} + \text{H}_2\text{O}$$

If more amino acids bind in a series by such a peptide bond, they will form a peptide chain that is known as a polypeptide. A polypeptide, like an amino acid, has either a carboxyl group or an amino group at each end of the peptide chain. The molecule can also be positively of negatively charged at different pH solutions. However, a difference should be noticed with respect to a single amino acid molecule. Polypeptides typically have many ionizable groups on the side chains of the amino residues. The acid–base properties of the molecule are determined not only by the carboxyl and amino groups at the ends but also by the ionizable groups on these side chains. As the polypeptide chain becomes longer, the number of ionizable groups on the side chains increases, and the groups will play a more important role in the determination of the acid–base behavior. Consequently, a short peptide, e.g., a dipeptide or tripeptide, shows characteristics similar to those of the corresponding amino acids, while a long peptide has characteristics mainly determined by the sort and number of the ionizable groups on the side chains.

3. Proteins

Proteins are macromolecules that are composed of one or more polypeptides linked by secondary bonds such as disulfide bonds, hydrogen bonds, and covalent bonds, in a specific spatial conformation. This spatial conformation that refers to the secondary, tertiary, and quaternary structure of proteins is determined by their primary structure (this primary structure of a protein refers to the sequence of the amino residues in the peptide chains). Most of the physical and chemical properties of a protein are similar to those of the long constituent polypeptides [5]. As will be clear by now, proteins have many sites that are able to bind with or release protons.

The acid–base behavior of proteins can reveal some important properties with respect to both their composition (selectivity) and their concentration (sensitivity). The most direct way to exploit these acid–base properties is to make use of acid–base titration. Titrant should be added somehow and the resulting change in pH should be measured. Since the ion-sensitive field-effect transistor (ISFET) is suitable for fast (and local) pH detection, an ISFET can be used for protein titration if the protein to be detected can be immobilized in a membrane, deposited on top of the device. Advantages are the small amount of protein necessary for the characterization owing to the small membrane volume, and the relatively short time needed to perform a full titration.

B. The ISFET

The ion sensitive field-effect transistor (ISFET) is a special member of the family of potentiometric chemical sensors [6,7]. Like the other members of this family, it transduces information from the chemical into the electrical domain. Unlike the common potentiometric sensors, however, the principle of operation of the ISFET cannot be listed on the usual table of operation principles of potentiometric sensors. These principles, e.g., the determination of the redox potential at an inert electrode, or of the electrode potential of an electrode immersed in a solution of its own ions (electrode of the first kind), all have in common that a galvanic contact exists between the electrode and the solution, allowing a faradaic current to flow, even when this is only a very small measuring current.

The working principle of an ISFET is essentially different, which is evident from the name of this transducer: information is transferred via an electric field. As is known, the source of any static electric field is charge. The nature of this charge in our case is concealed in the first two letters of the acronym ISFET: ions form the source of the charge, of which the resulting electric field controls the electronic behavior of the transistor. It is important to observe that in this case no galvanic contact exists between the solution and the conducting part of the sensor, so there is no faradaic current.

The ISFET is today a well-known transducing element for the development of chemical sensors and biosensors. The transducing principle of an ISFET is based on the dependence of the drain current of the transistor on the surface charge

The Acid–Base Behavior of Proteins

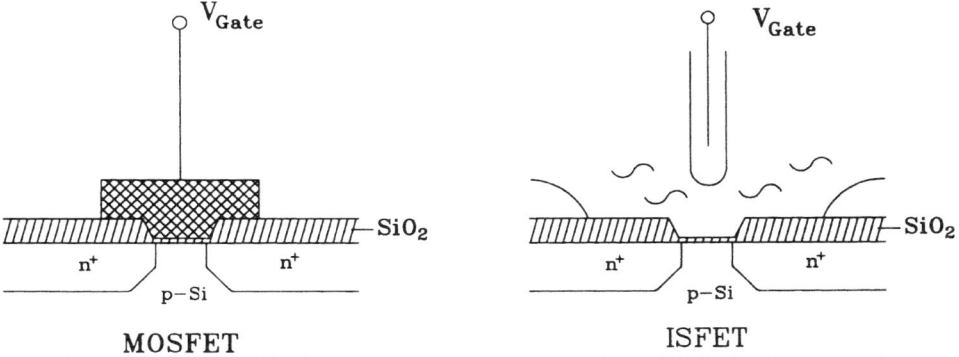

FIG. 1 Cross sections of a MOSFET and an ISFET.

of the inorganic gate insulator, which is in contact with an aqueous solution. The surface charge is determined by the pH of the solution in which the ISFET is immersed. The corresponding drain current is electronically converted to an output voltage of an amplifier [8].

The ISFET is deduced from the MOSFET, which is a well-known electronic device. Cross section of the MOSFET and the ISFET are shown in Fig. 1. A MOSFET consists of a p-type silicon substrate in which two n-type diffusions are realized, which are called the source and the drain. The structure is covered with an insulating layer (usually SiO_2), and a metal gate electrode is deposited over the area between the source and the drain.

In normal operation, the drain is maintained at a voltage V_D with respect to the source, which is normally at 0 V. When a positive voltage V_G is applied to the gate with respect to the substrate (which is also normally at 0 V), and this voltage exceeds the so-called threshold voltage V_T, electrons (the minority carriers in the substrate) are attracted to the surface and create a conducting channel between the source and the drain. The current I_D, which consequently flows between source and drain, is determined by the electrical resistance of the conducting channel (determined by V_G) and the magnitude of V_D. For the condition that $V_D < (V_G - V_T)$, the drain current is given by

$$I_D = \frac{C_{ox}\mu W}{L}\left(V_G - \frac{1}{2}V_D - V_T\right)V_D \tag{1}$$

where C_{ox} is the gate insulator capacitance per unit area, μ the electron mobility in the channel, and W/L the width-to-length ratio of the channel.

The expression for the drain current is also valid for the ISFET, which in fact is a MOSFET without a metal gate, immersed in an aqueous solution. The reference electrode, which defines the potential of the solution in which the ISFET is immersed, has the same function as the metal gate of a MOSFET. However, the threshold voltage V_T of a MOSFET is constant, whereas the threshold voltage of an ISFET is influenced by the surface potential ψ_0 that exists at the oxide surface–solution interface. The acid–base equilibria at the oxide surface makes the surface potential ψ_0 a function of the pH of the solution [7,9].

If an ISFET is connected to a so-called source-drain follower, a change in the threshold potential V_T, resulting from a change in the surface potential ψ_0 (which is caused by a pH change), is compensated by an equal change of the gate-source voltage, which is the output of the source-drain follower. The final result is an output voltage V_{GS} which has a Nernstian dependence on the pH of the solution:

$$V_{GS} = \alpha \frac{RT}{F} \ln a_{H^+} + \text{constant} \tag{2}$$

where a_{H^+} represents the activity of the protons in the solution and α the sensitivity factor of the ISFET ($0 < \alpha \leq 1$), which is 1 for maximum Nernstian behavior. The constant reflects, among other constants, the reference electrode potential.

The transducing principle of an ISFET makes it an interesting device for the development of chemical sensors. The most direct application is of course the direct use of the ISFET as a pH sensor. As described previously, the surface potential at the interface between the gate oxide and the solution is determined by the acid–base equilibria of the OH groups and is therefore a function of the pH [7,9]. For the development of pH-sensitive ISFETs, several gate oxides have been investigated, including Si_3N_4, Al_2O_3, and Ta_2O_5 [8]. These oxides were believed to have better properties then SiO_2, which was initially used, and were deposited on top of the initial SiO_2 layer. It appears that Ta_2O_5 has the best properties and thus gives a pH sensitivity factor α of almost 1 and thus a pH sensitivity of -59 mV/pH at $25°C$ [9]. Today, several pH meters, based on a pH-sensitive ISFET, are commercially available. Most of them use Al_2O_3 or Si_3N_4 as gate oxide.

II. METHODS OF PROTEIN DETERMINATION USING ISFETs

In this section, four quite different methods of protein determination are described. Some of these methods mainly give qualitative information, e.g., a protein-specific signature of the solution, while others directly yield quantitative data about the protein to be detected. All methods have in common that they somehow rely on the acid–base properties of proteins and use (modified) ISFETs as transducing elements.

A. External pH Step

A stepwise change in pH can be applied outside a protein-containing membrane applied on top of an ISFET, at the interface between the membrane and the electrolyte, using a flow-through system [10]. This pH step will lead to simultaneous diffusion and chemical reaction of protons and hydroxyl ions in the membrane. A theoretical description of these phenomena, elaborated in the next subsection, leads to the conclusion that the diffusion of protons in the membrane is delayed by a factor that depends linearly on the protein concentration. Consequently, the time needed to reach the end point in the obtained titration curve also depends linearly on the protein concentration. The effect of both the incubation time and the protein concentration will be simulated and experimentally verified.

1. Theory

After a stepwise change in pH at the interface between the protein-containing membrane and the surrounding electrolyte solution, both protons and hydroxyl ions will start to diffuse through the membrane in opposite directions. (We assume 1 M KNO_3 solutions of pH = 4.0 and 10.0 without addition of any pH-buffering components). These diffusion processes will be delayed as a result of protein dissociation reactions of immobilized protein molecules. For protons this delayed diffusion process can be described by a modified form of Fick's second law of diffusion [11]:

$$\frac{\partial c_{H^+}}{\partial t} = D_{H^+} \cdot \frac{\partial^2 c_{H^+}}{\partial x^2} - \frac{\partial c_{HProt}}{\partial t} \tag{3}$$

where c_{H^+} is the concentration of mobile protons in the membrane, D_{H^+} is the diffusion coefficient of these protons in the membrane, and c_{HProt} is the total concentration of protonated groups in the membrane. When only diffusion in one dimension is taken into account, all concentrations are a function of time t and distance x from the ISFET surface. According to Eq. (3), the change in the proton concentration at a location x in the membrane e is a function of both diffusion (first term on the right-hand side) and proton dissociation reactions of immobilized protein molecules (second term on the right-hand side). This last term can be written as

$$\frac{\partial c_{HProt}}{\partial t} = \frac{\partial c_{HProt}}{\partial pH} \cdot \frac{\partial pH}{\partial c_{H^+}} \cdot \frac{\partial c_{H^+}}{\partial t} = \beta_{Prot} \cdot \frac{1}{2.3 c_{H^+}} \cdot \frac{\partial c_{H^+}}{\partial t} \tag{4}$$

where β_{Prot} is the buffer capacity of the immobilized protein and $2.3 c_{H^+}$ that of the supporting electrolyte. For β_{Prot} we can write

$$\beta_{Prot} = c_{Prot} \cdot f(c_{H^+}) \tag{5}$$

where c_{Prot} represents the immobilized protein concentration and f is a complex function that is independent of this protein concentration but dependent on c_{H^+}

and on reaction constants of protein groups. If we assume that $\beta_{Prot} \gg 2.3 c_{H^+}$, then combination of Eqs. (3), (4), and (5) yields

$$\frac{\partial c_{H^+}}{\partial t} = \frac{1}{c_{Prot}} \cdot \frac{D_{H^+}}{f'(c_{H^+})} \cdot \frac{\partial^2 c_{H^+}}{\partial x^2} \tag{6}$$

Eq. 6 shows that the diffusion of protons in the membrane is delayed by a factor that depends linearly on c_{Prot}. Diffusion of hydroxyl ions through the membrane can be described in a similar way. Thus the time needed to obtain a certain change ΔpH at the ISFET surface always depends linearly on c_{Prot}.

2. Experimental

ISFETs with a Ta_2O_5 gate insulator, having a sensitivity of -58.5 mV/pH, were fabricated in the MESA clean room following the usual processing steps. After mounting the chips on a piece of printed circuit board, they were encapsulated using hysol epoxy, leaving a circular area around the gate with a diameter of 1.7 mm and a depth of 150 μm uncovered. The membranes, consisting of polystyrene beads ($\phi = 0.1 \mu$m) in an agarose gel with a thickness of 8–16 μm, were casted in this area [12]. Lysozyme (from chicken egg white, Sigma) was adsorbed into the membrane by incubation in solutions containing lysozyme (3E-7–3E-4 M) in 100 mM KNO_3, pH = 7.4. Except for the first series of experiments, the protein solution was stirred during the incubation, in order to reduce protein concentration gradients in the solution. After each incubation step the devices were rinsed for one minute and mounted in the wall-jet cell of a flow-through system in which a pH step was applied in 0.1 s as shown in Fig. 2 [13]. All experiments were performed in 1 M KNO_3, using HNO_3 and KOH to adjust the pH to respectively 4.0 and 10.0. After mounting the devices in the wall-jet cell, they were rinsed with the pH = 4 solution for 2 min before the pH step was applied. The ISFET response, measured as the output voltage V_S of a source-drain follower, was recorded with a Nicolet 310 digital oscilloscope.

3. Results and Discussion

(a) *Simulation Results.* A one-dimensional simulation model based on the Nernst–Planck and Poisson equations [14], in which all the acid–base reactions occurring in the membrane are taken into account, has been used to give a qualitative description of the pH step titration process. In these simulations, a pH step is applied outside a 2 mm thick stagnant layer, which is assumed to be present in front of an 8 mm thick membrane. Diffusion coefficients in the membrane are assumed to be 4/10 of those in water (this value is based on experience with ion step experiments). Lysozyme, used as a model protein, is assumed to contain 11 carboxylic groups ($pK_a = 4.4$), 2 imidazole groups ($pK_a = 6.0$), and 9 amino groups ($pK_a = 10.4$) per molecule. Concen-

The Acid–Base Behavior of Proteins

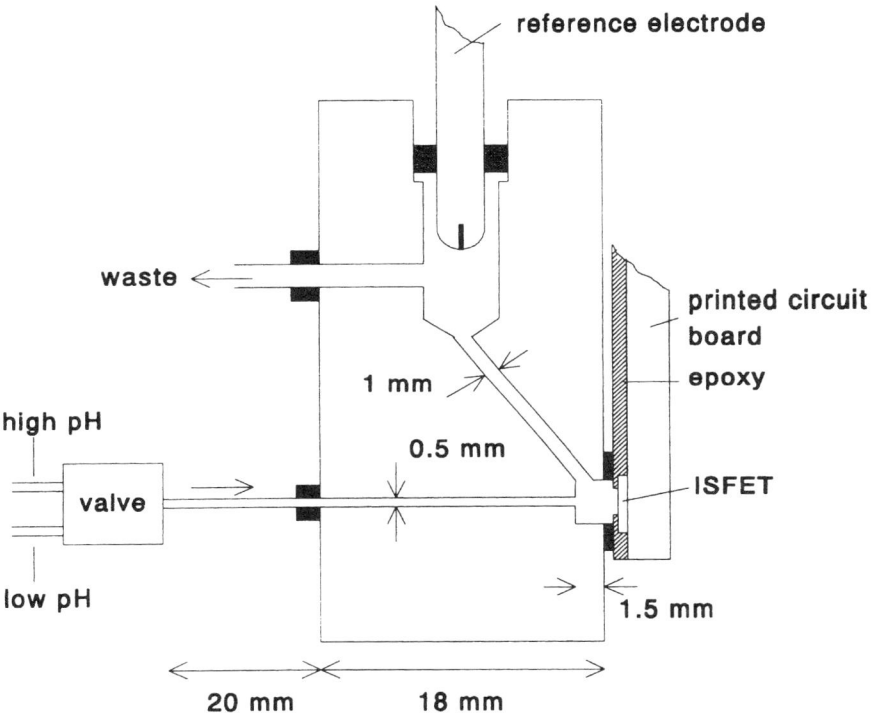

FIG. 2 Cross section of the wall-jet cell.

tration profiles of all species involved in the titration process at any moment in time as well as the pH at the ISFET surface as a function of time can be calculated.

As an example, the ISFET response to a pH step from 4 to 10 applied outside an 8 mm thick membrane containing lysozyme in equilibrium with a 1 M KNO_3 electrolyte solution is simulated. It should be noted that not only mobile protons react with hydroxyl ions but also protons that dissociate from acidic and basic lysozyme groups. These dissociation reactions strongly delay the titration process.

Figure 3 shows the ISFET response as a function of time for different concentrations of lysozyme present in the membrane. The dashed line indicates the ISFET response corresponding to pH = 7.0. Clearly, an increasing concentration of lysozyme immobilized in the membrane corresponds with a stronger delay in ISFET response. Figure 4 shows the linear relation between the time needed to reach pH = 7.0 at the ISFET surface, defined at t_{pH7}, and the concentration

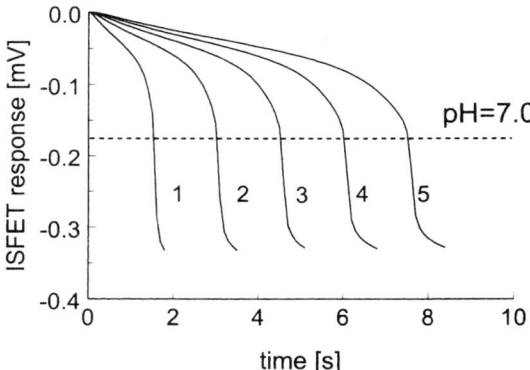

FIG. 3 Simulated ISFET response to a pH step from pH = 4 to 10 as a function of time for membranes containing 1.14 (1), 2.28 (2), 3.42 (3), 4.56 (4), and 5.7 mM (5) of lysozyme. The horizontal dashed line represents the ISFET response at pH = 7.0.

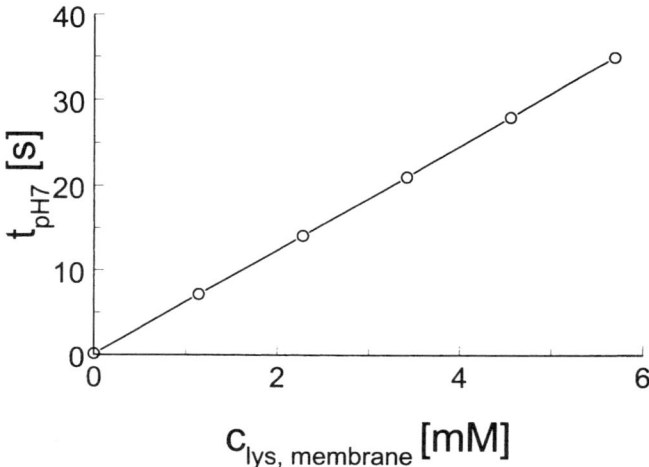

FIG. 4 Relation between t_{pH7} and the concentration of lysozyme immobilized in the membrane, found by simulations.

The Acid–Base Behavior of Proteins

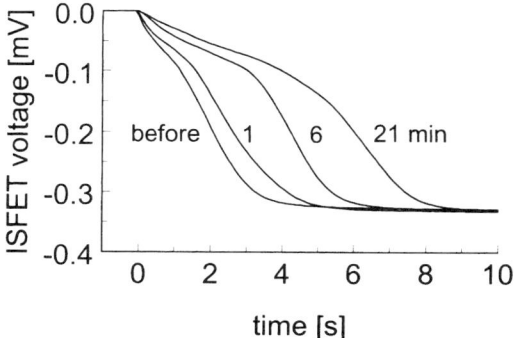

FIG. 5 Typical titration curves measured with the pH step method (from pH = 4 to 10) before and after the mentioned number of minutes of incubation in $3 \cdot 10^{-5}$ M lysozyme.

of immobilized lysozyme, c_{Prot}, that can be derived from these responses. This linear relation is in agreement with Eq. (6).

The simulation results presented show clearly the dependence of the ISFET response on the concentration of protein immobilized in a membrane on top of the device. Since the number and type of proton dissociating groups determine the protein buffer capacity (dc_{HProt}/d_{pH}), and since the ISFET response directly reflects the change in pH as a function of time, in principle the shape of an ISFET response also contains information on the type of protein that is immobilized.

Note that, since in practical experiments the concentration of immobilized protein after an incubation step is related to the protein concentration in the incubation solution through a complex adsorption process, the measured relation between t_{pH7} and protein concentration is not expected to be linear in that case.

(b) Experimental Results. Figure 5 shows typical ISFET responses to a step from pH = 4 to 10, as a function of the cumulative incubation time in a $3 \cdot 10^{-5}$ M lysozyme solution. A time shift, resulting in a clear dependence of t_{pH7} (intersection with dashed line) on the incubation time in a lysozyme solution is found. Since $t_{pH7} \gg 50$ ms for an ISFET without a membrane, the value $t_{pH7} = 1.85$ s found before incubation(curve 1) indicates that the bare membrane already contains a certain amount of buffering components. This is probably owing to the presence of carboxylic groups on the polystyrene beads. Since a measured value for t_{pH7} can be regarded as the sum of the time needed for the titration of the bare membrane plus the time needed for the titration of the immobile protein, we define a Δt_{pH7} as the shift in t_{pH7} as a result of incubation in a protein solution. Figure 6 shows Δt_{pH7} as a function of the cumulative incubation time for incubation in 3E-6, 3E-5, and 3E-4 M lysozyme solutions.

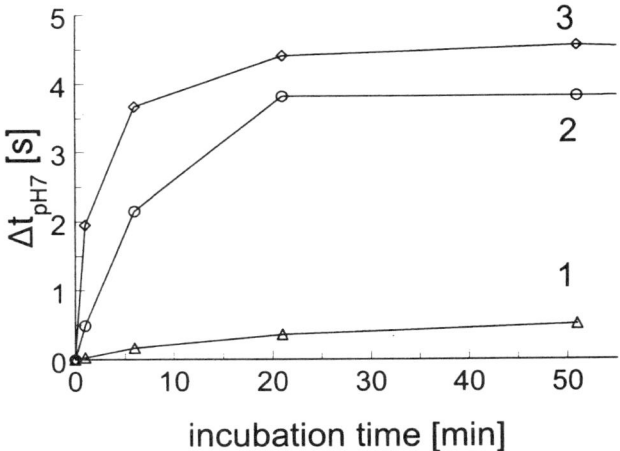

FIG. 6 Δt_{pH7} measured as a function of the cumulative incubation time in a 3E-6 (1), 3E-5 (2), and 3E-4 M (3) lysozyme solution.

After this first series of experiments we incubated six devices in six different lysozyme solutions for 30 min. The results shown in Fig. 7 show a clear dependence of Δt_{pH7} on the lysozyme concentration in the sample solution. We stirred during the 30 min incubation process in order to reduce protein concentration gradients in the solution. A different membrane composition may further im-

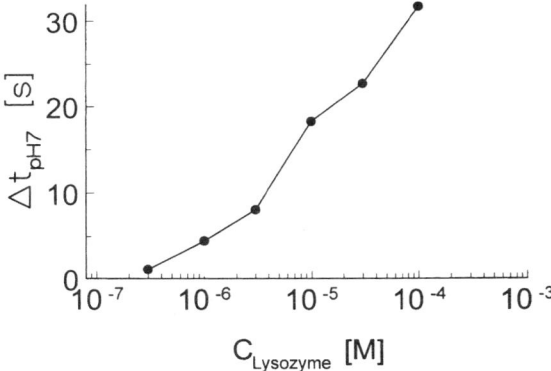

FIG. 7 The measured relation between Δt_{pH7} and concentration of lysozyme in the incubation solution for a pH step from pH = 4 to 9 and an incubation time of 30 min.

The Acid–Base Behavior of Proteins

prove the protein diffusion in the membrane and hence shift the response curve of Fig. 7 to smaller protein concentrations.

These results show that without any optimization of membrane composition, measurement of lysozyme in concentrations ranging from 3E-7 to 3E-4 M is possible with the pH step titration method. In order to lower the detection limit and to improve the device-to-device reproducibility, other membrane materials may have to be investigated. Especially for selective protein detection, requiring specific receptor molecules deposited in the membrane, more hydrophilic membrane materials are probably necessary, since the hydrophobic polystyrene beads are known to adsorb easily proteins aspecifically.

B. Constant Current Coulometry

As an alternative for the external application of a stepwise change in pH, as described above, titrant can be added by coulometry. Coulometric generation of protons or hydroxyl ions is possible by sending a current between a noble metal actuator electrode, situated closely around the ISFET gate, and a distant counter electrode [15].

The measured pH as a function of time results in a titration curve, of which the time t_{end} needed to reach the end point of this titration is related to the type and concentration of protein immobilized in the membrane, since the pH change in the membrane is retarded by the buffer action of the immobilized protein [16].

1. Theory

The material for the actuator has to be selected on the basis of the titrant to be generated. For the titration of an acid or base, a noble metal electrode is usually adopted. A constant current is often used in the coulometric titration, and the quantity of charge is then calculated by the time of the electrolysis. The analyte can either be an acid or a base. If there are no other interfering redox couples, the titrant generated at the electrodes depends on the direction of the applied current:

anode: $2H_2O \rightarrow 4H^+ + 4e + O_2$
cathode: $2H_2O + 2e \rightarrow 2OH^- + H_2$

The titration of an acid or base can be carried out by choosing a cathodic or anodic current, respectively. The ions produced cause a local change in the pH, which can easily be measured by the pH-sensitive ISFET, located in the direct vicinity of the actuator electrode [15]. This change in pH will lead to simultaneous diffusion and chemical reaction of protons and hydroxyl ions in the membrane. These diffusion processes will be delayed as a result of protein dissociation reactions of immobilized protein molecules [10]. The description of the diffusion and the effect of the concentration of immobilized protein on the

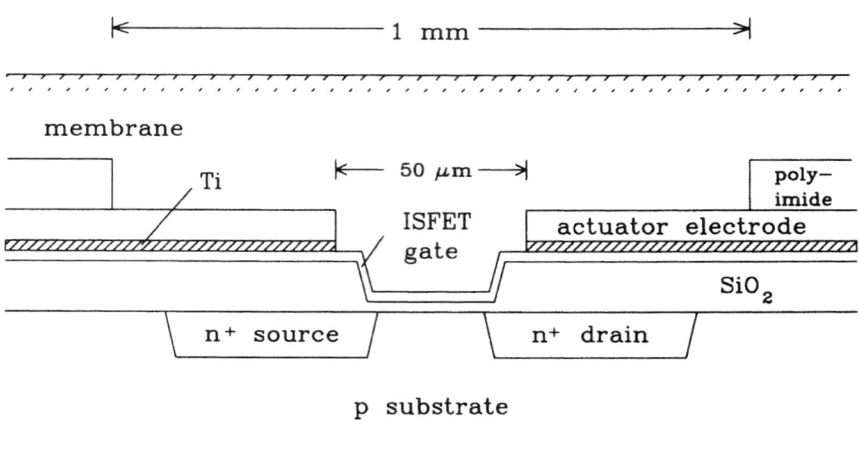

FIG. 8 Cross section of the measuring device with the actuator electrode deposited around the ISFET gate and the membrane covering both. The typical thickness of the polystyrene beads/agarose membrane is 8 μm. The gate is the pH-sensitive part of the ISFET.

delayed mass transport is comparable to the one given in the previous Section A.1 and is not repeated here.

2. Experimental

The processing of the devices and the casting of the membranes is identical to the steps as described in Section A.2, except for the additional deposition of a platinum actuator electrode around the gate of the ISFET. A cross-sectional view of the membrane-covered ISFET with actuator is shown in Fig. 8. Also in this case, the membrane consists of an 8 μm thick layer of polystyrene beads with a diameter of 0.1 μm and agarose.

The setup for these types of measurements is shown in Fig. 9. After an incubation step in a protein-containing solution, the actual titration takes place in a protein-free unbuffered solution of 1 M KNO_3 at pH = 4, in order to avoid migration effects. After inserting the ISFET in the solution, we waited 2 min before starting the titration in order to minimize convection. A current of 10 $\mu A/mm^2$ was sent between the actuator and the counter electrode at $t = 0$ for 10 to 25 s.

3. Results and Discussion

The measured pH as a function of time results in a titration curve of which the time t_{end} needed to reach the end point of this titration is related to the type and

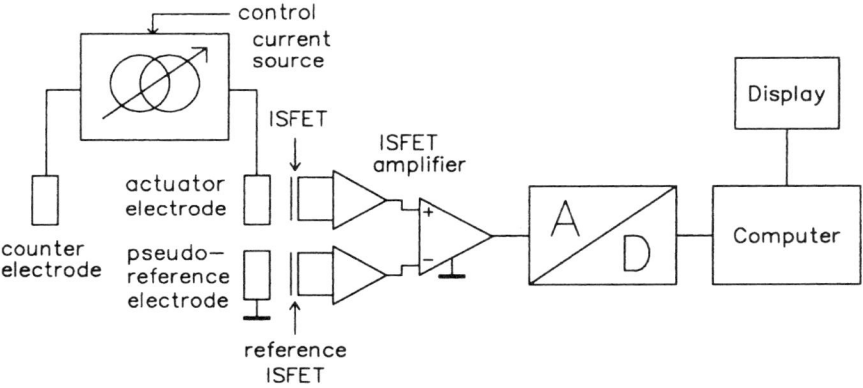

FIG. 9 Measurement setup. Note that by measuring differentially with respect to a second ISFET at which no coulometric titration is carried out, a conventional liquid-filled reference electrode is superfluous. The ISFET is provided with a membrane containing incubated protein.

concentration of protein immobilized in the membrane, since the pH change in the membrane is retarded by the buffer action of the immobilized protein [16]. Typical results showing the effect of the incubation time at a fixed concentration of lysozyme are shown in Fig. 10.

As expected, the end point in the titration curve shifts to higher values for longer incubation times, indicating a larger amount of lysozyme entering the membrane. The time needed to pass pH 7 as a function of the lysozyme concentration at fixed incubation time is shown in Fig. 11. In Fig. 11, the relation between this time and the lysozyme concentration is shown for two different values of the actuator current. The higher this current, the more titrant per second is generated, and consequently the shorter the time it takes to pass pH 7.

C. Sinusoidal Current Coulometry

The slope of the measured titration curve is inversely proportional to the buffer capacity of the protein, which contains specific information on the type of protein. After modification of the coulometric sensor-actuator device with a porous gold actuator electrode, the device is shown to be suitable for the determination of protein buffer capacity in solution [17]. A schematic representation of the device is shown in Fig. 12. The buffer capacity can be measured by sending a small alternating current through the porous actuator electrode, resulting in the alternating generation of protons and hydroxyl ions. This will result in a small pH perturbation in the free space of the porous electrode, which is detected by

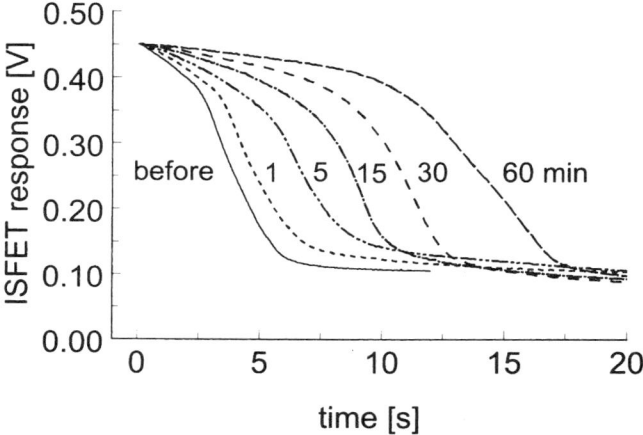

FIG. 10 Typical titration curves measured with a coulometric sensor-actuator device at 5 µA generating current before and after the mentioned number of minutes of incubation in $3 \cdot 10^{-5}$ M lysozyme.

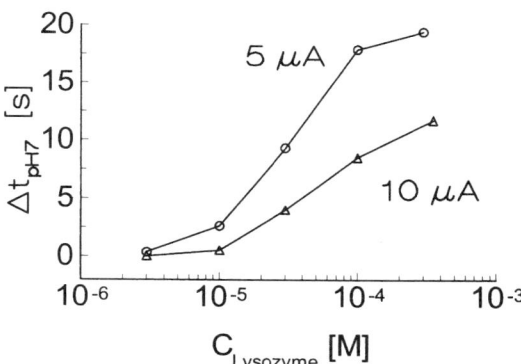

FIG. 11 The measured time needed to cross pH = 7 at the titration curve, t_{pH7}, minus t_{pH7} of the bare membrane as a function of the lysozyme concentration after 1 h incubation.

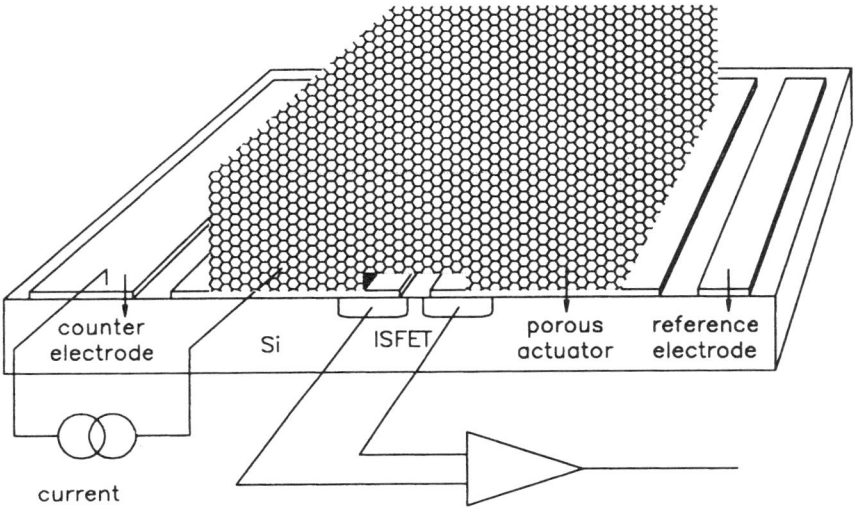

FIG. 12 Basic elements of the coulometric sensor-actuator system based on an ISFET with gate-covering porous actuator electrode. The thickness of the porous actuator is about 100 μm and the inner pore size is about 1 μm.

the underlying ISFET. The amplitude of the measured signal is a function of the buffer capacity of the solute, in which the protein is present. After a theoretical description of this process, experimental results will be shown.

1. Theory

Buffer capacity is by definition the reciprocal of the slope of the titration curve. Besides the conventional way that uses a complete titration, the buffer capacity can also be measured by applying a small perturbation of titrant at different pHs. This is illustrated in Fig. 13. The solid curved line shown in this figure is the titration curve of an acid. The abscissa represents the equivalents of the added base (left from zero) and acid (right from zero) to the original species. If a small amount of titrant is added alternatively as shown in this figure, the resulting pH changes will depend on the buffer capacity of the solution, i.e., the reciprocal of the slope of the titration curve at the pH concerned. The perturbation of titrant can be coulometrically generated and modulated pH can be detected by an ISFET. Both of these can be achieved by the ISFET-based coulometric sensor-actuator system with gate-covering porous actuator [18,19]. This operation of the system provides a dynamic way of measuring the buffer capacity.

(a) Mathematical Description. After the start of the coulometric generation of the titrant, the resulting diffusion is considered to be one-dimensional, which

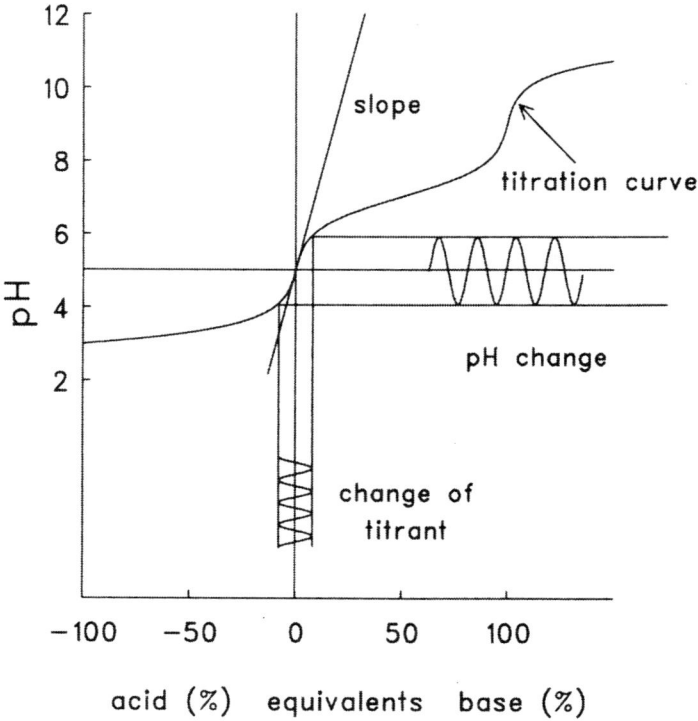

FIG. 13 Illustration of the buffer capacity measurement by means of an applied sinusoidal perturbation of the titrant. Basically, the method can be described by an electrochemo-electrical transfer function, parametrically dependent on the buffer capacity in the chemical domain.

is reasonable because the geometric area of the actuator is much larger than its thickness. With other assumptions stated elsewhere [18] the derivation of the model can proceed. First, a unit ρ_f, defined as the titrant generated per unit time and per unit volume, will be introduced for the description of the titrant source. For an applied sinusoidal current, the generation of titrant will exhibit a certain initial phase shift θ, because of the charge transfer and the mass transport from the actuator surface into the actuator pores. When this initial phase shift is taken into account, the applied sinusoidal current can be defined as

$$i = I \cos(\omega t - \theta) \tag{7}$$

where I is the amplitude and ω is the angle frequency of the applied current. Then the titrant production rate per unit volume, ρ_f, as has just been defined is

The Acid–Base Behavior of Proteins

$$\rho_f = \frac{I_f \cos[(\omega t - \theta) + \theta]}{F V_{\text{act}}} = \frac{I_f \cos \omega t}{FSl} \tag{8}$$

where I_f is the faradaic portion of the amplitude of the applied current I, F is the Faraday constant, V_{act} is the total volume of the porous actuator, and S and l are the geometric area and thickness of the porous actuator, respectively. Here the subscript f is used to denote the corresponding faradaic quantity, because only the faradaic current takes part in the electrode reaction.

The equations governing the mass transport of the electroactive species inside and outside the porous actuator with the corresponding initial and boundary condition are described and solved elsewhere [18]; they result in a general solution for $C_1(x, t)$, representing the concentration profile of the species inside the porous actuator. Since the detection is carried out be the gate of the ISFET, which is very closely located at the edge of the actuator ($x = 0$), only the solution of $C_1(0, t)$ is of interest. The steady-state response for $C_1(0, t)$ is

$$C_1(0, t) = C_0 + A \cos(\omega t + \varphi) \tag{9}$$

The response is sinusoidal with the same frequency ω as the applied current. Its amplitude A and phase shift φ are dependent on the frequency of the applied current, the diffusion coefficients of the active species, and the thickness of the porous actuator.

(b) *Buffer Capacity Dependent System Response.* Considering a buffer system consisting of a weak acid: $HA \Leftrightarrow H^+ + A^-$, the buffer capacity β is defined as a small amount of base $d[B]$ needed to change the pH by dpH. If the adding of titrant is coulometrically generated at the porous actuator, the effective change of titrant $d[B]$ should be taken as equivalent to the associated change of the concentration of the species that will be detected by the ISFET at $x = 0$, i.e., $d[B] = \Delta C_1(0, t)$ [see Eq. (9)]. The response to this pH change ΔpH at the gate of the ISFET is obtained by incorporating Eq. (9) in the expression for the buffer capacity [20]:

$$\Delta \text{pH}(0, t)|_{\text{ISFET}} \cong \frac{d\text{pH}}{d[B]} \Delta C_1(0, t) \tag{10}$$

$$= \frac{A \cos(\omega t + \varphi)}{2.3 \left\{ \frac{K_a C_0 [H^+]}{(K_a + [H^+])^2} + [H^+] + \frac{K_w}{[H^+]} \right\}}$$

The concentration $C_1(0, t)$ can be considered as constant and equal to the bulk concentration C_0 for a small titrant perturbation.

Equation (10) shows that the response of the ISFET with respect to a coulometrically generated titrant perturbation is inversely proportional to the buffer capacity of the analyte, if the titrant perturbation is small. It indicates that the buffer capacity can be determined. It can be seen that this dynamic way of measuring differs from the conventional derivative titration in that the titrant is locally generated by coulometry and the associated local pH changes are rapidly detected by the ISFET. The advantage is that the measurement can be carried out very rapidly without disturbing the bulk solution.

2. Experimental

(*a*) *Sensor-Actuator Device with Porous Actuator Electrode.* The procedure for making the coulometric sensor-actuator device has been extensively described elsewhere [21]. A flat ISFET is used as the pH sensor with an additional thick layer of Ta_2O_5 of about 150 nm as a barrier to prevent gold from migrating into the silicon. The starting layer of the actuator was a 0.5 μm layer of gold deposited on the ISFET chip by thermal evaporation. After patterning by photolithography, a *ca.* 3 μm SiO_2 layer was deposited by PECVD (plasma enhanced chemical vapor deposition) on the top of the gate as the sacrificial layer. Then a layer of gold thick-film paste (DuPont 9910) was deposited. After drying, the thick-film paste was sintered at 600°C in a nitrogen atmosphere. The etching of the sacrificial layer and the glass compound in the thick-film paste was carried out in buffered HF for 1.5 h under stirring condition. A typical thickness of the obtained porous actuator is *ca.* 100 μm. The chip was finally mounted on a printed circuit board and encapsulated by epoxy.

(*b*) *Measurement Setup.* The buffer capacity measuring setup is shown in Fig. 14. Since the current source is not floating, the grounded counter electrode works as a reference electrode as well. In this case, the current at low frequency will cause a certain polarization in spite of the very large area of the counter electrode. This polarization potential of the counter electrode will be superposed on the output of the ISFET amplifier and interfere with the measurement. Therefore an additional saturated calomel electrode (denoted S.C.E. in Fig. 14) is used to measure separately the polarization potential, and the signal is sent to the lock-in amplifier for subtraction. The measured current and voltage are presented in effective (root-mean-square or RMS) values.

(*c*) *Measurement Protocol.* Before dissolution of the proteins, the sample solution is initially bubbled with nitrogen for 15 min. Then the protein is added and the solution is stirred for several minutes until complete dissolution. The measurement vessel was purged with nitrogen throughout the measurement. As the supporting electrolyte, 0.1 M potassium nitrate was used. The pH was changed by adding 1.0 M nitric acid or 1.0 M potassium hydroxide to the sample solution and was monitored by a pH meter with a glass electrode (Radiometer PHM83). For all measurements the pH was changed from a high to a low value. A measurement for correction of the results for the buffer capacity of the blank solution without proteins was carried out in the supporting electrolyte.

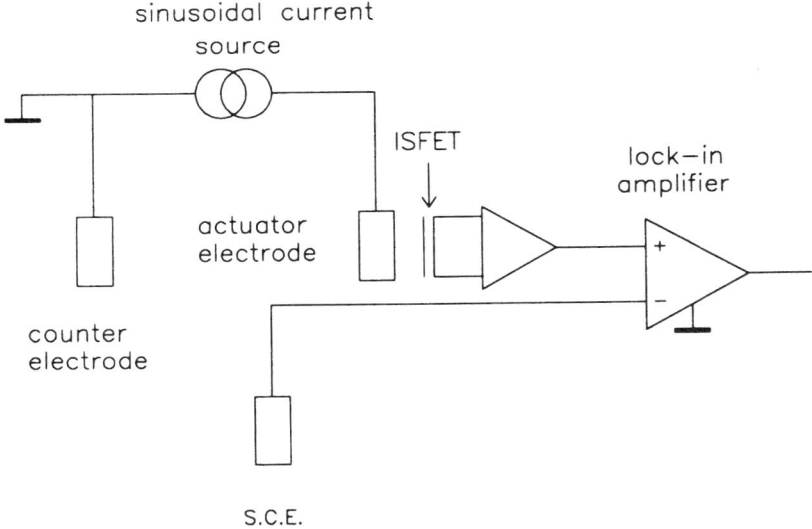

FIG. 14 Setup for the buffer capacity measurement.

3. Results and Discussion

(a) *Buffer Capacity of Acetic Acid.* To characterize and test the system, the buffering property of acetic acid at different pH values was measured, and the result is shown in Fig. 15 (triangles with solid line in upper part of the figure). The measurement results are presented as the reciprocal of the ISFET amplifier output voltage V_S. This reciprocal value of the measured results directly reflects the buffer capacity as a function of pH. For comparison, the theoretical curve was calculated and is also shown in Fig. 15 (lower part of the figure). At pH = pKa (= 4.77 for acetic acid), a maximum buffer capacity is expected. The measured results are in fairly good agreement with the theoretical description.

(b) *Buffer Capacity of Lysozyme and Ribonuclease.* The buffer capacity measurement of lysozyme (5 g/L, Sigma L-2879) and of ribonuclease (5 g/L, Sigma R-4875) after a simple correction of the data with the measurement results of a blank solution is presented in Fig. 16. The output voltage of the ISFET amplifier is presented reciprocally, which corresponds to the measured buffer capacity.

Lysozyme shows a high buffer capacity in the acidic region around pH 5 and has almost no buffer function above pH 10. It appears that this lowest point is around the isoelectric point of lysozyme, which is reported to be about pH 11 [22]. Ribonuclease has a good buffer function in the acidic region around pH 5

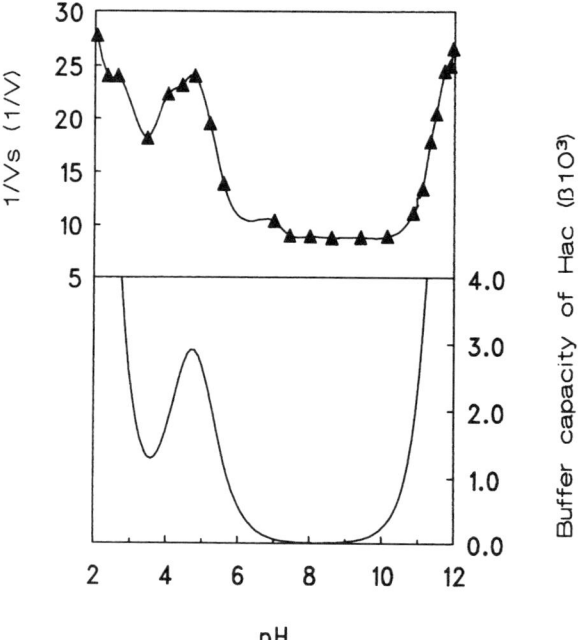

FIG. 15 Buffer capacity of 5 mM acetic acid as a function of pH. [KNO$_3$] = 0.1 M, $I = 10\ \mu A$, $f = 1$ Hz. Upper marks and fitting line are from the measurement; lower curve is calculated.

as well as in the basic region around pH 10, but it shows a low buffer capacity in the neutral solution around pH 8.

The interpretation of the experimental results has not taken into account any kinetic effects, which might be of importance in this particular case of dynamic operation. The results obtained here are therefore not exactly comparable to those carried out by a normal potentiometric titration. Nevertheless, the value of the buffer capacity at a certain pH contains quantitative information about the amount of protein that is present. Moreover, the curve as a function of the pH as shown in Fig. 16 reveals a protein signature that is specific for each different type of protein. This feature might be exploited in future biosensor research.

D. External Ion Step

In addition to the methods described above, the dependence of a protein titration curve on the salt concentration can also be exploited by making use of the ion step method. The essential aspect of this approach is, as with the previously de-

The Acid–Base Behavior of Proteins

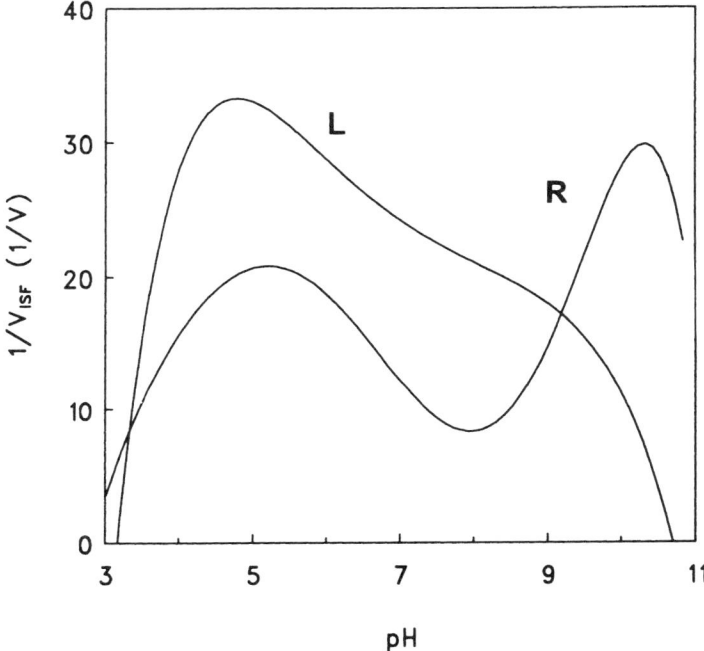

FIG. 16 Buffer capacity measurement of 5 g/L ribonuclease (R) and lysozyme (L) after correction for the buffer capacity of the blank solution.

scribed methods, that the measurement is not carried out during thermodynamic equilibrium, but rather the effect of a disturbance of this equilibrium is measured. The ISFET with protein membrane is incorporated in a flow-through system and thermodynamic equilibrium is established. The equilibrium is then disturbed by a sudden increase of the electrolyte concentration. The response of the ISFET to this ion step is a transient potential. The amplitude of this potential peak is a measure of the charge density, and with that of the concentration of protein immobilized in the membrane. It became clear that the response is caused by a temporary pH change in the membrane due to the release or uptake of protons by the protein molecules in the membrane. The release or uptake of protons by protein molecules is caused by the fact that the titration curves of proteins are a function of the ionic strength of the solution. If the ionic strength is changed, as occurs during an ion step, the amount of bound protons per protein molecule also changes. Two possible mechanisms will be described in more detail in the next subsection [23,24].

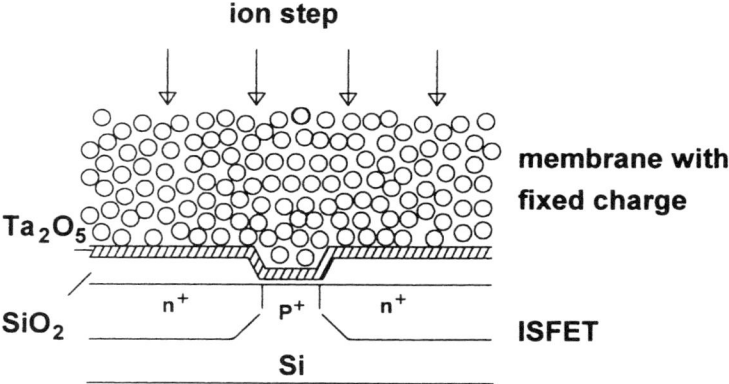

FIG. 17 The ion step measuring method. The ion step applied to the protein containing membrane on top of the ISFET.

1. Theory

(*a*) *Donnan Potential Model.* The device used in this measuring method consists of a porous membrane of polystyrene beads linked together with agarose, which is deposited on the gate of an ISFET (Fig. 17). A protein can be immobilized in the membrane via physical adsorption or covalent binding. The charged groups of the polystyrene beads, together with the charged groups of the protein, result in a net charge c_x in the membrane that is pH dependent in case the charged groups are titratable. If the sensor is actuated by a stepwise change of electrolyte concentration, as shown in Fig. 18, upper, the ISFET shows a transient response of which the amplitude ΔV is a function of the net charge density in the membrane (Fig. 18, lower). The ISFET response is positive if the net charge in the membrane is negative and vice versa. The origin of the response investigated in detail by Eijkel et al. [25]. Here we would like to confine ourselves to a short description of the mechanism.

When a layer of protein molecules is in thermodynamical equilibrium with a solution with a certain pH and ion concentration, the ions in the solution and in the protein layer will be distributed according to the Donnan ratio (the ratio of the ion activities in the two phases). As a results, there will exist an electrical potential difference between the protein layer and the bulk solution, which is known as the Donnan potential.

The dissociation of titratable groups of a protein molecule is influenced by the electrolyte concentration. A positively charged protein will contain more bound protons per molecule at higher electrolyte concentrations, and a negatively charged protein will contain fewer bound protons per molecule at higher concentrations.

The Acid–Base Behavior of Proteins

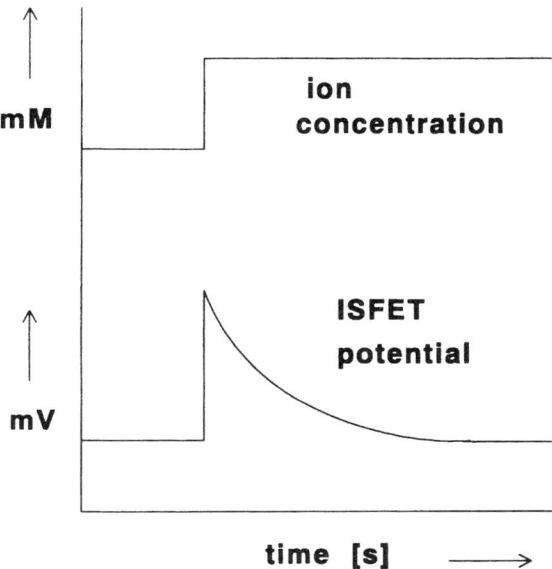

FIG. 18 The ion step measuring method. The schematic ISFET potential response (lower) on an ion concentration stimulus (upper).

When the membrane of polystyrene beads is in thermodynamic equilibrium, Donnan potentials exist at the pore walls of the porous membrane, where the protein is immobilized. At high charge densities, the resulting fields can extend to the pore centers, and consequently the membrane itself will have a potential with respect to the bulk solution.

As a result of a stepwise increase in electrolyte concentration, the Donnan potentials of the protein layer will diminish, thereby also diminishing the potential of the membrane itself versus the bulk solution. Furthermore, protons will be released by negatively charged protein molecules or taken up by positively charged protein molecules because the dissociation is changed with the ion concentration. The underlying ISFET measures the change in the potential of the membrane with respect to the bulk solution as well as the pH effect of the protons released or taken up. The pH effect of the protons is responsible for almost the entire response on an ion step.

(*b*) *Double-Layer Model.* In the description of the charged membrane in the previous subsection, the charged groups were considered to be homogeneously distributed in the membrane. The Donnan theory was used to describe the membranes, and in this Donnan model, an electrical double layer only exists at the membrane–solution interface. However, the membranes that are actually used with the ion step measuring method consist of polystyrene beads in an agarose gel. The charged groups are not homogeneously distributed but are concentrated on the surface of the beads. Charged surfaces are usually described by a double-layer theory, and Eijkel has described the implication of using a double-layer model for the ion step responses of a charged membrane [14].

The double-layer model of the membrane consists of many particles (assume a diameter of 0.1 μm) that are impenetrable for solution and carry a surface charge. Electrical double layers exist around each particle, and because the dimensions of the membrane pores are of the same order as the double layers around the particles, double layers exist throughout the membrane pores. The potential measured by the underlying ISFET, with respect to the bulk potential, is on one hand determined by the mean pore potential, which is the net result of the contribution of all surface potentials of the charged particles, and on the other hand by the pH at the membrane–ISFET interface. The measured ISFET response in equilibrium is therefore the same as that of an ISFET without a membrane, because the distribution of the protons between membrane and solution results in a pH difference, which compensates the mean membrane potential (this is the same mechanism as in the Donnan model). The relation between the surface charge on the particles σ (C/cm^2) and the surface potential ψ of each particle is given by

$$\sigma = C_{dl}\psi \tag{11}$$

where C_{dl} is the double-layer capacitance (F/cm^2).

First, assume that the surface charge on the membrane particles does not interact with the mobile protons (no proton release or uptake). An ion step will result in an increase in the double-layer capacitances of the particles and consequently in a decrease of the surface potentials ψ, because the charge densities remain constant. The ISFET will measure a transient change in the mean pore potential. As a result of the potential changes, an ion redistribution will take place and the equilibrium situation is re-established. The theoretical maximum ion step response is the change in the mean pore potential. This is comparable with the Donnan model where the theoretical maximum is determined by the change in the Donnan potential at the membrane solution interface.

If the surface groups at the particles are mildly acidic or basic, the change in surface potential after an ion step additionally results in a release or uptake of protons, depending on the pK$_a$ of the groups. In this case the theoretical maximum of an ion step response will not be determined by the change in the mean pore potential but by the change in the surface potentials at the particles, which will change more then the mean pore potential [14]. The buffer capacity of the surfaces is much higher than the mean buffer capacity of the pores, and therefore the release or uptake of protons resulting from the change in surface potential will now fully determine the pH in the membrane pores, which is measured by the underlying ISFET.

Eijkel found that, using the double-layer model, measured amplitudes of ion step responses better matched calculated values than using a Donnan model. Especially when larger beads are used (up to 1 μm), the double-layer model seems to be more realistic.

2. Experimental

(*a*) *Measurement Devices.* ISFETs with a Ta_2O_5 gate insulator were fabricated in the MESA clean room laboratory following the usual ISFET processing steps [8]. The ISFETs showed a response of about -58 mV/pH. The ISFET chips were mounted on a piece of printed circuit board and encapsulated with Hysol epoxy. On some of these devices, a subsequent 1 μm layer of silver was deposited around the gate. Using a polyimide mask, a square surface of 1 mm^2 of Ag is left uncovered. Then, the Ag electrode is chloridized by immersing it for 20 min in a 1% $FeCl_3$ solution, resulting in a *ca.* 300 nm thick layer of AgCl, acting as reference electrode under the membrane. Around the gate a circular area with a diameter of 2.5 mm and a depth of about 150 μm was left uncovered. The devices were mounted on a small (1×10 cm^2) piece of printed circuit board, contacted by wire bonding and covered with a protecting layer of Hysol epoxy resin. Thus, a dipsticklike measuring device was obtained. The membranes were made by 1:1 mixing and ultrasonication of a 0.25% agarose solution with a suspension of polystyrene beads at 40–50°C and subsequently casting portions of 3 μL on top of the gate area of the ISFETs. After overnight drying at 4°C and a temperature step of 55°C, during one hour, membranes with a thickness of about 10–15 μm (in dry condition) resulted.

The devices were incubated in different concentrations of the protein lysozyme for different intervals of time. The lysozyme was dissolved in a solution containing 10 mM KCl and 2 mM HEPES buffer to maintain the pH at 7.6.

(*b*) *Measurement Setup.* The measurement setup for the ion step measurements with membrane-covered ISFETs consists of a computer-controlled flow-through system in which a peristaltic pump ensures a flow of 2.7 mL/min. The ISFET is mounted in a wall-jet cell in which the liquid flow is perpendicular to the ISFET surface. A cross section of the wall-jet cell is shown in Fig. 2 (instead of high and low pH, the input is now high and low electrolyte concentration). A saturated calomel electrode, placed downstream, is used to define the potential of the solution. Two bottles containing the two ion step solutions are connected via valves to the measurement cell in such a way that the electrolyte concentration at the ISFET surface can be increased with a rise time (to 90% of the final value) of about 200 ms.

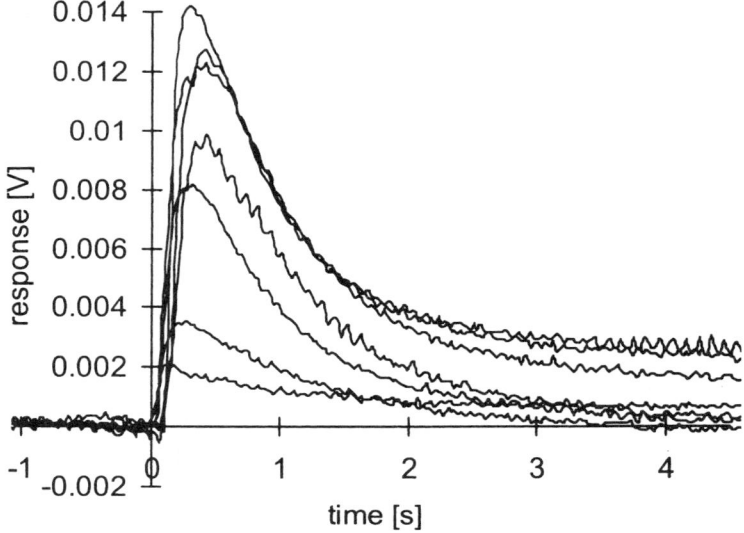

FIG. 19 Measured response on a membrane-covered ISFET to an ion step as a function of incubation time in $3 \cdot 10^{-5}$ M lysozyme. The incubation times were (from high to low transient response) 0, 1, 5, 15, 30, 60, and 120 min, respectively.

The devices with internal Ag/AgCl reference electrodes were used to perform manually an ion step by transferring the device from a vessel with a low-concentration electrolyte to a vessel with a high-concentration electrolyte.

The ISFETs are connected to a source-drain follower, and the output of this amplifier is connected to a digital oscilloscope. For presentation purposes, the curves are filtered with a software low-pass filter using a cutoff frequency of 40 Hz for elimination of the 50 Hz main supply interference, which has a top–top amplitude of typically 0.2 mV.

3. Measurement Results

Results are presented as obtained with the dipstick device, manually transferred between the two bottles of electrolyte. The dipstick device is meant for the detection of a protein concentration in a sample solution. During incubation in a protein-containing solution, a certain amount of protein will be adsorbed on the beads of the membrane. This amount is a function of both the incubation time and the concentration of protein present in the sample solution. The effect of the different incubation times at a fixed concentration of lysozyme ($3 \cdot 10^{-5}$ M) is shown in Fig. 19.

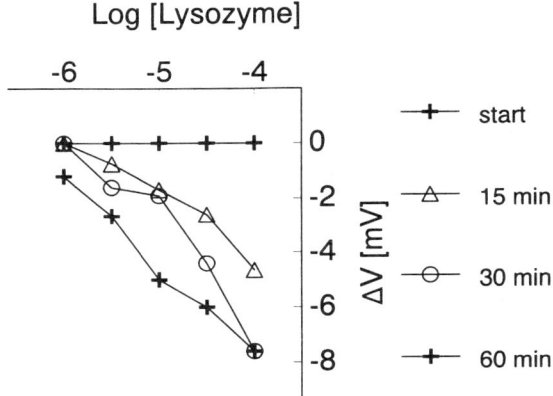

FIG. 20 Measured ion step response (with respect to the response of a bare membrane) as a function of the lysozyme concentration for different incubation times.

Since the point of zero charge of lysozyme is at pH 11, the protein is positively charged at pH 7.6. However, the bare membrane is charged strongly negatively, resulting in a net negative charge. As is shown in Fig. 19, the response decreases, while the incubation time increases with respect to the bare-membrane response, because of the addition of positive charge.

The effect on the amplitude of the responses as a function of the lysozyme concentration at a fixed incubation time is shown in Fig. 20 for several fixed incubation times [26].

III. CONCLUSIONS

In this chapter, we have proposed to use the acid–base properties of proteins as the transducing parameter in a biosensor. The acid–base behavior of proteins can reveal some important properties with respect to both their composition (selectivity) and their concentration (sensitivity). A change in this intrinsic parameter of the protein, when used as binding ligand, must be adequately determined. The classical method of acid–base determination is by volumetric titration. Successful application in a sensor requires another approach. Since the ion-sensitive field-effect transistor (ISFET) is suitable for fast (and local) pH detection, an ISFET can be used for protein titration.

We have presented four alternative methods for volumetric protein titration, based on a membrane-covered ISFET, in which proteins can be incubated. These methods are

1. External pH step application
2. Internal constant current titrant generation
3. Internal sinusoidal titrant generation by coulometry
4. External ion step application

These methods require a protein-containing volume of <50 nL with a response time of the order of 10 seconds after incubation.

ACKNOWLEDGMENTS

The authors are greatly indebted to J. Kruise, J. C. T. Eijkel, J. C. van Kerkhof, J. Luo, and F. Burlando for the work they carried out, as comprised in this chapter.

REFERENCES

1. F. Scheller and F. Schubert, *Biosensoren*, Birkhäuser Verlag, Basel, 1989.
2. A. P. F. Turner, I. Karube, and G. S. Wilson, *Biosensors—Fundamentals and Applications*, Oxford University Press, Oxford, 1987.
3. R. B. M. Schasfoort, P. Bergveld, R. P. H. Kooyman, and J. Greve. Anal. Chim. Acta *238* (1990): 323–329.
4. B. H. van der Schoot and P. Bergveld. Biosensors *2* (1988): 161–186.
5. L. Stryer, *Biochemistry*, 3d ed., W. H. Freeman, New York, 1988.
6. P. Bergveld. IEEE Trans. Biomed. Eng., BME-17 (1970): 70–71.
7. L. B. Bousse, N. F. de Rooij, and P. Bergveld, IEEE Transactions on Electron Devices *30* (1983): 1263.
8. P. Bergveld and A. Sibbald, in G. Svehla, ed., *Comprehensive Analytical Chemistry*, Vol. 23: *Analytical and Biomedical Applications of Ion-Selective Field-Effect Transistors*, Elsevier, Amsterdam, 1988.
9. R. E. G. van Hal, J. C. T. Eijkel, and P. Bergveld. Adv. Colloid Interface Sci. *68* (1996): 31–62.
10. J. Kruise, J. C. T. Eijkel, and P. Bergveld. Sensors and Actuators, in press.
11. J. Crank, *The Mathematics of Diffusion*, Oxford University Press, London, 1975, p. 326.
12. R. B. M. Schasfoort, R. P. H. Kooyman, P. Bergveld, and J. Greve. Biosensors and Bioelectronics *5* (1990): 103–124.
13. J. C. van Kerkhof, P. Bergveld, and R. B. M. Schasfoort. Biosensors and Bioelectronics *10* (1995): 269–282.
14. J. C. T. Eijkel, Potentiometric detection and characterization of adsorbed protein using stimulus-response measurement techniques, thesis, University of Twente, Enschede, The Netherlands, ISBN 90-9008615-3, 1995.
15. W. Olthuis, J. Luo, B. H. v. d. Schoot, P. Bergveld, M. Bos, and W. E. v. d. Linden. Anal. Chim. Acta. *229* (1990): 71–81.
16. J. Kruise, J. C. T. Eijkel, and P. Bergveld, Proc. 6th Int. Meeting Chem. Sens., Gaithersburg, MD, USA, July 22–25, 1996, p. 125.

17. W. Olthuis, J. Luo, and P. Bergveld. Biosensors and Bioelectronics 9 (1994): 743–751.
18. J. Luo, W. Olthuis, P. Bergveld, M. Bos, and W. E. van der Linden. Anal. Chim. Acta 274 (1993): 7–23.
19. J. Luo, W. Olthuis, P. Bergveld, M. Bos, and W. E. van der Linden. Sensors and Actuators B, 20 (1994): 7.
20. D. D. Perrin and B. Dempsey, *Buffers for pH and Metal Ion Control*, Chapman and Hall, London, 1974.
21. J. Luo, W. Olthuis, P. Bergveld, M. Bos, and W. E. van der Linden, Proc. Transducers '91, The 6th Int. Conf. on Solid-State Sensors and Actuators, San Francisco, 1991, pp. 229–232.
22. C. Tanford and M. Wagner. J. Am. Chem. Soc. 76 (1954): 3331–3336.
23. J. C. van Kerkhof, The development of an ISFET-based heparin sensor, thesis, University of Twente, Enschede, The Netherlands, ISBN 90-9007514-3, 1994.
24. J. C. van Kerkhof, P. Bergveld, and R. B. M. Schasfoort. Biosensors and Bioelectronics 8 (1993): 463–472.
25. J. C. T. Eijkel, W. Olthuis, P. Bergveld, and J. Greve, Ext. Abstr., Eurosensors VI, San Sebastian, Spain, 1992, p. 336.
26. F. Burlando, An ISFET-based dipstick device for protein detection with the ion-step method, Research Report, University of Twente, 1996.

Index

Acid–Base
  behavior
    of amino acids, 374
    of peptides, 374
    of proteins, 373 *et seq*
  dissociation constants, 118
  equilibria, 378, 393
  interactions, 114
Adhesion
  force of, 11, 12
  work of, 9, 11, 21, 67, 70
Adsorption (*see also* Polymer
    adsorption; Protein
    immobilization)
  of anions, 28
  -desorption behavior, 352
  equilibrium constant, 350
  isotherm
    BET, 15
    Gibbs, 14
    Langmuir, 348, 351, 361
  kinetics, 352, 355, 369
Aerogels, 340
Albumin
  bovine serum, 101
  human serum, 133, 362
Aluminium, 8

[Aluminium]
  oxide
    in AFM force studies, 98
    in electrokinetic studies, 136
    heat of immersion of, 17
    as polishing agent, 78
    as semiconductor gate, 378
    spreading pressure upon, 15
Amino acids, 374
3-Aminopropyltriethoxysilane, 127, 130
Antigen-antibody
  association, 345 *et seq*, 373
  E. Coli β-galactosidase, 359
Atomic Force Microscope (AFM), 12,
    87 *et seq*, 265
  cantilevers, 13, 25, 89, 90
  colloid probe equipped, 12, 91, 93, 96
  force measurement, 13, 25, 91, 96
    electrical double-layer forces, 93,
      97, 98
    force curve construction, 91
    polyelectrolyte bridging force, 101
    polyelectrolyte depletion force, 103
    van der Waals forces, 87, 98, 103
  surface stress measurement, 25
Attenuated total reflectance spectroscopy
    (ATR), 175, 183

Auto correlation function, 187, 188, 193, 194, 203, 210, 212, 216, 218, 220, 225, 231, 253

Bending plate measurement, 26
Biochemical sensors, 373
  enzymes, 374
Boltzman distribution
  of ions, 93, 117
Bragg diffraction patterns, 8, 311
Brownian motion, 212, 224, 239, 253, 269, 274
Buffer capacity, 389, 393

Cadmium sulfide, 239
Calcium
  fluoride, 20
  hydroxide, 18
  oxide, 18
  sulfate, 20
Calorimetry, 16
Capillary
  penetration, 65
  rise, 43, 59
Carbon, 17, 25
Cholesterol acetate, 47
Chromatography, 14, 69, 126, 345 *et seq*
  column, 349, 350, 357, 358, 359, 362
  gas adsorption, 347
  high performance liquid (HPLC), 346
  immunochromatography, 346, 352
  instrumentation and methodology, 356, 366
  inverse gas, 14
  repeated pulse injection, 365
  theoretical model for, 347
Chromophore, 292, 306
Circular dichroism (CD), 292
Colloid size distribution, 200, 208, 213, 219
Contact Angle
  experimental values for various materials, 58(t)
  hysteresis, 38, 73
  measurement of, 21, 37 *et seq*

[Contact Angle, measurement of]
  adhering bubble technique, 48
  axisymmetric drop technique, 52
  capillary rise, 43, 59
  flat surfaces, 39, 58
  nonflat surfaces, 58
  sessile drop method, 48
  Wilhelmy plate method, 28, 40, 47, 59
Correlator, 208, 220
Coulometry
  constant current, 385
  sinusoidal current, 387
  theory of, 389
Critical micellization concentration, 289

Debye length, 94, 99, 228
Depletion force, 101, 103
Derjaguin approximation, 94
Derjaguin–Landau–Verwey–Overbeek (DLVO) theory, 88, 93, 97
Diffuse layer potential, 95, 98, 119
Diffuse reflection infrared Fourier transform spectroscopy (DRIFT), 128
Diffusion
  coefficient, 189, 192, 212, 239, 253, 274, 348, 379, 280
  equation, 212
Dimethyldichlorosilane, 77
Donnan effect, 103, 396
Dynamic light scattering (DLS), 199 *et seq*, 201, 250
  data analysis methods, 216
    binning, 221
    Laplace inversion, 220
    summarized, 217(t)
  depolarized light, 227
  dual beam arrangement, 231 *et seq*
  of evanescent waves, 187
  theory of, 209, 253

Elastic strain, 5
Electrical double-layer, 88, 93, 100, 106, 115, 119, 127, 280, 397

Index

Electrocapillarity, 27, 47
Electrode, 27, 385
  gold, 99
  mercury, 27
  pH, 392
  platinum, 119, 121, 123
  saturated calomel, 392
  semitransparent, 185
  silver/silver chloride, 119, 399
  tin-indium oxide, 280
Electron
  binding energy of, 145
  energy in XPS, 146
  exchange, 294
  mobility, 377
Electronic
  absorption spectra, 292, 315, 322
    energy level diagram, 294
  emission spectra, 322
Electron microscopy, 8, 219, 222, 257, 286
  scanning, 299, 323, 331
  transmission, 219, 298, 326
Electron spectroscopy for chemical analysis (ESCA), 134, 136
Electro-osmosis, 113 *et seq*, 280
  measurement of, 119
    cylindrical cell, 120
    rectangular cell, 123
Electrophoresis, 98, 114, 121, 126, 199, 200, 201, 280
Electrophoretic light scattering, 228
  experimental methods, 237
  laser Doppler method, 237, 238, 280
  phase analysis light scattering, 239
Elution
  frontal mode, 347
  peak splitting, 354
End group analysis, of polymers, 163
Enthalpy
  of aggregation, 297, 309
  of immersion, 16
  of surface, 4, 16, 18
Entropy
  of aggregation, 297, 309

[Entropy]
  of surface, 19
  of tethered polymers, 192
Evanescent wave, 173 *et seq*
  penetration depth, 174, 178
  scattering of, 174
    description of apparatus, 177
    dynamic, 187
    from particular ensemble, 181, 187
    from polymer molecules, 187
    static, 179
Exciton band, 307
  splitting, 307

Fibrinogen, 134
Ficks second law, 379
Fluoresence, 286, 296
Fluorescent probes, 282
Forces
  adhesive, 11, 12
  capillary, 287, 299
  electrical double-layer, 87, 93 *et seq*, 127, 200
  polymer mediated, 101, 103, 200
  van der Waals, 6, 67, 87, 106, 200, 288
Free energy of aggregation, 288, 297

Gel permeation chromatography, 164, 170
Gelation, 286 *et seq*
  temperature, 310
Gel-liquid crystal phase transition, 262
Gel-sol (*see* Sol-gel)
Gibbs–Duhem equation, 4, 5, 14, 16
Glass
  bearing adsorbed polymer layer, 185, 194
  chromatography beads, 360
  evanescent wave cell, 177
  nylon coated, 195
  siliconized, 47
  -water interface, 190
Globulin
  gamma, 259
  human immuno, 133

Gold
  in AFM force measurements, 99, 100, 106
  with alkanethiol monolayers, 76
  electrodes, 27, 100
  Faraday sols, 100
  Surface energy of, 7
Gouy–Chapman theory, 93, 115, 116
Gouy–Chapman–Stern–Grahame theory, 119
Greens function, 176

Hagen–Poiseuile equation, 65
Hamaker constant, 67, 99, 104
Heat of immersion, 16
  values for various materials, 17
Helmholtz–Smoluchowski equation, 229
Henry equation, 228
Heterodyne detection, 90, 190, 216, 231
Homodyne detection, 90, 190, 209, 216, 231
Hooke's law, 93
Hückel equation, 228

Image threshold analysis, 270
Immunoglobulin, human, 361
Insulin, bovine, 362
Internal surface energy, 16
Ion mediated forces
  polyelectrolyte species, 101, 103
  simple electrolytes (*see* Electrical double-layer forces)
Ion sensitive field effect transistor (ISFET), 373, 376
  enzyme sensor, 374
Ionic strength, 93, 125, 395

Jellies (*see* Organogels)

Kelvin Equation, 19

Laminar flow, 65, 124
Langmuir
  adsorption rate equation, 348, 351
  isotherm, 348, 351, 360, 361

Langmuir–Adam surface balance, 75
Langmuir–Blodgett films, 75, 185, 193
Laplace
  equation, 43, 50, 51, 65, 287
  pressure, 19
  transform, 182, 219, 220, 254
    summarized, 223(t)
Laser
  Doppler electrophoresis, 238
  pressure of, 283
Lattice constant, 8
Lifshitz theory, 104
Line tension, 73
Liposomes, 190, 250, 262
Lippmann equation, 27
London theory, 87, 103
Luminescence, 286 *et seq*, 292
Lumophore, 293, 306
Lysozome, 380, 387, 393
  adsorption on antilysozome, 360

Magic angle, 308
Magnesium oxide, 18, 136
Mass spectrometer, 148, 149
Maximum entropy, 222
Membrane, 376, 379, 380
  calcium-silicate, 98
  thickness, 257
3-Mercaptopropyltriethoxysilane, 130
Mercury electrode, 27
Metal oxide semiconducting field effect transistor (MOSFET), 377
Metastability, 298
Mica, 9, 10, 11, 25, 75, 98, 101, 106
Micelle, 250, 288, 289
  living polymer, 291
Microcapsules, 249 *et seq*
  poly(lysine)-alt-(terephthalic acid), 263
  poly(urea-urethane), 257
Microemulsion, 262
Microparticles, 157, 269
Mie theory, 215
Modulus
  elastic, 13, 25, 295
  loss, 295, 309 312

Index

[Modulus]
  shear, 295
  storage, 295, 309, 312, 313, 338
Molecular ion species
  designation of, 149, 152, 154
Multiple scattering, 227

Navier–Stokes equation, 114, 117, 120, 122, 124
Nernst equation, 378
Nernst–Planck equation, 380
Nickel (II) oxide, 17

Organogels, 286 *et seq*
  phase transitions of
    compared to jellies, 291
    nucleated crystal growth, 310
  physico-chemical properties of
    anthryl derivatives, 308
    sorbitol based, 338
    steroid based, 313
Osmotic
  pressure, 94, 101, 103, 263
  swelling, 262

Paraffin, 25
Particles
  in AFM studies, 91, 93 *et seq*
  in contact angle measurement, 58, 63, 69, 79
  iron, 278
  scattering factors, 251
    summarized, 255(t)
  size distribution, 200, 208, 213, 216, 219, 221, 225
  solubility of, 19
Peptide, 374
Permittivity, of dielectric, 205
Phase analysis light scattering (PALS), 239 *et seq*
Phase
  diagram, 296, 299
  transition, 47, 291, 296
pH electrode, 378–380, 386
Phosphatidylcholine vesicles, 262

Phosphorescence, 286, 308
  spectra, 325
Photoionization, 146
Photon correlation spectroscopy (PCS)
  (*see* Dynamic light scattering)
Piezoelectric scanner, 90, 92, 271, 278
Plastic strain, 25
Platinum
  electrodes, 121, 123
Poisson–Boltzman equation, 93, 115, 119
Poisson equation, 93, 117, 380
Poisson ratio, 25
Poly(acrylic acid) (PAA), 101, 103
  crosslinked surface, 131
Poly(α-methylstyrene) (PαMS), 161
Poly(carbonate), 161
Poly(dimethylsiloxane) (PDMS), 151, 324
Polyelectrolytes, 100
  and surface forces, 101
Poly(ethylene), 25, 74
Poly(ethylene imine), 130
Poly(ethylene oxide) (PEO), 168, 184
  oligomeric (PEG), 129, 161
  with perfluorodecanoyl termination, 168
  tetrameric (PTMO), 151
Poly(fluoroethylene propylene) (FEP), 58, 74
Poly(hydroxybutyrate), 170
Poly(isobutylene), 151
Poly(lactic acid), 147, 163
  -*co*-(glycolic acid) block copolymer, 157
Poly(lysine), 98, 101
  -alt-(terephthalic acid) microcapsule, 263
Polymer
  adsorption and grafting to surfaces, 102, 129, 161, 185, 187, 192, 194, 195, 260
  desorption, 170
  layer thickness, 260
  surface engineering, 145
Poly(methyl methacrylate) (PMMA), 25, 123, 185

[Poly(methyl methacrylate) (PMMA)]
  copolymer, with (PEG) methacrylate
    macromonomer, 155
  surface tension of, 25
Poly(n-butyl methacrylate), 56
Poly(propylene), 74
Poly(propylene glycol), 152
Poly(sebacic anhydride), 163
Poly(siloxane), 132
Poly(styrene) (PS), 25, 161
  block copolymer with
    latex, 122, 259, 380
      PEG grafted, 161
    perfluoroalkyl terminated, 164
    poly(ethylene oxide), 185, 195
    poly(methyl methacrylate), 185
Poly(styrene sulfonate), sodium salt, 103
Poly(sulfone), 74
Poly(tetrafluoroethylene) (PTFE), 25
  in AFM force studies 106
  contact angle with various liquids,
    58(t)
  surface preparation, 77
Poly(urethane)
  XPS and SIMS of copolymer
    derivatives, 151
Poly(vinyl acetate)-co-(vinyl alcohol)
  copolymer, 257
Poly(vinyl pyridine), 101
Potassium nitrate, 99, 107, 379, 386
Potential energy between sphere and flat,
  183
Power spectrum, 240, 261
  Lorentzian, 259
Pluronic, 157, 258
Protein (*see also* Lysozyme albumin),
  375
  adsorption/immobilization, 145, 259,
    359, 379, 380, 383
  recognition, 378
  titration of, 385 *et seq*

Quartz
  deposition of Langmuir–Blodgett films
    upon, 75

[Quartz]
  electrophoretic cell, 120
  silanol functionalities upon, 126
  spreading pressure upon, 15
  surface grafting of, 130
  vapor deposition onto, 77
Quasielastic light scattering (QELS)
  (*see* Dynamic and Static light
  scattering)

Rayleigh–Gans–Debye approximation,
  215
Redox couples, 385
Refractive index, 173, 269
  in calculation of Hamaker constants,
    104
  of continuous phase, 204, 205
  of scattering centers, 188
Reynolds number, 276
Rheology, 287, 295
  of organogels, 286
ribonuclease, 393

Scanning tunneling microscope, 89, 265
Scattering vector, 179, 186, 204, 235,
  251, 300
Self assembly
  of monolayers, 76, 100
  of organogels, 285 *et seq*
Sessile drop, 48
Shuttleworth equation, 5
Siegert relationship, 211, 213, 221
Silanization (*see* Siliconization)
Silica (*see* Quartz; Silicon dioxide)
Silicon, 13
  carbide, 136
  dioxide
    aerogels, 339
    in AFM force studies, 107, 97 *et seq*
    dissolution of, 136
    with grafted alkyl layers, 24
    heat of immersion, 17
    in HPLC columns, 358, 359, 362
    insulation on semiconductor devices,
      377, 378, 392

Index

[Silicon, dioxide]
    particles, 192, 219, 224, 272, 280
    spreading pressure upon, 15
    nitride, 13, 106, 136, 378
Silicone oil, 278, 280
Siliconization, 47, 77, 128, 130
Small angle scattering
    neutron (SANS), 261, 300, 310, 326, 328
    x-ray (SAXS), 257, 326, 328
Sodium
    chloride, 124, 126, 128
    dodecyl sulphate, 170
Solidification front, 69
Solubility of solids, 19
Sol-gel, and gel-sol phase transition, 291, 297, 309, 321
Spherical particles
    in atomic force microscopy, 92, 97–99, 102
    in chromatography, 348, 360
    as light scattering centers, 180, 187, 188, 206, 219
    tracking of, 271 *et seq*
Spinodal decomposition, 287, 297
Spreading pressure, 15
Staphylococcus epidermidis, 63
Static secondary ion mass spectroscopy (SSIMS), 143 *et seq*, 148
Static light scattering (SLS), 201, 250, 251, 257
Steric layer thickness, 183
Stokes–Einstein equation, 213, 228, 239
Stokes law, 254, 275, 276
Stokes radius, 259
Stoneys formula, 24
Streaming potential, 98, 115
Structure factor
    in scattering studies, 180, 181, 188, 211
Surface
    charge development, 88, 118
    cleavage, 9
    energy, 288
    enrichment of polymeric species, 151

[Surface]
    modification (*see also* Polymer adsorption; Protein immobilization; Self assembled monolayers; Siliconization), 127
    with 3-Aminopropyltriethoxysilane, 127
    with 3-Mercaptopropyltriethoxy-silane, 130
    by plasma discharge, 77, 130
    with poly(ethylene glycol), 129, 161
    with poly(ethylene imine), 130
    plasmon resonance, 370
    potential, 93, 119, 378, 398
    preparation
        for contact angle measurement, 74
    roughness and contact angle measurement, 60
    segregation of polymer blends, 161
    stress, 2 *et seq*
        defined, 2, 5
        measurement of, 7, 8, 9, 20, 26
        summary of experimental techniques, 29
    tension
        defined, 3
        measurement of, 7, 12, 53
        relationship to van der Waals forces, 22, 67
        values of
            of liquids, 23(t)
            of solids, 24(t), 25(t), 37
Surface forces apparatus, 11, 107

Tantalum oxide, 378, 380
Thiobacillus ferrooxidans, 62
Thixotrophy, 324, 326
Titanium dioxide, 17, 98, 106
Total internal reflection, 174
Tungsten, 90, 92, 106

Ultra high vacuum, 8, 10, 146, 149

van der Waals interactions, 6, 87 *et seq*, 114, 127, 200, 288

[van der Waals interactions]
   AFM studies, 98, 103
   attractive, 67, 104
   between planar surfaces, 6, 67
   Lifshitz theory, 104
   London theory, 87, 103
   relationship to surface tension, 22, 67
   repulsive, 67, 71
   retardation of, 88, 105
Vesicles (*see* Liposomes)
Viscoelasticity of gels, 291

Washburn equation, 65
Wide angle x-ray scattering (WAXS), 330, 336
Wilhelmy plate, 28, 38, 40, 47, 59

X-ray photoelectron spectroscopy (XPS), 143

Young contact angle, 38, 56
Young equation, 2, 21, 38, 67, 73
Youngs' modulus, 10, 295

Zeta potential, 98, 99, 119, 136, 201, 228, 269, 281, 282
Zimm plot, 225
Zinc oxide, 17
Zinc sulfide
   in AFM force studies, 98, 101
   in evanescent wave studies, 184
Zirconium oxide, 98, 101